# Computational Molecular Science and Engineering Forum 2017



Core Programming Area at the 2017 AIChE Annual Meeting

Minneapolis, Minnesota, USA
29 October – 3 November 2017

ISBN: 978-1-5108-5807-7

**Printed from e-media with permission by:**

Curran Associates, Inc.
57 Morehouse Lane
Red Hook, NY 12571

**Some format issues inherent in the e-media version may also appear in this print version.**



Printed by Curran Associates, Inc. (2018)

For permission requests, please contact AIChE
at the address below.

AIChE
120 Wall Street, FL 23
New York, NY 10005-4020

Phone: (800) 242-4363
Fax: (203) 775-5177

www.aiche.org

**Additional copies of this publication are available from:**

Curran Associates, Inc.
57 Morehouse Lane
Red Hook, NY 12571 USA
Phone: 845-758-0400
Fax: 845-758-2633
Email: curran@proceedings.com
Web: www.proceedings.com

# Computational Molecular Science and Engineering Forum 2017

Core Programming Area at the 2017 AIChE Annual Meeting

Minneapolis, Minnesota, USA
29 October – 3 November 2017

AIChE
The Global Home of Chemical Engineers

# TABLE OF CONTENTS

**Some abstracts in this book refer to images that were not made available by the publisher**

# 504302 On-Boarding Computational Scientists with Bash, Git, and Python

**Abstract:**

New researchers in the computational sciences face common activation barriers to authentically participating in distributed project development. In this workshop we discuss strategies for efficiently overcoming these barriers informed by the pedagogical literature for how students learn to programming and how scientists manage code. This hands-on session focuses specifically on examples with command-line-interfaces (bash), version control (git), and python that will be used throughout the day's workshop. This introduction is targeted at graduate students new to the computational sciences and will be equally useful to faculty and postdocs looking for a refresher or who serve in computational mentoring roles.

# 499166 Mosdef: Molecular Simulation and Design Framework

**Abstract:**

We present a collection of tools designed to enable the facile setup, initialization, atom-typing, and molecular dynamics (MD) simulations, termed MoSDeF [1]. In particular, MoSDeF has been developed to enable computational MD-based screening of soft materials [2]. The core tool, mBuild [2,3], generates starting configurations by minimizing or even eliminating the need to explicitly translate and orient components when building systems - users simply state which components to connect. mBuild supports parameterized structures through generative modeling. This allows for declaratively expressing repetitive structures, e.g. polymer chains, crystal structures, planar or spatial tiling, as well as for parameterizing the affine transformations applied to subcomponents of a composite component. For end users, this minimizes and often eliminates the need to explicitly rotate and translate components when assembling systems. Additionally, the approach enables users to programmatically vary parameters for a family of systems (e.g., polymer chain length) or interchange individual components (e.g., polymer type), while still employing the same general framework. mBuild integrates with the Foyer tool for cataloging and applying force fields to molecular systems (i.e., atom-typing). Foyer provides a force field and simulator agnostic method for defining parameter usage that relies upon SMARTS [5] based annotations of chemical context, providing both human and machine readable documentation of parameter usage, which also aids in the dissemination of force fields. These tools are coupled with the Signac-flow workflow manager [6], allowing large scale screening over parameter space. Building upon prior work [2], we demonstrate the utility and flexibility of our tool chain by screening the frictional properties of several families of monolayers.

[1] https://github.com/mosdef-hub

[2] C. Klein, J. Sallai, T. J. Jones, C. R. Iacovella, C. McCabe, and P. T. Cummings, "A Hierarchical, Component Based Approach to Screening Properties of Soft Matter," in Foundations of Molecular Modeling and Simulation, 2016, pp. 79–92.

[3] https://github.com/mosdef-hub/mbuild

[4] https://github.com/mosdef-hub/foyer

[5] http://www.daylight.com/dayhtml/doc/theory/theory.smarts.html

[6] C. S. Adorf and P. Dodd, "Signac-Flow." [Online]. Available: https://bitbucket.org/glotzer/signac-flow.

# 504309 Managing Data Spaces, Performing MD, and Analyzing Trajectories with Signac, HOOMD-Blue, and FreudÂ

**Abstract:**

We demonstrate how to efficiently develop a complete, integrated, and reproducible molecular-dynamics workflow covering everything from the setup of a parameter space to the execution of simulations and the post-processing of output data for analysis and visualization. Participants are introduced to the basics of managing data spaces with signac [1]. That includes the management and searching of data and metadata. We show how to setup and execute restartable molecular-dynamics simulations with the GPU-accelerated particle simulation toolkit HOOMD-blue [2] with emphasis on the definition of initial configurations, boundary conditions, and simulation protocols. We present the analysis of trajectory data using the parallelized high-performance analysis code freud [3]. The individual steps will be integrated with signac-flow, which allows us to chain and execute data space operations and easily submit workflows to high-performance clusters.

All tutorial material is presented in the form of published Python Jupyter Notebooks. This enables participants to easily follow all examples, and follow-up on the presented workflows.

[1] https://glotzerlab.engin.umich.edu/signac
[2] https://glotzerlab.engin.umich.edu/hoomd-blue/
[3] https://glotzerlab.engin.umich.edu/freud/

# 504310 How to Recognize Garbage

**Abstract:**

One of the most difficult problems when starting to perform atomistic simulations is identifying when the output of a simulation corresponds to what you actually put in. In this workshop, we will cover a number of ways to examine and diagnose what can go wrong with then simulations, so that the results properly reflect the model you intend to simulate, rather than something nonphysical that keeps you and your advisor tearing their hair out for weeks instead of generating cool science. Examples will be presented in LAMMPS and GROMACS, but techniques and principles will be generally applicable to any software.

# 504303 How to Use the Atomistic Monte Carlo Package Cassandra: Liquid Phase Properties and Vapor-Liquid Phase Equilibria

**Abstract:**

Cassandra is an open source atomistic Monte Carlo simulation package that is used by over 300 researchers worldwide. Cassandra is capable of simulating any number of molecules composed of rings, chains, or both. It can be used it to simulate small organic molecules, oligomers, and ionic liquids. It handles a standard "Class I"-type force field having fixed bond lengths, harmonic bond angles and improper angles, a CHARMM or OPLS-style dihedral potential, a Lennard-Jones 12-6 potential or Mie potential and fixed partial charges. Cassandra can simulate the following ensembles: canonical (NVT), isothermal-isobaric (NPT), grand (muVT), osmotic (muPT), and Gibbs (NVT and NPT versions).

In this brief hands-on tutorial, we will cover the essentials of how Cassandra works and the types of properties is can compute. We will provide step-by-step instructions for creating input files, how to interpret the output from Cassandra and how to avoid common mistakes. At the end of this tutorial, users will be able to simulate a pure liquid phase at constant pressure and temperature as well as vapor-liquid equilibria using the Gibbs ensemble. Extensions to more complex systems will be discussed if time permits.

Who should attend: undergraduate students, graduate students or other researchers who are new to molecular simulation and are interested in learning how to carry out Monte Carlo simulations of fluids.

# 504301 Using Python to Standardize and Format Input to Popular Molecular Simulation Software

**Abstract:**

High performance simulation software often requires the construction of highly formatted and sometime cryptic input that, out of context, can have little or no meaning. The aim of this workshop is to illustrate ways in which the Python scripting language can be used to create a common interface to independent molecular simulation software. User scripts can then be more informative when read as stand alone code. For example, two interfaces to LAMMPS and CASSANDRA, implemented in the open-source Python package *pysimm* will be demonstrated from an end-user perspective. These working examples will show the utility of this type of coding practice.

Hands-on examples will include building polymer systems in LAMMPS with control over molecular weight distribution, and adsorbing gas molecules in porous polymeric materials using CASSANDRA.

# 504308 Quantifying Uncertainty in Molecular Simulations

**Abstract:**

The collection of averages by molecular simulation is a stochastic process, and consequently it yields results that are not perfectly reproducible. To use these averages or to compare them to other quantities it is essential to understand the uncertainty associated with them, i.e., by how much might they differ if the calculation were repeated with different initial conditions? The means to quantify this uncertainty is well established and cuts across all fields of science, but the reporting of uncertainties with molecular simulation data is still not a universal practice. Moreover, there are pitfalls to avoid in applying error analysis to the results of a molecular simulation, as it is frequently necessary to manipulate and combine the direct simulation results to obtain the property of interest. Proper attention is needed to correlation both across averages and within subaverages to ensure that the simulation uncertainty is correctly propagated to the final result.

In this workshop we review these issues and provide several instances demonstrating the need for attention to uncertainty calculation. We guide the participants through example calculations demonstrating techniques for computing and propagating uncertainties.

# 499033 Contact Freezing Vs. Surface Freezing in Water

**Abstract:**

Contact freezing is a phenomenon in which heterogeneous ice nucleation occurs on a surface close to a vapor-liquid interface. In general, contact freezing is considerably faster than immersion freezing, i.e., heterogeneous nucleation on a surface fully immersed within the bulk [1-2]. Surface freezing, however, refers to the enhancement of homogeneous ice nucleation at or near a vapor-liquid interface [3]. Unlike contact freezing that is supported by direct experimental evidence, surface freezing is a more subtle phenomenon that is far more difficult to prove and characterize directly, and even though strong indications exist in its support, no direct experimental evidence for facilitation of homogeneous ice nucleation at vapor-liquid interfaces is available. Computational studies of surface freezing have also been inconclusive [4-5]. Contact freezing and surface freezing are conceptually related to one another as both involve nucleation close to a vapor-liquid interface. It has therefore been argued that the ability of a free interface to enhance heterogeneous nucleation (contact freezing) would also imply its propensity to facilitate homogeneous nucleation (surface freezing). This is, however, a speculation that has never been tested. The focus of this work is to test this hypothesis, by computing heterogeneous nucleation rates in several model tetrahedral liquids with known surface freezing propensities.

1- R. A. Shaw, et al., J. Phys. Chem. B, 109, 9865 (2005).

2- A. J. Durant, et al., Geophys. Res. Lett., 32, L20814 (2005).

3- A. Tabazadeh, et al., PNAS, 99, 15873 (2002).

4- A. Haji-Akbari, et al., Phys. Chem. Chem. Phys., 16, 25916 (2014).

5- A. Haji-Akbari, et al., PNAS, 114, 3316 (2017).

# 492411 Modeling the Homogeneous and Heterogeneous Nucleation of Crystals of Ionic Liquids in Different Environments

**Abstract:**

Solidification of ionic liquids (ILs) is used in the synthesis of optically-active and magnetic nanomaterials based on ILs. However, very little is understood about the mechanism of nucleation of solid phases of ILs, mainly because most of their applications involve liquid phases. Here we studied the nucleation of crystals of two common ILs, 1,3-dimethylimidazolium chloride [dmim$^+$][Cl$^-$], and 1-butyl-3-methylimidazolium chloride [bmim$^+$][Cl$^-$], from their supercooled liquid phases in the bulk, in contact with graphitic surfaces, and inside slit-like graphitic nanopores. The string method in collective variables[1,2] was used in combination with Markovian milestoning with Voronoi tessellations[2,3] and order parameters for molecular crystals[4] to sketch a minimum free energy path connecting the supercooled liquid and the crystal phases, and to determine the rates involved in the nucleation process. This particular combination of methodologies is well suited to study nucleation of ILs, as these systems exhibit very slow dynamics and very large simulation box sizes are required in order to avoid finite-size effects. We will present and discuss results[5,6] obtained for the free energy profiles, mechanisms and rates involved in the nucleation processes of these systems, aiming at understanding the nucleation and growth of crystals of organic salts in the bulk, near surfaces and inside nanopores.

[1] L. Maragliano, A. Fischer, E. Vanden-Eijnden and G. Ciccotti, "String Method in Collective Variables: Minimum Free Energy Paths and Isocommittor Surfaces", *J. Chem. Phys.* **2006**, *125*, 024106

[2] V. Ovchinnikov, M. Karplus and E. Vanden-Eijnden, "Free Energy of Conformational Transition Paths in Biomolecules: The String Method and Its Application to Myosin VI", *J. Chem. Phys.* **2011**, *134*, 085103

[3] L. Maragliano, E. Vanden-Eijnden and B. Roux, "Free Energy and Kinetics of Conformational Transitions from Voronoi Tesellated Milestoning with Restraining Potentials", *J. Chem. Theory Comput.*, **2009**, *5*, 2589

[4] E. E. Santiso and B. L. Trout, "A General Set of Order Parameters for Molecular Crystals", *J. Chem. Phys.* **2011**, *134*, 064109

[5] X. He, Y. Shen, F. R. Hung and E. E. Santiso, "Molecular Simulation of Homogeneous Nucleation of Crystals of an Ionic Liquid from the Melt", *J. Chem. Phys.* **2015**, *143*, 124506

[6] X. He, Y. Shen, F. R. Hung and E. E. Santiso, "Heterogeneous Nucleation from a Supercooled Ionic Liquid on a Carbon Surface", *J. Chem. Phys.* **2016**, *145*, 211919

# 486893 Using Transition Path Sampling to Determine the Mechanism of the Fcc-Hcp Phase Transformation in Solid Ar

**Abstract:**

We present an atomistic description of the fcc-to-hcp transformation mechanism in solid argon (Ar) obtained from transition path sampling molecular dynamics simulation. The phase transition pathways collected during the sampling for an 8000--particle system reveal three transition types according to the lattice deformation and relaxation details. In all three transition types, we see a critical accumulation of defects and uniform growth of a less ordered transition state, followed by a homogeneous growth of an ordered phase. Stacking disorder is discussed to describe the transition process and the cooperative motions of atoms in 111 planes. We investigate the nucleation with larger system. In a system of 18000--particles, the collective movements of atoms required for this transition are facilitated by the formation and growth of stacking faults. However the enthalpy barrier is still far beyond the thermal fluctuation. The high barrier explains previous experimental observations of the inaccessibility of the bulk transition at low pressure and its sluggishness even at extremely high pressure. The transition mechanism in bulk Ar is different from Ar nanoclusters as the orthorhombic intermediate structure proposed for the latter is not observed in any of our simulations.

# 492854 Using Metadynamics to Resolve and Characterize Complex Reactions at the Molecular Scale

**Abstract:**

While molecular dynamics simulations can be a powerful tool for characterizing the thermodynamics and kinetics of reaction mechanisms, they often come at a significant computational cost. Specifically, analyzing reaction networks with many pathways and states dramatically increases the computational expense of the simulations due to the vast landscape of the system. Enhanced sampling methods, such as the metadynamics (MetaD) family of methods,[1] are often employed to help mitigate this issue, but typically require a high level of insight about the chemistry involved in a system, making it difficult to use a tool for studying poorly understood systems. MetaD, specifically, relies on biasing a few degrees of freedom, or collective variables (CVs), that dictate the transitions between stable states. For complex reacting systems the number of relevant CVs can be very high, constraining the effectiveness of the bias, or difficult to identify due to lack of knowledge about the system.

In this study we discuss the use of generic CVs, i.e., those CVs that do not rely on a priori knowledge of a chemical transformation, to help ease the challenge of discovering both complex chemical reaction networks and the associated kinetics of pathways in the networks. We present how two different generic CVs, namely Social Permutation Invariant (SPRINT) coordinates[2] and collective variable-driven hyperdynamics (CVHD),[3] require a relatively low level of system knowledge to construct, but act as suitable CVs for estimating transition rates from biased simulations.[4] We biased these generic CVs, as well as typical ones derived from chemical intuition, following the infrequent MetaD method[5] to calculate the transition rates of two reaction systems: an SN2 reaction and a Diels Alder reaction. For both systems, we show that regardless of the biased CVs, consistent transition rates and energy barriers are recovered an ensemble of biased simulations, while reducing the simulation time. In addition, we demonstrate an approach for exploring large, complex reaction networks with high dimensionality by biasing the SPRINT coordinates of the atoms in a given system following the Parallel Bias MetaD method.[6] Using only knowledge of the initial state, we explore the reaction pathways of the decomposition of g-ketohydroperoxide described by the PM6 Hamiltonian. From analyzing an ensemble of biased simulations at various temperatures, we are able to construct a network of species that encompasses pathways consistent with the Korcek reaction mechanism,[7] as well as typical hydrocarbon chemistries. This presentation will close with notes on how these methods can be broadly applied to resolve and characterize the relevant pathways involved in poorly understood, complex reacting systems.

References:

(1) Valsson, O.; Tiwary, P.; Parrinello, M. Enhancing Important Fluctuations: Rare Events and Metadynamics from a Conceptual Viewpoint. *Annu. Rev. Phys. Chem.* **2016**, *67*, 159–184.

(2) Pietrucci, F.; Andreoni, W. Graph Theory Meets Ab Initio Molecular Dynamics: Atomic Structures and Transformations at the Nanoscale. *Phys. Rev. Lett.* **2011**, *107* (8), 85504.

(3) Bal, K. M.; Neyts, E. C. Merging Metadynamics into Hyperdynamics: Accelerated Molecular Simulations Reaching Time Scales from Microseconds to Seconds. *J. Chem. Theory Comput.* **2015**, *11* (10), 4545–4554.

(4) Fu, C. D.; Oliveira, L. F. L.; Pfaendtner, J. Assessing Generic Collective Variables for Determining Reaction Rates in Metadynamics Simulations. *J. Chem. Theory Comput.* **2017**, acs.jctc.7b00038.

(5) Tiwary, P.; Parrinello, M. From Metadynamics to Dynamics. *Phys. Rev. Lett.* **2013**, *111* (23), 230602.

(6) Pfaendtner, J.; Bonomi, M. Efficient Sampling of High-Dimensional Free-Energy Landscapes with Parallel Bias Metadynamics. *J. Chem. Theory Comput.* **2015**, *11* (11), 5062–5067.

(7) Jalan, A.; Alecu, I. M.; Aguilera-Iparraguirre, J.; Merchant, S. S.; Yang, K. R.; Merhcant, S. S.; Truhlar, D. G.; Green, W. H. New Pathways for Formation of Acids and Carbonyl Products in Low Temperature Oxidation. *J. Am. Chem. Soc.* **2013**, *135*, 11100–11114.

# 501792 Improving Ab Initio Molecular Dynamics Simulations of Water with Experimental Data

**Abstract:**

Accounting for electrons and nuclei simultaneously is a powerful capability of ab initio molecular dynamics (AIMD). However, AIMD is often unable to accurately reproduce properties of systems such as water due to inaccuracies in the underlying electronic density functionals. This shortcoming is often addressed by added empirical corrections and/or increasing the simulation temperature. We present here a maximum-entropy approach to directly incorporate limited experimental data via a minimal bias. Biased AIMD simulations of water and an excess proton in water are shown to give significantly improved properties both for observables which were biased to match experimental data and for unbiased observables. This approach also yields new physical insight into inaccuracies in the underlying density functional theory as utilized in the unbiased AIMD. The method is general and can be applied to simulations of other liquids with DFT. This talk will specificilly explore recent advances in maximum-entropy biasing of water density, a significant weakness of DFT water at 300K.

# 500284 Predicting Hydrogen-Deuterium Exchange Rates in Proteins Using Molecular Dynamics Simulations

**Abstract:**

Hydrogen-deuterium exchange (HDX) is a novel experimental method that uses exchange of hydrogen for deuterium as a probe for studying dynamics in proteins because certain hydrogen atoms in proteins, especially in highly flexible regions, are more prone to exchange by deuterium in solution than those in rigid structural motifs. Although HDX combined with mass spectrometry (HDX-MS) analysis can provide a resolution of deuterium incorporation at the level of peptide fragments, there is no direct route to obtaining residue-level exchange rates and hence their contribution to overall flexibility. Among many modeling approaches to predicting such residue-level exchange rates, molecular dynamics (MD) simulations are emerging as a highly useful tool. However, a large number of studies in the literature have used significantly variable criteria (solvent-accessible surface area, number of water molecules near proteins, hydrogen-bonds in the vicinity of water molecules or polar atoms) for predicting exchange rates and no generally applicable criteria have been discovered yet. In this work, we present a comprehensive analysis on prediction of HDX rates in three closely-related signaling proteins by carrying out multiple microsecond-long classical MD simulations. Specifically, we discover novel correlations between known criteria and the protection factor of each amino-acid by optimizing parameters using experimentally measured rates. These methods reveal dramatically different patterns of flexibility in these proteins that cannot inferred by their sequences or structures alone and suggest new ways of targeting them using small molecules. Our results have wide implications for studies on dynamics of other proteins, especially when unique modes of flexibility can be exploited to gain control of protein function via direct or allosteric inhibition.

# Using Semidefinite Programming to Calculate Bounds on Stochastic Chemical Kinetic Systems

**Garrett R. Dowdy and Paul I. Barton**

**Process Systems Engineering Laboratory**
**Department of Chemical Engineering**
**Massachusetts Institute of Technology**

**29-Oct-2017**

## Stochastic Chemical Kinetics

Summary:
Describes the time-evolution of reacting chemical systems.

Applies when the number of reacting molecules is small.

- On the order of 10s to 1000s rather than moles ($10^{23}$).
- Ex: single biological cell

Models reactions as discrete random events.

- Markov Process

Explains phenomena such as binary cell "decisions".

- Can't be explained through deterministic equations.

# Stochastic Chemical Kinetics

The state of the system is specified in a **state vector** listing the count of molecules of each species.

Ex: Reversible Dimerization

$$2A \xrightarrow{c_1} B$$

$$B \xrightarrow{c_2} 2A$$

The state may be

$$\mathbf{x} = \begin{bmatrix} x_1 \\ x_2 \end{bmatrix} = \begin{bmatrix} 4 \\ 1 \end{bmatrix}$$

4 molecules of $A$

1 molecules of $B$

# Stochastic Chemical Kinetics

Knowing

1. the initial state and
2. the reaction stoichiometry,

we can enumerate all **reachable states**.

Ex: For the system,

$$2A \xrightarrow{c_1} B$$

$$B \xrightarrow{c_2} 2A$$

with the initial condition

$$\mathbf{x} = \begin{bmatrix} 6 \\ 0 \end{bmatrix}$$

the reachable states are

| $x_1$ | 6 | 4 | 2 | 0 |
|---|---|---|---|---|
| $x_2$ | 0 | 1 | 2 | 3 |

The **set** of all reachable states:

$$X = \left\{ \begin{bmatrix} 6 \\ 0 \end{bmatrix}, \begin{bmatrix} 4 \\ 1 \end{bmatrix}, \begin{bmatrix} 2 \\ 2 \end{bmatrix}, \begin{bmatrix} 0 \\ 3 \end{bmatrix} \right\}$$

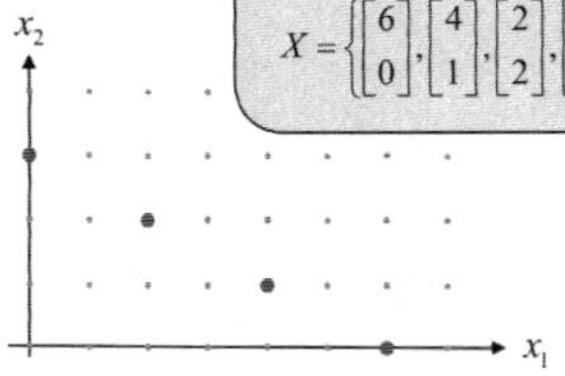

## Stochastic Chemical Kinetics

Each reaction is a random event.

⇒ Beyond the initial time, we don't know the state exactly.

⇒ We can only assign **probabilities** to each state $\mathbf{x} \in X$.

Ex: For our system, at time $t$, we might have

| $x_1$ | 6 | 4 | 2 | 0 |
|---|---|---|---|---|
| $x_2$ | 0 | 1 | 2 | 3 |
| $P(\mathbf{x},t)$ | 0.2 | 0.4 | 0.3 | 0.1 |

Graphically,

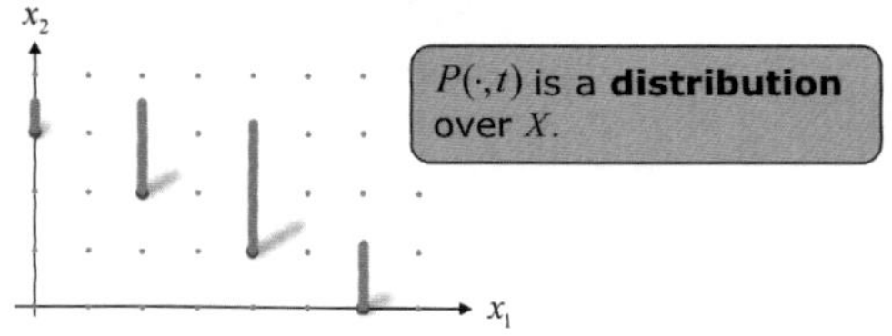


## Stochastic Chemical Kinetics

The probability distribution changes over time according to the **chemical master equation** (CME):

$$\frac{\mathrm{d}P}{\mathrm{d}t}(\mathbf{x},t) = \sum_{r=1}^{R} \left[ P(\mathbf{x}-\mathbf{s}_r,t)a_r(\mathbf{x}-\mathbf{s}_r) - P(\mathbf{x},t)a_r(\mathbf{x}) \right], \qquad \forall \mathbf{x} \in X$$

$$\Downarrow$$

Infinitesimal Generator Matrix

Linear Time-Invariant ODE

$$\frac{\mathrm{d}\mathbf{p}}{\mathrm{d}t}(t) = \mathbf{Q}\mathbf{p}(t)$$

Vector of probabilities for each state, $\mathbf{x} \in X$

Problem: There is often a huge (even infinite) number of states, $\mathbf{x} \in X$.

⇒ The dimensions of $\mathbf{p}$ and $\mathbf{Q}$ are too large.

⇒ We can't solve the problem.

## The Moment Approach

Often, we only need summary descriptions of the distribution:

- Mean molecular counts
- Variances

Alternative: model only the **moments** of the distribution:

$$y_{\mathbf{i}} = \sum_{\mathbf{x}\in X} \mathbf{x}^{\mathbf{i}} P(\mathbf{x},t)$$

where $\mathbf{i} = (i_1,\ldots,i_N)$ is a multi-index, and $\mathbf{x}^{\mathbf{i}} = x_1^{i_1}\ldots x_N^{i_N}$.

The **order** of $y_{\mathbf{i}}$ is the sum $|\mathbf{i}| = i_1 + \ldots + i_N$.

## The Moment Approach

Advantages:

Low-order moments provide a summary description of the distribution:

$y_{\mathbf{e}_j}(t) = \langle x_j(t) \rangle =$ Mean count of molecules of species $j$ at time $t$.

$y_{2\mathbf{e}_j}(t) - y_{\mathbf{e}_j}^2(t) = \sigma_j^2(t) =$ Variance in the count of molecules of species $j$ at time $t$.

The number of low-order moments scales better than the number of states.

## The Closure Problem

One can use the CME to derive an ODE describing how the moments change over time:

$$\frac{\mathrm{d}P}{\mathrm{d}t}(\mathbf{x},t)=\sum_{r=1}^{R}\left[P(\mathbf{x}-\mathbf{s}_r,t)a_r(\mathbf{x}-\mathbf{s}_r)-P(\mathbf{x},t)a_r(\mathbf{x})\right],\qquad \forall\mathbf{x}\in X$$

$$\Downarrow$$

$$\frac{\mathrm{d}\mathbf{y}_L}{\mathrm{d}t}=\mathbf{A}_L\mathbf{y}_L+\mathbf{A}_H\mathbf{y}_H$$

Constant matrices

A vector of all moments up to order $m$

Finitely many moments with order > $m$

We have the **closure problem**.

- Time derivatives of low order moments depend on higher order moments. $\Rightarrow$ Can't solve system.
- Occurs whenever we have bimolecular (or higher order) reactions.

Smadbeck, P., and Y. N. Kaznessis. "Efficient moment matrix generation for arbitrary chemical networks." *Chemical engineering science* 84 (2012): 612-618.

## The Closure Problem

The problem persists at steady state, where

$$\mathbf{0}=\mathbf{A}_L\mathbf{y}_L+\mathbf{A}_H\mathbf{y}_H$$

We'd like to solve this system to obtain $\mathbf{y}_L$ and extract

$$y_{\mathbf{e}_i}=\langle x_i\rangle_{\mathrm{ss}}\qquad y_{2\mathbf{e}_i}-y_{\mathbf{e}_i}^2=(\sigma_j^2)_{\mathrm{ss}}$$

Fewer equations than unknowns

$\Rightarrow$ We can't solve for the steady-state $\mathbf{y}$.

The equation $\mathbf{0}=\mathbf{A}_L\mathbf{y}_L+\mathbf{A}_H\mathbf{y}_H$ is only a **necessary condition**.

## The Bounding Paradigm

Instead of calculating $\langle x_i \rangle_{ss}$, let's calculate bounds:

$$\langle x_i \rangle_{ss}^{L} \leq \langle x_i \rangle_{ss} \leq \langle x_i \rangle_{ss}^{U}$$

Solve the optimization problem:

$$\langle x_i \rangle_{ss}^{U} \equiv \max_{\mathbf{y}} \; y_{\mathbf{e}_i}$$

$$\text{s.t. } \mathbf{y} \text{ satisfies necessary steady-state moment conditions.}$$

True $\mathbf{y}_{ss}$ is a feasible point.

$$\Rightarrow \langle x_i \rangle_{ss} \leq \langle x_i \rangle_{ss}^{U}$$

## Necessary Conditions

Dynamics:

$$\mathbf{0} = \mathbf{A}_L \mathbf{y}_L + \mathbf{A}_H \mathbf{y}_H$$

Total Probability:

$$y_{\mathbf{0}} = 1$$

Invariants:

$$\mathbf{B}\mathbf{y} = \mathbf{0}$$

Nonnegative Probability:

$$\mathbf{M}_n^{\mathbf{0}}(\mathbf{y}) \succeq \mathbf{0}$$

Nonnegative Molecular Counts:

$$\mathbf{M}_n^{\mathbf{e}_k}(\mathbf{y}) \succeq \mathbf{0}, \quad \forall k \in \{1,\ldots,N\}$$

## Bounding Steady-State Moments

Instead of calculating $\langle x_i \rangle_{ss}$, let's calculate bounds:

$$\langle x_i \rangle_{ss}^L \leq \langle x_i \rangle_{ss} \leq \langle x_i \rangle_{ss}^U$$

Solve the optimization problem:

$$\langle x_i \rangle_{ss}^U \equiv \max_{\mathbf{y}} \ y_{\mathbf{e}_i}$$
$$\text{s.t. } \mathbf{y} \text{ satisfies necessary steady-state moment conditions.}$$

## Bounding Steady-State Moments

Instead of calculating $\langle x_i \rangle_{ss}$, let's calculate bounds:

$$\langle x_i \rangle_{ss}^L \leq \langle x_i \rangle_{ss} \leq \langle x_i \rangle_{ss}^U$$

Solve the optimization problem:

$$\begin{aligned} \langle x_i \rangle_{ss}^U \equiv \max_{\mathbf{y}} \ & y_{\mathbf{e}_i} \\ \text{s.t. } & \mathbf{A}_L \mathbf{y}_L + \mathbf{A}_H \mathbf{y}_H = \mathbf{0}, \quad y_{\mathbf{0}} = 1, \\ & \mathbf{B}\mathbf{y} = \mathbf{0}, \\ & \mathbf{M}_n^{\mathbf{0}}(\mathbf{y}) \succeq \mathbf{0} \\ & \mathbf{M}_n^{\mathbf{e}_k}(\mathbf{y}) \succeq \mathbf{0}, \quad \forall k \in \{1, \ldots, N\}. \end{aligned}$$

Similar SDPs provide $\langle x_i \rangle_{ss}^L$ and $(\sigma_i^2)_{ss}^U$

This is a **semidefinite program** (SDP)

Lasserre, Jean Bernard. *Moments, positive polynomials and their applications*. Vol. 1. World Scientific, 2009.

## Mean Bounds Example

Ex:

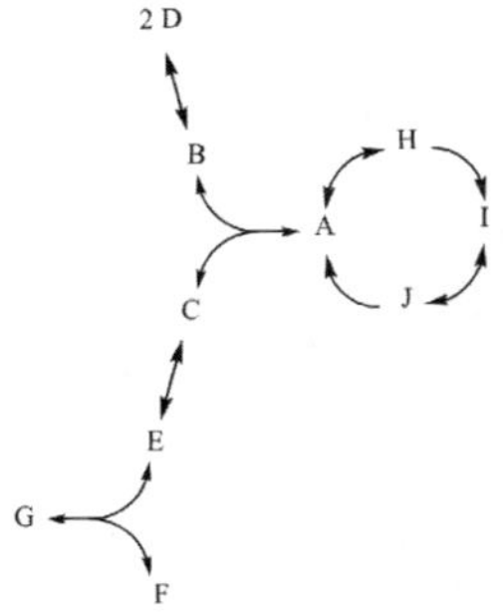


$A_0 = 100, \quad F_0 = 100,$

All other initial counts 0.

74,816,108,146 States

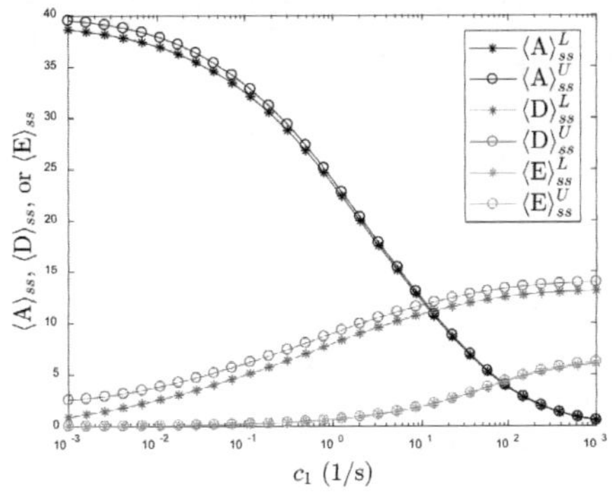


## Variance Bounds Example

Ex:

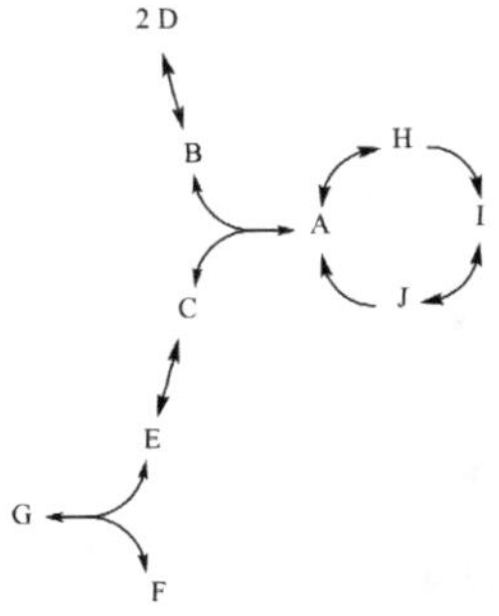


$A_0 = 100, \quad F_0 = 100,$

All other initial counts 0.

74,816,108,146 States

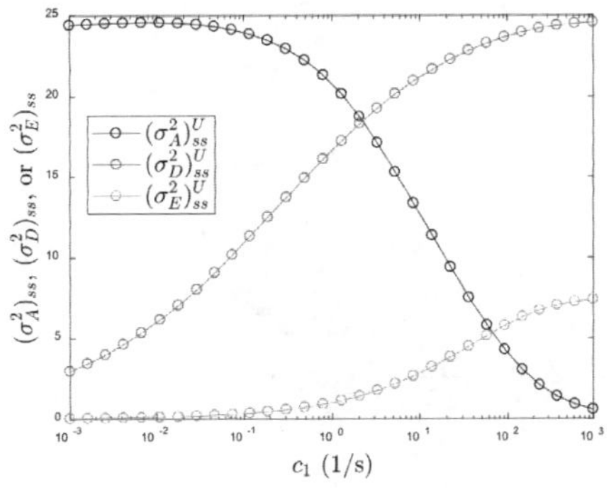

# Histogram Bounds Example

Ex:

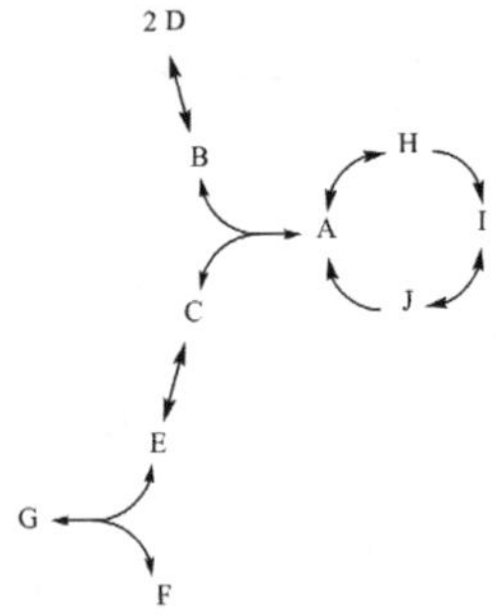


$A_0 = 100, \quad F_0 = 100,$

All other initial counts 0.

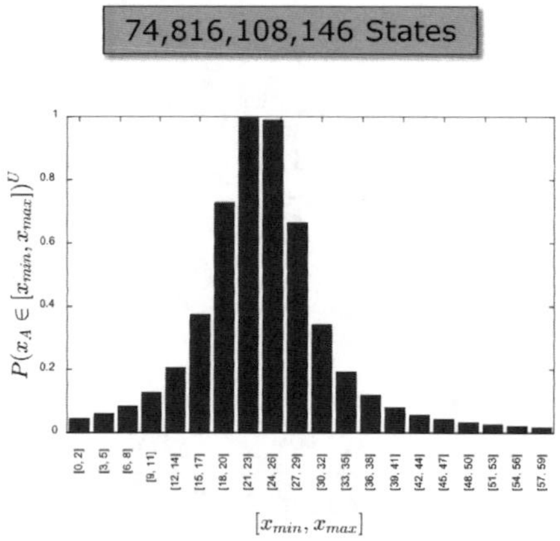


# Dynamic Bounds

$P(\cdot,t)$ or $\lim_{t\to+\infty} P(\cdot,t)$

Distribution over state space

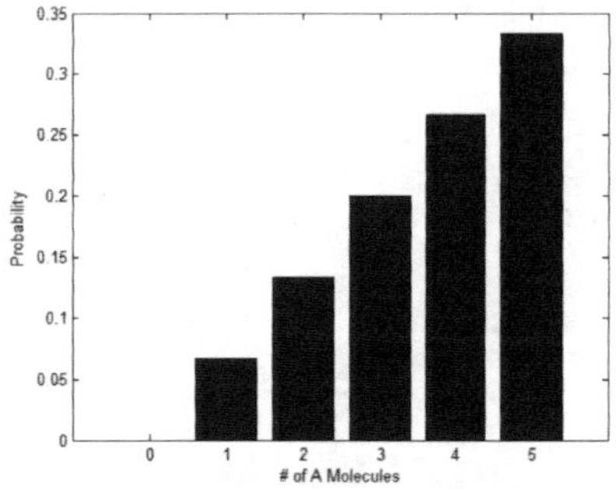


$P(\cdot,\cdot)$

"Distribution" over state space and time

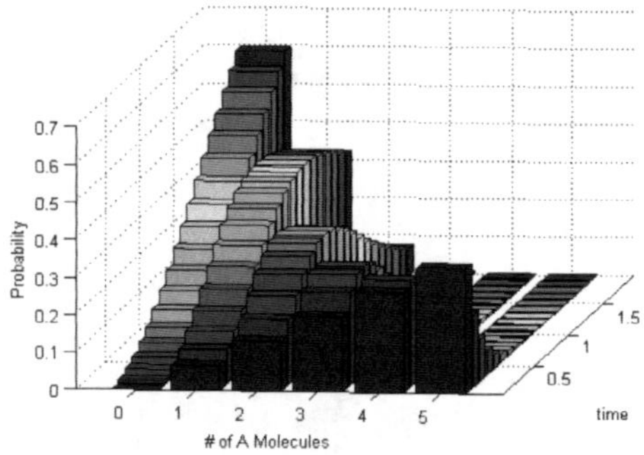


Bertsimas, Dimitris, and Constantine Caramanis. "Bounds on linear PDEs via semidefinite optimization." *Mathematical programming* 108.1 (2006): 135-158.

## Dynamic Bounds

Define the moments of $P(\cdot,\cdot)$ as

$$z_{\mathbf{i}}^{(\rho)} = \int_0^t e^{\rho(t-\tau)} \sum_{\mathbf{x}\in X} \mathbf{x}^{\mathbf{i}} P(\mathbf{x},\tau) d\tau$$

for all $\mathbf{i} = (i_1,\ldots,i_N)$ and $\rho \in \{\rho_1,\ldots,\rho_d\}$.

Necessary conditions:

| | |
|---|---|
| Dynamics: | $\mathbf{y}_L(t) - e^{\rho t}\mathbf{y}_L(0) = (\mathbf{A}_L - \rho\mathbf{I})\mathbf{z}_L^{(\rho)} + \mathbf{A}_H\mathbf{z}_H^{(\rho)}$ |
| Invariants: | $\tilde{\mathbf{B}}\mathbf{z} = \mathbf{0}$ |
| nonnegative: | $\mathbf{M}_n^{(0,\mathbf{0})}(\mathbf{z}) \succeq \mathbf{0}$ |
| Nonnegative molecular counts: | $\mathbf{M}_n^{(0,\mathbf{e}_k)}(\mathbf{z}) \succeq \mathbf{0}, \ \forall k \in \{1, \ldots, N\}$ |

Bertsimas, Dimitris, and Constantine Caramanis. "Bounds on linear PDEs via semidefinite optimization." *Mathematical programming* 108.1 (2006): 135-158.

## Dynamic Bounds

$$\begin{aligned}
\langle x_i(t)\rangle^U \equiv & \max_{\mathbf{y}(t),\mathbf{z}} y_{\mathbf{e}_i}(t) \\
& \text{s.t. } \mathbf{y}_L(t) - e^{\rho t}\mathbf{y}_L(0) = (\mathbf{A}_L - \rho\mathbf{I})\mathbf{z}_L^{(\rho)} + \mathbf{A}_H\mathbf{z}_H^{(\rho)}, \ \forall \rho \in \{\rho_1,\ldots,\rho_d\}, \\
& \tilde{\mathbf{B}}\mathbf{z} = \mathbf{0}, \ \mathbf{B}\mathbf{y}(t) = \mathbf{0} \\
& \mathbf{M}_n^{(0,\mathbf{0})}(\mathbf{z}) \succeq \mathbf{0}, \\
& \mathbf{M}_n^{(0,\mathbf{e}_k)}(\mathbf{z}) \succeq \mathbf{0}, \ \forall k \in \{1, \ldots, N\}, \\
& \mathbf{M}_n^{\mathbf{0}}(\mathbf{y}(t)) \succeq \mathbf{0}, \\
& \mathbf{M}_n^{\mathbf{e}_k}(\mathbf{y}(t)) \succeq \mathbf{0}, \ \forall k \in \{1, \ldots, N\},
\end{aligned}$$

Similar SDPs provide $\langle x_i(t)\rangle^L$ and $(\sigma_i^2(t))^U$

MIT

PSEL

# Dynamic Bounds

Ex:

$$A+B \xrightarrow{c_1} C \xrightarrow{c_2} D, \qquad \rho \in \{0, -c_1, -c_2\}$$

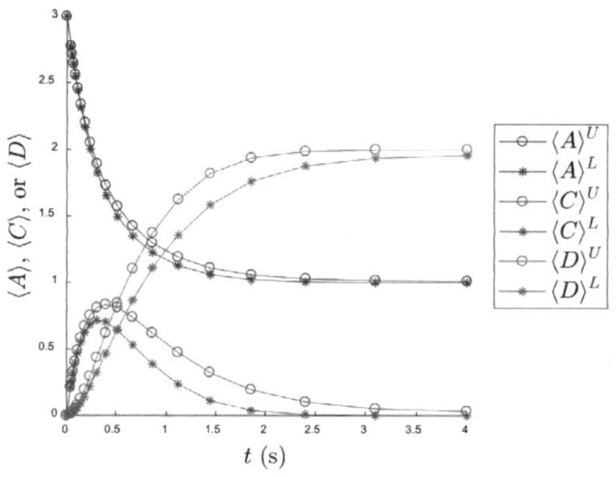


MIT

PSEL

# Summary

Calculated **bounds** on probability distributions in **stochastic chemical kinetics**.

Used **semidefinite programming**, with moments as decision variables.

Analyzed both **steady-state** and **dynamic** problems.

Applied the bounding method to a **nontrivial** model.

The method is a **work in progress**.

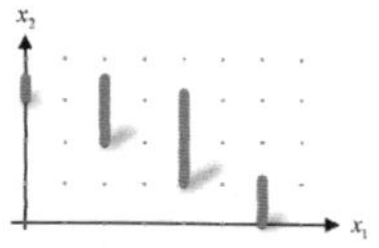


$$\begin{aligned} \langle x_i \rangle_{ss}^U \equiv \max_{\mathbf{y}} \ & y_{\mathbf{e}_i} \\ \text{s.t.} \ & \mathbf{Ay} = \mathbf{0}, \quad y_{\mathbf{0}} = 1, \\ & \mathbf{By} = \mathbf{0}, \\ & \mathbf{M}_n^{\mathbf{0}}(\mathbf{y}) \geq \mathbf{0} \\ & \mathbf{M}_n^{\mathbf{e}_k}(\mathbf{y}) \geq \mathbf{0}, \quad \forall k \in \{1, \ldots, N\}. \end{aligned}$$

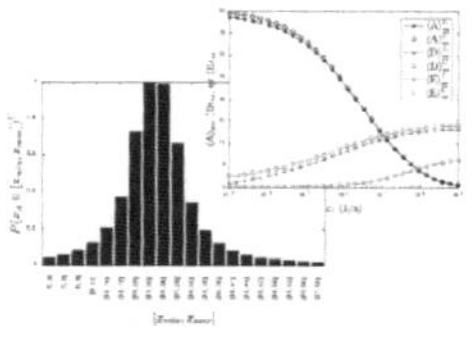

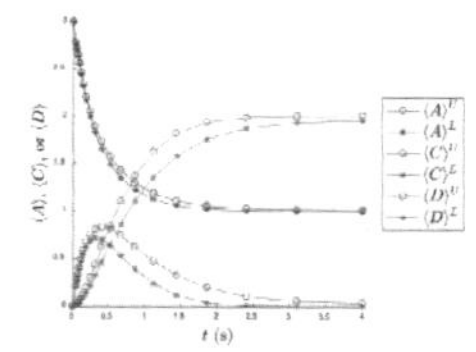

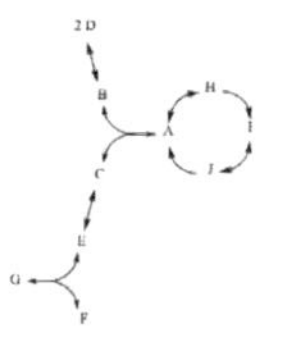

$7.4 \times 10^{10}$ States

## Acknowledgements

Advisor: Paul Barton

Jean Bernard Lasserre

Funding

NOVARTIS


---

## Questions?

**Email: gdowdy3@mit.edu**

# 500087 Automated Discovery of Reaction Pathways for the Combustion of Alternative Fuel Candidates

**Abstract:**

A primary culprit of emission-related health issues is the soot produced via the combustion of common fuels. This soot, often formed from polyaromatic compounds, has been observed to irritate tissue, and can also be a carrier for mutagenic and carcinogenic organics.[4,7] Because of the severity of this problem, accurate and detailed analysis of these combustion byproducts and how they are formed has become increasingly necessary.

We use reactive molecular dynamics to examine soot formation pathways starting with anisole, $C_7H_8O$, and toluene, $C_7H_8$.[2,3] Fuel and oxygen molecules are studied at different ratios, temperatures between 2000 K and 2400 K, and pressures between 187 atm and 224 atm. The system is simulated using LAMMPS with the ReaxFF force field developed for hydrocarbons by Chenoweth et al.[1] Thermodynamic and positional data is recorded throughout the length of the trials, and can then be passed to the ChemTraYzer software package for analysis of reaction pathways and pertinent reaction data.[3] The output of this analysis is a reaction network that can be visualized from the parent fuel through various radical pathways to the final product of the reaction, thereby allowing for accounting of all molecules in the reaction chain through various physical conditions and constraints.

References

(1) Chenoweth, K.; van Duin, A.C.T.; Goddard III, A.; ReaxFF Reactive Force Field for Molecular Dynamics Simulations of Hydrocarbon Oxidation. *J. Phys. Chem.* **2008** 112 (5), pp 1040-1053

(2) Döntgen, M.; ChemTraYzer – Reaction Models from Molecular Dynamics Simulations.

(3) Döntgen, M.; et al,; Automated Discovery of Reaction Pathways, Rate Constants, and Transition States Using Reactive Molecular Dynamics Simulations. *J. Chem. Theory Comput* **2015**, *11* (6), pp 2517–2524

(4) Lighty, J.; Veranth, J.; Sarofim, A.; Combustion Aerosols: Factors Governing Their Size and Composition and Implications to Human Health. *J. Air and Waste Management Association* **2000**, 50 (9), pp 1565-1618

(5) Plimpton, S.; Fast Parallel Algorithms for Short-Range Molecular Dynamics. *J Comp Phys* **1995** 117, pp 1-19 (http://lammps.sandia.gov)

(6) Richter, H.; Howard, J.B.; Formation of polycyclic aromatic hydrocarbons and their growth to soot—a review of chemical reaction pathways. *Progress in Energy and Combustion Science* **2000**, 26, pp 565–608

# 501706 A Generic Coarse-Grained Model of Influenza Budding: What Can We Learn?

**Abstract:**

Viral release of influenza is believed to occur independently of the usual cellular machinery of endosomal sorting complexes required for transport (ESCRT); instead, scission of the extruding bud is facilitated by the M2 protein. [1] The shape of the M2 transmembrane domain is distinctly conical and the assumed mechanism by which scission proceeds is that M2 aggregates in the budding neck to cause constriction that eventually pinches off the enveloped virus.

Constructing a physiologically inspired, minimal physical model of viral budding that can elucidate molecular mechanisms is nontrivial. In the spirit of Einsteinâ€™s famous quote: â€œEverything should be made as simple as possible, but no simpler,â€ we will discuss model design principles, appropriate modelistic idealizations, technical challenges, and validation exemplified by the chosen application in virology.

REFERENCES

[1] Rossman et al. Cell 141, pp. 902-913, 2010

# 492019 Allosteric Effects of Gold Nanoparticles on Human Serum Albumin

**Abstract:**

The ability of nanoparticles to alter protein structure and dynamics plays an important role in their medical and biological applications. We investigate allosteric effects of gold nanoparticles on human serum albumin protein using molecular simulations. The extent to which bound nanoparticles influence the structure and dynamics of residues distant from the binding site is analyzed. The root mean square deviation, root mean square fluctuation and variation in the secondary structure of individual residues on a human serum albumin protein are calculated for four protein-gold nanoparticle binding complexes. The complexes are identified in a brute-force search process using an implicit-solvent coarse-grained model for proteins and nanoparticles. They are then converted to atomic resolution and their structural and dynamic properties are investigated using explicit-solvent atomistic molecular dynamics simulations. The results show that even though the albumin protein remains in a folded structure, the presence of a gold nanoparticle can cause more than 50% of the residues to decrease their flexibility significantly, and approximately 10% of the residues to change their secondary structure. These affected residues are distributed on the whole protein, even on regions that are distant from the nanoparticle binding site. Our simulations suggest that allosteric effects must be considered when designing and deploying nanoparticles in medical and biological applications that depend on protein-nanoparticle interactions.

# 500931 Self-Assembly of Proteins: The Role of Shape and Specific Interaction

**Abstract:**

Crystallization is routinely exploited for understanding protein structure, but the molecular details of how proteins crystallize remain poorly understood. Using computer simulations, we investigate the role of patchy interactions and protein shape in self-assembly. We demonstrate that for green fluorescent protein with its beta-barrel shape, a patchy sphere model is not sufficient to explain the experimentally observed unit cell, but a model that accounts for anisotropic shape can reproduce protein crystal growth. We explore the role of rigid molecular shape in simulations of hard proteins represented solely by their solvent excluded surfaces. Hard convex polyhedra are known to crystallize into a multitude of close-packed structures, but for concave shapes such as proteins much less is known. In our simulations we observe metastable order, which could shed light on the role of shape in protein crystal nucleation.

# 497188 Predictive Design of Next-Generation Nanomaterials and Devices Via Bottom-up Approaches

**Abstract:**

Nanomaterials and devices that resemble biological matter in their ability to reconfigure and adapt on demand have captured increasingly growing interest over the past decades. Towards this end, bottom-up approaches including self- and directed-assembly techniques have been shown as a promising means for engineering the underlying nanostructure of this exciting class of materials and devices. In these approaches, the underlying nanostructures are obtained from the self- or directed-organization of nano building blocks such as block copolymers, nanoparticles and colloids. The fundamental challenges to the bottom-up techniques are to design the optimal assembling units, to tailor their effective interactions and to find efficient assembly pathways. In this contribution, I will address the design rules for the building blocksâ€™ geometry and their interactions targeting several hierarchically assembled structures including terminal supraparticles, helical ribbons and various columnar morphologies. I will also demonstrate numerous unconventional pathways to assemble these bio-mimicking and reconfigurable nanostructures including interaction tuning to shape shifting. The ultimate goals of my research program are to design of adaptive, programmable nanomaterials that are of potential use in a host of nanotechnology applications including, but not limited to, optical devices, drug delivery, biosensing and energy storage and conversion. Also presented are the tools and methods I have been developing to improve the efficiency of the computational studies of interest, ranging from large-scale GPU-accelerated codes to enhanced sampling methods.

# 492812 Solvation of Self-Assembled Complexes: Using Molecular Simulations to ProbeÂ Energetics, Structure, and Dynamics

**Abstract:**

Single walled-carbon nanotubes (SWCNTs) possess unique properties that make them attractive for a number of applications. However, the difficulty in obtaining uniform samples in terms of size, chirality, and handedness inhibits their widespread use. Various separation techniques address this problem through use of surfactants in order to suspend the tubes in aqueous media and then apply other approaches such as aqueous two phase extraction (ATPE) or ion-exchange chromatography (IEX) to sort the dispersed SWCNTs by their physiochemical properties. Single-stranded DNA (ssDNA) is a very effective dispersant that shows sequence-specific behavior during separation. Previous experimental work has shown that this specificity can be harnessed to tune the separation in favor of particular nanotube chiralities. Currently, the nature of this specificity is not yet well understood and optimal ssDNA/SWCNT pairs must be searched by trial and error. In this study, we use replica exchange molecular dynamics (REMD) simulations to assemble the ssDNA-SWCNT complexes and then apply free energy perturbation methods to obtain an estimate for the solvation energy of the complex. Such systems require the construction of a novel free energy pathway to access the quantity of interest. Additionally, the structure and dynamics of the hydrating water has been analyzed to better understand the solvation behavior on a molecular level and to provide insight into the effects of ssDNA sequence. These investigations work towards our goal to provide better insight into the sequence/chirality specific separation mechanism, and to eventually develop a model that allows for the prediction of other pairs leading to the efficient sorting of a desired nanotube chirality.

# 495168 Level Set Strategy for Self Consistent Field Theory

**Abstract:**

We introduce level set methods to design new algorithms for self-consistent field theory (SCFT) in polymer physics. SCFT computes the structure and energy of inhomogeneous self-assembling polymers at thermodynamic equilibrium. Level set methods enable to represent the surface of an enclosing domain with an implicit function which in turn provides a powerful way to evolve a free surface. We introduce the concept of *shape derivative* into SCFT and rigorously derive expressions for the change of energy of a diblock copolymer melt with respect to its enclosing shape. The shape derivative is then used to embed SCFT into a variable shape simulation where the internal structure and the enclosing shape are coupled and evolve in tandem in order to reduce the energy. Finally an algorithm for the inverse geometric problem is presented. The algorithm finds a shape in order to obtain a desired internal structure of the confined polymeric material.

# 501391 Simulations of Nonlinear Flows in Nonequilibrium Complex Liquids

**Abstract:**

Complex liquids, such as active biofluids and colloidal suspensions, are important for biophysical and other industrial applications. Here we focus on dense suspensions of semi flexible bioplomers, such as microtubule and actin filaments. When the biopolymer concentration is high, it exhibits nematic phase. And if motor protein clusters and adenosine triphosphate (ATP) are added to the suspension, biopolymers are able to slide on each other due to the action of motors - such collective motion leads to a spontaneous, cohesive flow in the system. The nature of such flows is not fully elucidated. Here we rely on a continuum hydrodynamic model of active nematics to characterize such nonequilibrium liquids. The model is realized via a hybrid lattice Boltzmann solver. We first validate the model by comparing it to benchmark problems of passive nematic liquid crystal flows and active nematics experiments. We then extend it to include unequal elasticities and local concentration variations. The model is able to describe actin-based active nematics with variable activity and elasticity. We further apply it to explain the ordered flow emerged from confinements and anisotropic boundaries. Based on the above results, we propose that active nematics can be the candidate for active microfluidic devices. Finally, we investigate the transient flows in dense suspensions. Discrete element method is used to simulate a suspension of bi-disperse non-Brownian spherical particles. As the system is subject to a sudden simple shear, dynamic jamming happens as there is a propagating wave moving with a constant speed. We focus on how the inter particle frictional force and the short-range repulsive force determine the formation of the front and its velocity. Our simulations provide detailed structures of the suspension at transient state and therefore helpful for elucidating the transient dynamics of dense suspensions.

# 491508 Employing a Multipole Approximation in a Hybrid Fluid Via Relative Resolution

**Abstract:**

Recently, we presented a novel framework for multiscale simulations, termed "Relative Resolution" (RelRes) [1]. Inspired by another hybrid approach [2], our algorithm also contains both Fine-Grained (FG) and Coarse-Grained (CG) models in asingle fluid system. However, a unique feature of RelRes is that it switches between molecular resolution in terms of relative separation: While nearest neighbors are characterized by a FG (geometrically detailed) model, other neighbors are characterized by a CG (isotropically simplified) model. We in turn make important connections with classical theories in statistical mechanics [3]. Instead of performing an iterative procedure for relating the FG and CG models [4], we formulate for this task an analytical expression which is based on a multipole approximation at appropriate distances: This algebraic equation stems in the energy conservation between a molecular pair in an infinite limit, and it substantially enhances the computational efficiency of RelRes. With this hybrid formalism for fluid mixtures, we consequently examine several systems in molecular simulations (e.g., a liquid-liquid of oxygens-nitrogens, a liquid-gas of tetrachloromethanes-thiophene, etc.). We primarily focus on structural correlations and on thermal properties, correctly retrieving their corresponding behavior across state space. As a next step, we also proceed by implementing RelRes for polymers. In summary, we show here for fluids, that molecular resolution is inherently hybrid in terms of relative separation.

[1] A. Chaimovich, C. Peter, and K. Kremer. Relative resolution: A hybrid formalism for fluid mixtures. The Journal of Chemical Physics 143:243107, 2015.

[2] M. Praprotnik, L. Delle Site, and K. Kremer. Adaptive resolution molecular-dynamics simulation: Changing the degrees of freedom on the fly. The Journal of Chemical Physics 123:224106, 2005.
[3] B. Widom. Intermolecular forces and the nature of the liquid state. Science 157:375-382, 1967.
[4] M. G. Saunders and G. A. Voth. Coarse-graining methods for computational biology. Annual Review of Biophysics 42:73-93, 2013.

# 489615 Quantitatively Reliable Molecular Modeling and Simulation of Vapor-Liquid Equilibria

**Abstract:**

For a variety of low-molecular fluids, two-center Lennard-Jones plus point dipole (2CLJD) and two-center Lennard-Jones plus point quadrupole (2CLJQ) models, which describe bulk properties of pure-component vapor-liquid equilibria with a good accuracy, are available from the literature [1]. In the present work, these models are applied for simulating the vapor-liquid surface tension. Thereby, the literature models, which were adjusted to bulk properties, but not to interfacial properties, are validated against the surface tension of the real fluids.

By molecular dynamics simulation, employing a recently developed efficient long-range correction scheme [2, 3], it is found that on average, the literature models overestimate the vapor-liquid surface tension by 15 to 20% [3, 4]. To improve the overall accuracy of the models, optimization criteria based on the vapor pressure and the saturated liquid density, i.e., bulk properties which were previously taken into account during model parameterization, are combined with the vapor-liquid surface tension, which was not considered when the literature models were developed. This multicriteria optimization problem is addressed by constructing the Pareto front, i.e., the set of rational compromises between the competing objectives, from which models tailored to special needs can subsequently be chosen [5].

[1] J. Vrabec, J. Stoll, H. Hasse, J. Phys. Chem. B 105(48), 12126-12133, 2001.
[2] S. Werth, G. Rutkai, J. Vrabec, M. Horsch, H. Hasse, Mol. Phys. 112(17), 2227-2234, 2014.
[3] S. Werth, M. Horsch, H. Hasse, Mol. Phys. 113(23), 3750-3756, 2015.
[4] S. Werth, M. Horsch, H. Hasse, J. Chem. Phys. 144, 054702, 2016.
[5] K. Stöbener, P. Klein, M. Horsch, K. Küfer, H. Hasse, Fluid Phase Equilib. 411, 33-42, 2016.

# 505008 Atomistic Simulations of Ordered and Disordered Carbons: "Mimicking" Versus "Targeting"

**Abstract:**

Atomistic simulation techniques have become extremely valuable tools for the elucidation of the structure of disordered or (partially) ordered materials. These techniques not only allow for the production of large-scale three-dimensional representations of the materials from the atomic scale, but also to predict most of their properties, and in some cases, to assess their formation mechanisms. In this talk, I will focus on carbon materials that can be ordered as well as disordered. There two routes to obtain realistic molecular-scale reconstruction of porous carbons: mimicking their formation process and targeting a â€œfinalâ€ state by constraining/guiding the simulation using experimental texture information. Obviously the first is a priori more predictive than the second but requires dealing upfront with chemistry (through reactive force fields) than can be rather complex with kinetically hindered processes.

To illustrate the mimicking route, I will present (i) the Grand-Canonical numerical synthesis of ordered carbon phases by carbon vapor deposition on zeolites and MCM-type mesoporous oxides, (ii) the Replica Exchange MD simulation of the geological degradation of the main components of wood (lignin and cellulose that are both organic molecules) into disordered nanoporous carbons and kerogen phases with the production of a complex liquid phase. The so-called targeting route will be illustrated with the example of the kerogen phases using the Hybrid-Reverse Monte-Carlo approach with an extension towards their mechanical properties from elasticity to failure (fracture).

# 505007 Graphene Oxide Membranes: A Molecular Simulation Approach

**Abstract:**

Graphene oxide membranes have attracted a lot of attention from the scientific community due to their potential application in challenging separations. Nevertheless, to optimise a material for an engineering application, it is necessary to understand their structures and the features that can improve their performance. Following on Keith Gubbins philosophy that realistic models of materials at a molecular level can be obtained from mimicking some aspects of their synthesis using molecular simulations, we have developed a protocol to create model graphene oxide sheets and mimic the membrane formation procedure to obtain model materials, with properties comparable to those measured experimentally.

# 495702 Polyethylene Oxide (PEO) in a Polyethylene (PE) Framework: A Simple Model for Simulation Studies of Scaling and Solvent Effects on Polymers in an Open Framework

**Abstract:**

We report molecular dynamics simulations studies of a polyethylene oxide polymer in a cubic polyethylene framework. Radius of gyration, shape, and other structural parameters are computed. Self-diffusion and displacement relative to the framework are also computed to investigate mobility and transport behavior. To investigate scaling effects, polyethylene oxide polymers of three different lengths and framework of three different sizes are simulated. Explicit water molecules are introduced to investigate solvation effects in different dimensions of polymer and framework.

Confinement of the polymer in smaller framework lead to closer interactin of polymer and framework and a more elongated configuration. The elongated configuration is favorable for diffusion through small windows of the framework. A larger mean square displacement is observed for the smaller framework, particularly in the presence of water. Adjusting the framework interaction potential sees a critical threshold dividing normal diffusion behavivoir and a regime dominated by hopping between cavities of the framework.

Although the model framework contains only simple van der Waalsl groups, confinement and solvent effects on polymer observed agree qualitatively with those in experiments reported recently for polyeletrolyte threaded in metal organic frameworks [1,2]

[1] L. Gao, C.Y.V. Li, and K.Y. Chan, *Chem. Mater.* 27(10) (2015) 3601-3608.

[2] L. Gao, C.Y. V. Li, K.Y. Chan, and Z.N. Chen, *J. Am. Chem. Soc.* 136 (2014) 7209-7212.

# 505006 Molecular Understanding and Design of Zwitterionic Materials


**Abstract:**

An important challenge in many applications, ranging from medical devices to ship hulls, is the prevention of nonspecific protein adsorption on surfaces. To address this challenge, our goals are twofold. First, we strive to provide a fundamental understanding of nonfouling mechanisms at the molecular level using an integrated experimental and simulation approach. Both simulation and experimental results show that the strong hydration of zwitterionic materials is responsible for their excellent nonfouling properties (JACS, **127**, 14473, 2005). Further studies examine zwitterionic structures and reveal the role of these structures in nonfouling properties beyond hydration (Advanced Materials, **27**, 15, 2015). Second, we aim to develop biocompatible and environmentally benign ultra low fouling materials based on the molecular principles learned. We have demonstrated that zwitterionic and mixed charge materials and surfaces are highly resistant to nonspecific protein adsorption, cell adhesion and bacteria adhesion/biofilm formation from complex media. Recent results show that zwitterionic materials induce no capsule formation upon implantation (Nature Biotechnology, **31**, 553, 2013) and no immunological response in blood circulation (PNAS, **112**, 12046, 2015). Furthermore, these materials are shown to preserve protein (Nature Chemistry, 4, 59, 2012) and cell (Angewandte Chemie, **53**, 12729, 2014) bioactivity. Because of these unique properties, zwitterionic materials are excellent alternatives to widely used poly(ethylene glycol) (PEG)-based materials.

# 504992 Interfacial Transport of Protons on a 2-Dimensional Functionalized Graphane Surface

**Abstract:**

Proton transport is crucial for many industrial and biological processes. The movement of protons is most often facilitated by the presence of water because water forms an extended hydrogen bonded network allowing for fast hopping of protons from one molecule to another. We have developed a novel 2-dimensional material that is capable of facilitating fast proton transport in the complete absence of water. Anhydrous proton transport is important for applications such as high- and intermediate-temperature fuel cell membranes. Our material was designed and tested using first principles density functional theory calculations. The material is based on graphane, which is an sp3 analogue of graphene. We have functionalized the graphane surface with hydroxyl groups and have used ab initio molecular dynamics and transition state calculations to study the mechanism and barriers of proton transport. We examine the limit of 1-dimensional proton hopping along a line of OH groups and the limit of fully hydroxylated graphane, forming 2-dimensional hydrogen bonding networks. We report diffusivities and transition state theory calculations. We find that anhydrous functionalized graphane has higher transport rates than hydrated Nafion.

# 504993 Gas Adsorption Behavior in Ionic Polyimide Composite Membranes

**Abstract:**

Industrial gas separation and storage processes present many technical and economic challenges to the natural gas industry and to power plants. In particular, palliating $CO_2$ from emission sources is a critical need in industry, and it is necessary in order to meet current and future environmental regulations. Polyimides are at the forefront of advanced membrane materials for $CO_2$ capture and gas purification processes. Recently, â€œionic polyimidesâ€ (i-PIs) have been reported as a new class of condensation polymers which combine structural components of both ionic liquids (ILs) and polyimides through covalent linkages. In this study, we report $CO_2$ and $CH_4$ adsorption and structural analyses of an i-PI and an i-PI + IL composite containing [$C_4$mim][$Tf_2$N]. The combination of molecular dynamics (MD) and grand canonical Monte Carlo (GCMC) simulations are used to compute the gas solubility and the adsorption performance with respect to the density, fractional free volume (FFV), and surface area of the materials.

Our results highlight the polymer relaxation process, and its correlation to the gas solubility. In particular, the surface area can provide meaningful guidance with respect to the gas solubility, and it tends to be a more sensitive indicator of the adsorption behavior versus only considering the system density and FFV. For instance, as the polymer continues to relax, the density, FFV, and pore-size distribution remain relatively constant, while the surface area can continue to increase, enabling more adsorption. Structural analyses are also conducted to identify the nature of the gas adsorption once the ionic liquid is added to the polymer. The presence of the IL significantly displaces the $CO_2$ molecules from the ligand nitrogen sites in the neat i-PI to the imidazolium rings in the i-PI + IL composite. Whereas, the $CH_4$ molecules move from the imidazolium ring sites in the neat i-PI to the ligand nitrogen atoms in the i-PI + IL composite. These molecular details can provide critical information for the experimental design of highly selective i-PI materials, as well as provide additional guidance for the interpretation of the simulated adsorption systems.

# 493614 Computational Simulation of Supported Nanocatalysts Under Realistic Conditions

**Abstract:**

Supported noble metal catalysts have played an important role in the conversion of energy and resources. Computational simulations ranging from quantum chemistry density functional calculations, molecular dynamic have been widely used to investigate the reaction mechanisms and the relationship between structure and properties of nanocatalysts. However, there are still huge gaps between computational simulations and realistic experiments. During the recent years, we try to computational simulate the nanocatalysts by consideration of realistic conditions in order to reduce the pressure and materials gap.

# 498780 Metastable Phase Transitions

**Abstract:**

Several model systems exhibit a low-temperature first-order transition between two disordered phases that are metastable with respect to an ordered phase. The preponderance of experimental evidence is consistent with the possibility that water may behave in this way. The confluence of structural, kinetic and thermodynamic factors that make this interesting phase behavior possible is an active area of inquiry. I will present a theoretical and computational overview of metastable phase transitions, including discussions of finite size effects, the interplay of nucleation and relaxation times, the stability of interfaces, and metastable criticality.

# 498824 Computational Investigation of Multipolar Colloidal Particle

**Abstract:**

Colloids with anisotropic charge distributions hold promise for creating a number of useful new materials including optic materials with novel symmetries, materials for information storage, and dampers for controlling vibrations in structures. Experimentally manipulating properties of anisotropic particles in a controlled fashion can sometimes be difficult and is time consuming. The search for novel anisotropic colloidal materials can be enhanced and even guided by simulations of colloidal system assembly. We report the results of two projects. In the first we use discontinuous molecular dynamics (DMD) simulation to investigate how the internal charge separation of anisotropic, dipolar colloidal rods with 4:1 aspect ratio affects their phase behavior. The differences between the head-tail and side-side ordering displayed in these simulations could be exploited to design gels whose properties (especially those related to percolation: viscosity, conductivity, etc.) are responsive to external stimuli. In the second project we perform quasi-2D Monte Carlo simulations of systems of colloidal spheres with off-center, extended dipoles. These colloidal spheres are found to aggregate into cyclical structures whose interior angle is equal to the angle between the charges on a single dipolar sphere.

# 498803 Interfacial Tensions from SAFT: Connecting Equations of State to Molecular Simulations

**Abstract:**

A description of fluid systems with molecular-based algebraic equations of state (EoSs) and by direct molecular simulation is common practice in chemical engineering and the physical sciences, but the two approaches are rarely closely coupled. The key for an integrated representation is through a well-defined force field and Hamiltonian at the molecular level. We discuss the latest developments in a top-down representation of fluids, where we use an accurate EoS to link the macroscopic properties of the fluid and the force-field parameters, with a particular focus on a coarse-grained formulation of the statistical associating fluid theory (SAFT-Î³ Mie). The EoS is used to estimate the parameters of the Mie force field, which can then be used with confidence in direct molecular simulations to obtain structural and dynamical properties that are otherwise inaccessible from the EoS, including the interfacial tension, which is a property not included in the original parametrization.

A further theoretical framework is developed based on scaling arguments applied to the influence parameter of the van der Waals theory of inhomogeneous fluids. The molecular model stems from the application of the square gradient theory to the SAFT-Î³ Mie equation of state. The theory is validated against computer simulation results for homonuclear pearl-necklace linear chains made up to six Mie (Î»-6) beads with repulsive exponents spanning Î» = 8 to 44 by combining the theory with a corresponding states correlation to determine the intermolecular potential parameters. We show how this can be employed to provide a predictive tool to determine interfacial tensions for a wide range of molecules including hydrocarbons, fluorocarbons, polar molecules, among others. The proposed methodology is tested against comparable existing correlations in the literature, proving to be vastly superior, exhibiting an average absolute deviation of 2.2%

# 493586 Generalized Gibbs Free Energy of Confined Nanoparticles

**Abstract:**

The nanoparticles generally show abnormal properties compared to those in the bulk phase, and they exhibit significant potential in various applications such as catalysis and energy conversion. However, the theoretical work for describing the properties of nanoparticles is limited with poor prediction capacity. In this work, the Gibbs free energy was studied, from both marcoscope and microscope, predictive models were proposed to study the thermodynamic properties of nanoparticles with a generalized description of the Gibbs free energy considering the effects of surface-energy and the substrate contacted. The proposed model from the microscope was based on the corresponding states theory to describe the effect of the substrate on the Gibbs free energy of nanoparticles, in which the molecular parameter with a generalized parameter was obtained from the melting point of metals due to sufficient information. The comparison with the new measured experimental results proves the reliability of the model prediction.

# 498818 Molecular Modeling of Polymeric Systems

**Abstract:**

In this talk, I will present the results of three exciting opportunities to comprehensively investigate nano- materials through the use of molecular simulations. Tailoring and optimization of these materials in- clude: 1) shape-directed packing of rigid 2D-like polymer chains (PIM-TMN-Trip), that provides ul- trapermable polymers with highly interconnected micropores facilitating rapid and selective gas transport of polymers, 2) ultra-thin molecular layer-by-layer polyamide membranes which can play a crucial role in the design and understanding of the next generation of reverse osmosis membranes, and 3) bio-conjugates, protein-polymer systems, which will benefit from a predictive knowledge of how an attached polymer will interact with a protein and its environment.
In all cases, it is anticipated that experimentalists will avoid the peril of performing a large number of difficult synthetic chemistry that would result in no enhancement of the target property. Concurrent to the above goals, is the generation of large-data sets and the extraction of critical information from that data (e.g., structure factors to understand the intrinsic correlations between structure and properties/ behavior) to catalyze materials breakthroughs.

# 492444 Effect and Regulation of Surface Wettability on Molecular Transport and Reaction

**Abstract:**

Whereas massive experimental works have been contributed into the investigation of surface wettability effect on fluid transport and reaction, theoretical study on molecular scale is still few. The obstacle mainly originates from the lack of both microscopic description for the surface wettability and accurate theoretical method for quantifying the effect. Here, we demonstrate that the surface wettability effect on molecular transport and reaction can be quantitatively accessed from molecular perspective. The surface wettability is characterized with the ratio of fluid-fluid interaction to fluid-substrate interaction, whose relation to contact angle is given by an empirical yet analytical equation. This mesoscale relation, connecting macroscopic measurement and microscopic modeling, allows for seeking regulation method by identifying the thermodynamic and/or dynamic properties of fluid nearby substrate, and these properties can be well predicted with the help of modern statistical theories including classical density functional theories and molecular simulation. By discussing two specific examples associated with the flux through nanopores and conversion rate of a surface catalytic reaction, we show that this route is generic and has directive and practical importance to the design of efficient processes.

# 503670 Digital Alchemy for Assembly Engineeringc

**Abstract:**

Back in the day, Stuart Churchill was a pioneer in the use of computers for modeling chemical and thermal processes. Since that time, the role of computing in chemical engineering has become indispensable. In this talk, we present new computational methods we call "digital alchemy" that allow for predictive modeling of thermodynamic assembly, and permit the design of building blocks optimized for assembly into target structures with desired properties. In this approach, standard thermodynamic ensembles are generalized to include alchemical variables and their thermodynamic conjugates, allowing for rapid, real-time computational investigation of large, complex design spaces.

# 503671 Improved Algebraic, Numerical, and Graphical Representations in Fluid Mechanics

**Abstract:**

The concept of equivalents is applied to fully developed flows through straight channels, and two applications are introduced: the equivalents that produce equal rates of flow through channels of different cross-section, and the diminishment that occurs between the regimes of laminar and turbulent flow. The validity of the eddy diffusivity and the invalidity of the mixing-length as measures of the shear stress due to turbulence have perhaps never before been demonstrated so clearly. The possible choices of primary variables and the consequences are also examined. The objective of these explorations is an improved understanding of the basic fluid mechanics by teachers, students, and industrial practitioners.

# 503672 The Scaling of Turbulence Near the Wall and the Churchill Turbulent Flux Correlation: Insights with Lagrangian Simulations

**Abstract:**

The evaluation of mean velocity and mean temperature in wall-bounded turbulent flows has been based on the "classical" theory of turbulence and the law of the wall. The viscous wall parameters, namely the friction velocity, the friction temperature, and the kinematic viscosity of the fluid have been used to scale not only the mean values of velocity, but also other mean turbulent quantities close to the wall, including the Reynolds stresses and the turbulent velocity and temperature fluctuations. Stuart Churchill, as he questioned the validity of the Reynolds analogy between momentum and heat or mass transfer in his works [1], he realized that extending the scaling of momentum to the scaling of scalar quantities in turbulence was not a sound approach. He proposed a scaling of turbulence that depends on the turbulent momentum flux and the turbulent heat flux. In his AIChE Institute Lecture (in Miami, 1998), Churchill presented a summary of his work on this topic, and closed form solutions for the mean velocity and temperature [2]. The turbulent Prandtl number, the most significant parameter for the modeling of turbulent transport not only in theory but also in modern CFD commercial software packages, was also interpreted physically based on these ideas [3]. We have used direct numerical simulation results and results from Lagrangian simulations [4] for the transport of passive scalars in turbulent channel flows to explore Churchill's ideas on turbulent flow scaling and to develop an updated theory for the scaling of heat transport in turbulent flows [5-7]. His passion to remove empiricisms and approximations from the development of predictive correlations in turbulent transport has been an inspiration. In this talk, we will revisit the Churchill scaling in view of our simulation results, and will discuss the counterintuitive finding that wall turbulence can be used to separate rather than mix substances under special circumstances [8].

1. Churchill, S. W., 1996, "A Critique of Predictive and Correlative Models for Turbulent Flow and Convection," Ind. Eng. Chem. Res., 35, pp. 3122–3140.
2. Churchill, S.W., 2000, "Progress in the Thermal Sciences: AICHE Institute Lecture," AIChE Journal, 46(9), pp. 1704-1722.
3. Churchill, S. W., 2002, "A reinterpretation of the Turbulent Prandtl Number," Ind. Eng. Chem. Res., 41, pp. 6393–6401.
4. Papavassiliou, D.V, 2002, "Turbulent transport from continuous sources at the wall of a channel," Int. J. Heat Mass Transfer, 45(17), pp. 3571-3583.
5. Le, P.M., and Papavassiliou, D.V., 2008, "On the scaling of heat transfer using thermal flux gradients for fully developed turbulent channel and Couette flows," Int. Commun. Heat Mass Transfer, 35(4), pp. 404-412.
6. Srinivasan, C., and Papavassiliou, D. V., 2011, "Prediction of Turbulent Prandtl Number in Wall Flows With Lagrangian Simulations," Ind. Eng. Chem. Res., 50(15), pp. 8881–8891.
7. Srinivasan, C., and Papavassiliou, D.V., 2013, "Heat transfer scaling for wall bounded turbulent flows," Applied Mechanics Review 65(3), Art. 031002.

8. Nguyen, Q., Srinivasan, C., and Papavassiliou, D.V., 2015, "Flow induced separation in wall turbulence", Phys Rev E, 91, Art. 033019.

# 503674 Flow Boiling Using a Piranha Pin Fin Heat Sink

**Abstract:**

**Flow Boiling using a Piranha Pin Fin Heat Sink**

Corey Woodcock, Xiangfei Yu, Yoav Peles[1] and Joel Plawsky

Rensselaer Polytechnic Institute; [1]University of Central Florida

The *Piranha Pin Fin* (*PPF*) microdevice is a heat exchanger engineered to dissipate extreme thermal loads using either dielectric fluids such as HFE7000 or water. Fabrication of the device is CMOS compatible and the first generations of the device have been fashioned from silicon substrates. The PPF operates on the principle that heat transfer would be more efficient if heated fluids could be removed nearly as fast as they are created. This process would keep the driving force as high as possible. Thus the heat exchanger is built as two layers; a primary substrate that dissipates the bulk of the heat and shunts the heated fluids to a booster section that continues to heat the fluid and extends the total surface area of the device. Three generations of devices were fabricated. The first two generations were single-layer systems designed to help understand the operation of the primary layer and optimize the shape and location of the pin fins. The third generation was the first, two-layer system to be fabricated and tested.

Figure 1 Representative PPF array

Correlations for the heat transfer coefficient, $h_{tp}$, based on the Boiling Number and the Weber Number in the PPF design was developed and validated with Gen II. The discrepancy of heat transfer coefficient between predicted and measured values was within 20%. The ratio of two-phase to single-phase pressure drop was correlated with the Boiling Number (Bo) and the Jakob Number (Ja). The discrepancy of the two-phase multiplier between prediction and experimental measurements were within 12%.

The third generation *PPF* heat sinks were the first series to provide both a primary substrate populated with *PPF*s, as well as a booster substrate populated with solid micro pin fins. The substrates are micro-aligned and fusion-bonded to provide a full three dimensional (3D) architecture with bulk silicon properties. Experimental results from the Gen III *PPF* microdevices provide a proof of concept of the true *PPF* heat sink concept with fluid venting into multiple substrates. At low flowrates ($G$ = 670 [kg/s/m$^2$]) the Gen III *PPF* heat sink has been shown to dissipate heat loads approaching 1.1 [kW/cm$^2$] at an exit quality of 95%, while at higher flowrates ($G$ = 3,300 [kg/s/m$^2$]) the thermal limit of the heat sink remains unknown due to exceeding the available power supply at >1.4 [kW/cm$^2$].

# 503673 Transport Problems in the Spirit of Stuart Churchill for Teaching and Research at the University of Michigan

**Abstract:**

Stuart Churchillâ€™s solutions to important heat transport problems have greatly influence both pedagogy and research. The correlation of Churchill and Chu for the natural convection heat transfer to or from a vertical plate, over an astonishing range of Rayleigh numbers is based partly on a similarity solution of a type common in heat, mass, and momentum transfer for problems in chemical engineering. One such problem is asphaltene or wax transport and deposition onto the wall of an oil pipeline, which can be solved using similar methods, as will be discussed. In addition, such problems, and the wealth of their applications make for excellent motivations and examples in the teaching of graduate or undergraduate transport, as will be discussed in the talk.

# 509054 Solve this! Fundamental Approach to Problem Solving in Industrial Processes I (Invited Talks)

**Abstract:**

Solving complex plant problems is an everyday exercise that affects all projects and budgets, and it is very rarely introduced to students at any level. In fact, by the time young engineers are few years into working in industry, most have lost the engineering skills that they developed in school. We hope to encourage future engineers to retain those skills and learn the importance of plant problem-solving.

# 490835 Comparison of PRISM Theory and Molecular Dynamics Simulations for Studying Assembly in Block Copolymer Solutions of Varying Sequences and Composition

**Abstract:**

In this poster, we will present our recent work directly comparing Polymer Reference Interaction Site Model (PRISM) theory and molecular dynamics (MD) simulations for studying assembly of amphiphilic block copolymers in solution. In both methods, we use a generic coarse-grained model to represent amphiphilic AB block copolymers in implicit solvent with the solvophobicity of the B block captured using effective B-B attractions modeled using Lennard Jones potential. We study the assembly of the amphiphilic AB block copolymer as a function of solvophobicity for varying copolymer sequences (diblock and triblock) and composition (solvophobic-rich or solvophilic-rich). We then solve the PRISM theory equation along with Percus Yevick closures for these block copolymer solutions and obtain the inter-molecular pair correlations in real space, g(r), and structure factors in Fourier space, S(k), at increasing values of solvophobicity. We directly compare the real-space intermolecular pair correlation functions and structure factors from PRISM theory and MD simulations. These structural results show excellent agreement between PRISM predictions and MD simulations at low solvophobicities. While PRISM theory captures the concentration fluctuations at low solvophobicities well, the intrinsically isotropic PRISM theory fails to converge to a numerical solution at higher solvophobicities where we see evolution of the assembled structures in MD simulations. Despite this drawback, PRISM theory is a valuable tool as we can use the low solvophobicity results from PRISM to predict all or most of the thermodynamic signatures of the polymer solutions at higher solvophobocities. For example, the inverse microphase peak 1/S(k*) at low solvophobicity from PRISM theory when extrapolated to zero quantifies the disorder-order transition solvophobicity value in good quantitative agreement with MD simulations. Additionally, in the systems where MD simulations predict aggregation of the micelles/clusters at high solvophobicity, the structure factors from PRISM theory results at low solvophobicity also present an up-turn in the zero-wave vector structure indicating tendency for macrophase separation at higher solvophobicity. These findings provide evidence of the capability of PRISM theory for predicting copolymer assembly in solution, where a wide range of design parameters can be explored much faster than MD simulations.

# 502532 Novel Computational/Experimental Apporaches to DNA/Proteins Interactions

**Abstract:**

Studying the interaction of Rad51 and ssDNA is very challenging since, the fact that the relevant supramolecular assemblies are often very stable, the intrinsic flexibility of ssDNA and the geometrical/energetic features of resulting interface concur to increase the complexity of the system under investigation. Yet, in this context, multiscale molecular simulations appear as a unique instrument for studying DNAâ€“protein interactions, owing to its atomistic resolution and capability of mimicking the considered biological environment (i.e., explicit solvent, temperature, and physiological/imposed ionic strength). In order to get a comprehensive structural and thermodynamic perspective of Rad51/ssDNA interaction, in this work we propose to apply large scale, parallel atomistic and coarse-grained (CG) simulations, coupled with high resolution imaging techniques and biophysical methods to i) dissect the docking mode of Rad51 to ssDNA, ii) deriving the energetics of the interface in the protein/nucleic acid interface, iii) uncovering the role of DNA and protein flexibility in their assembly formation, and iv) determining the mechanism of Rad51 binding to ssDNA.

# 502538 Wiggling, Crowding, Self-Assembling. Synthesis and Activity of Computer-Designed Nanovectors for Gene and Drug Delivery

**Abstract:**

Due to the relative easy synthesis and commercial availability, nanovectors based on self-assembling units are among the most utilized non-viral vectors for gene/drug transfer. Contextually, recent advances in molecular simulations and computer architectures, coupled with biophysical and molecular biology experiments, allow for 1) accurate predictions of many structural, energetic, and eventual self-assembly features of these nanocarriers *per se* and in complex with their genetic cargoes., and 2) thorough assessment of thier biological performace. In the present work, we aim at reviewing our own efforts in the field of computer-assisted design, synthesis, *in vitro* and in vivo activity of these (nano)biomaterials.

# 502566 New Anti-Mycobacterium AgentsÂ in Combination with Pgp Inhibitors: A Multidisciplinary Approach to Face an Old Re-Emerging Disease with New Tools

**Abstract:**

Tuberculosis (TB) is one of the most common infectious diseases worldwide, with about one-third of world population infected with Mycobacterium tuberculosis (M.Tb.). More frightening is the recent emergence of multi-drug and extensively drug resistant M.Tb. strains (MDR-M.Tb. and XDR-M.Tb., respectively). In the setting of drug resistance or intolerance to first-line agents (e.g. isoniazid, rifampinâ€¦), second-line agents may be used. Indeed, fluoroquinolones have been classified as second-line antituberculous drugs since they are active on isolated M. Tb. expressing resistance to both isoniazid and rifampin. Recently we synthesized a new series of quinolones derivatives, endowed with a good
anti-mycobacterial activity (against H37Rv and H37Ra and further clinical isolates of MDR-TB/XDR-TB strains) coupled with a low cytotoxicity.Â
In this work we present a combined in vitro/in silico approach with the aim (i) to improve knowledge on theÂ structure-activity relationship (SAR) of the new compounds, (ii) to identify a pharmacophoric map and (iii) enhance the biological activity against the biological target. According to the computational results, new derivatives will be designed and synthesized to obtain compounds able to inhibit replication in H37Rv and clinically isolated M.Tb. strains. The identification of a new action mechanism by our derivatives compared to classical quinolones could be used to generate a new class of drugs, with the aim to create an innovative treatment plan able to reduce possible pharmacological resistances.

In parallel a series of new quinoxaline derivatives, previously identified as P-glycoprotein (Pgp) inhibitors, will be investigate on M.Tb. strains, in combination with new compounds and classical drugs. In fact, efflux pump inhibitors could enhance a latent biological activity and implement the DNA-gyrase inhibitors activity against resistant M. Tb. strains. Even in this topic the computational and the experimental tecniques will be applied in order to improve the level of knowledge on the mechanism of action of these molecules againt their biological target. All obtained data will be analyzed and promising compounds will be optimized to obtain selected gold compounds.

# 497764 Qsars for Predicting Adipose:Blood Partitioning of Industrial Chemicals


**Abstract:**

In recent years, there is an increasing interest in the development of Physiologically Based Toxicokinetic (PBTK) models, which provide quantitative descriptors of Absorption, Distribution, Metabolism and Excretion (ADME) of environmental or pharmaceutical chemicals. However, their application in toxicity testing and health risk assessment is limited due to the lack of input parameters required for their development, especially when these systems comprise comprehensive mathematical descriptions of human physiology. Proper parameterization of PBTK models can be achieved using advanced Quantitative Structure-Activity Relationships (QSARs). QSARs are widely used for the estimation of physicochemical and biochemical properties of "data poor" chemical compounds, biological effects as well as understanding the physicochemical features governing a biological response. These relationships are described as regression or classification models, which connect the biological effects and chemistry of each chemical compound and comprise the activity data to be modeled, the data with which to model and a method to formulate the model.

Several approaches incorporating QSARs have been proposed for the prediction of partition coefficients for PBTK modeling, including

(a) an algorithm based on the fractional content of cells, interstitial fluid in tissue, plasma and erythrocyte in blood, tissue lipids and the lipophilicity of chemical compounds and

(b) the Linear Free Energy Relationship (LFER), proposed by Abraham and co-workers, for estimating biological properties, which takes into account excess molar refractivity, polarizability, solute effective or summation hydrogen-bond acidity and basicity, as well as the McGowan characteristic volume. The latter is a measure of the lipophilicity for chemical compounds.

Adipose/blood partition coefficient is considered as one of the most significant input parameters of PBTK models. The partitioning of chemical compounds into adipose tissue and blood provide information regarding distribution and toxicological effects of these substances.

The methodological approach presented in this study is based on the development of QSAR models to predict adipose/blood partition coefficient for environmental chemical compounds. The necessary input data for the models to be trained consisted of the experimental values of adipose/blood partition coefficient and two sets of molecular descriptors for 67 environmental chemicals of the initial set; a) the descriptors from Linear Free Energy Relationship (LFER) and b) the PaDEL descriptors. The PaDEL dataset included 1D, 2D and 3D descriptors, which are related to the molecular structure of the chemicals and are characterized as constitutional, topological, geometrical or electronic. The reduction process was followed for the derived molecular descriptors in order to avoid the semi-constant (>80%) and intercorrelated (>95%) ones. Principal Component Analysis (PCA) was used for further reduction of PaDEL inputs, as well as for the categorization of chemical compounds.

The datasets were randomly divided into the training set, containing 70% of the total data, the validation and the test set, each one representing 15% of the total. Then, they were analysed using two statistical methods; Genetic Algorithm based Multiple Linear Regression (GA-MLR) and Artificial Neural Networks (ANN). GA was implemented for the selection of the optimal set of descriptors for the models. The GA-MLR technique was implemented in the QSARINS software (Gramatica, Chirico et al. 2013, Gramatica, Cassani et al. 2014), while the ANN technique was implemented in MATLAB® (version R2016a, Mathworks Inc) using the Neural Network Toolbox.

The developed models with the LFER and PaDEL descriptors, coupled with ANN, produced excellent performance results. The fitting performance ($R^2$) of the models, using LFER and PaDEL descriptors, was 0.94 and 0.96, respectively. Cross Validation (CV) indicated that the predictive performance of the models ($Q_{cv}^2$) was equal to 0.96 and 0.95, while the external validation value ($R_{ext}^2$) was found to be 0.98 and 0.94 for LFER and PaDEL model, respectively. The Applicability Domain (AD), which defined the limitations regarding the structural and target domains of the models with the best performance, was determined using several approaches, including bounding box, bounding box on PCs, convex hull, leverage, distance to centroid, k Nearest Neighbors (kNN) approach with fixed k, k Nearest Neighbors (kNN) approach with variable k, Probability Density Function (PDF) based methods. Finally, the developed models were applied to a large number of chemical compounds with unknown values of adipose/blood partition coefficient.

The proposed models for the estimation of adipose/blood partition coefficients were checked for their fitting performance, validity and applicability. It was found that they are stable, reliable and capable to predict physicochemical parameters of "data poor" chemical compounds that fall within the applicability domain. The developed predictive models could serve as a tool to fill in data gaps of environmental chemicals with unknown values of adipose/blood partitioning. In this way, animal testing could be reduced significantly and the wide use of PBTK models could be reinforced. Finally, the "safe-by-design" concept for environmental chemicals is supported, by allowing the successful prediction of toxicokinetic behavior based on molecular parameters, promoting green chemistry and cost saving of product development.

# 498265 Pharmacometabonomics Approach for Early Prediction of Neuropathy


**Abstract:**

Vincristine (VCR) is a core chemotherapeutic drug administered to pediatric Acute Lymphoblastic Leukemia (ALL) patients. In a subgroup of population, it leads to Vincristine Induced Peripheral Neuropathy (VIPN), which is the dose-limiting toxicity. For few patients, VIPN is severe with long-term effects. Even though VCR has been used in chemotherapy for more than 50 years now, predictors and mechanism of VIPN induction are unclear. Historically, an empirical cap of 2 mg has been kept for dosage. Due to this, a population not susceptible to VIPN receives sub-therapeutic treatment, while another population may experience severe neuropathy. It is of interest to find predictors/biomarkers which can discriminate populations based on VIPN severity, before the treatment starts. A predictive model with the selected set of biomarkers will aid clinicians in identifying patients susceptible to neuropathy, and in adjusting dosage accordingly.

We are using pharmacometabonomics approach to find metabolites as biomarkers that can predict VIPN severity in pediatric ALL patients. In a retrospective pilot study done on 12 patients, metabolites were found to be differentially expressed in high and low VIPN patients. Presently, retrospective non-fasting plasma samples of 36 pediatric ALL patients are being used. These samples were collected at three time points: day 8, day 29 of induction period, and after around 6 months from the start of the therapy. A preliminary analysis (feature selection using elastic net logistic regression) showed that a set of 17 metabolites in +6 months samples could accurately identify patients with high and low VIPN. A logistic regression model built with these 17 metabolites had an area under the Receiver Operating Characteristics curve of 0.99. Currently, extensive analysis of metabolomics data at all the time points is being done. We aim to find metabolites that are becoming differentially expressed over time. Longitudinal feature selection algorithms are being explored for this purpose. To further explore the biomarkers found, we intend to identify the structure of these metabolites and find their relevant pathways.

# 490552 A Theoretical Study of the Activation of Hydrogen and Methane By Frustrated Lewis Pairs

**Abstract:**

In this work, we provide a systematic investigation of the activation of the H-H bond of molecular hydrogen and the C-H bond of methane mediated by intermolecular Frustated Lewis Pairs. To be specific, we use the classical sterical encumbered P(t-Me)$_3$ phosphine as Lewis bases and two Lewis acids: $B(C_6F_5)_3$ and $Al(C_6F_5)_3$. Calculations were performed by means of DFT calculations using the dispersion-corrected ωB97X-D functional in conjunction with the 6-311G(2d,2p), including toluene as solvent through the PCM-SMD implicit solvent scheme. In all cases, the activation barriers are larger for methane than for hydrogen. This is due to a large distortion of the C-H bond to reach the transition state, and additionally, the distortion of the Lewis acid attains a better interaction with the s-bond in methane. Finally, we found that the Lewis acid $Al(C_6F_5)_3$ has a considerable reduction in the free energy activation barrier for both hydrogen and methane activation. Thus, this work demonstrates that, if properly designed, the use of alanes as Lewis acid can reach high degree of reactivity to methane activation.

# 490800 Differences in Relative Free Energy Versus Temperature Curves for Small Organic Molecules Â between Quantum Mechanical and Classical Potentials.

**Abstract:**

The crystalline packing or polymorphic state of a material can have an effect on many of the physical properties of that material. Traditional crystal structure prediction methods rely on the potential energies of the 0K structures to make determinations about the most stable polymorph at higher temperatures. However, because of entropic effects, this approach may not always yield the correct room temperature structure. In this project, we show the effect of using a quantum mechanical potential on the estimation of the entropic effects in the temperature dependent polymorph stability.

In previous work, we have shown that in some small molecule organic crystals, the entropic effects are large enough to change the ordering of stability in crystal polymorphs between the ordering determined by the 0K structure and by molecular dynamics or the quasiharmonic approximation at room temperature. We have also shown that the potential used has an effect on the ordering of the stability of the polymorphs and the temperature of the temperature mediated transformation of the polymorph. We have determined the free energy versus temperature curves for full molecular dynamics in point charge and polarizable potentials, and their comparison to the quasiharmonic approximation in point charge potentials. To further examine the effect of changing potentials on the temperature mediated transformation of polymorphs, this project uses the quasiharmonic approximation in quantum mechanical potentials and the non-equilibrium work implementation of the Jarzynski equation for to obtain dG versus T stability curves.

In this project, two crystals were studied , benzene and aripiprazole. Benzene is a rigid molecule and aripiprazole has more flexibility. The cp2k package was used, with DFT-D3 corrections and the PBE functional. The vibrational analysis module in cp2k was used to obtain the normal mode frequencies and the corresponding eigenvectors of the two systems, and the Gruneisen parameter approach was used to construct the dG vs T curves for the free energy difference between polymorphs. the We can compute the free energy versus temperature for ab initio quantum mechanical approach by computing the free energy correction from a classical force field to a classical one, and adding this correction to previously computed classical results. We examined the length of switching simulations necessary to efficiently estimate the free energy of switching between potentials.

# 492690 Density Functional Theory Screening of Metal Catecholates for Adsorption of Toxic Pnictogen Hydride Gases

**Abstract:**

Ammonia ($NH_3$), phosphine ($PH_3$), and arsine ($AsH_3$) are toxic gases that are commonly used in industry, which places workers at risk of accidental exposure. Most widely used filtration methods rely on activated carbon, which has low capacity and selectivity for these gases. However, in recent years, researchers have looked to other porous materials, such as metal-organic frameworks (MOFs), as sorbents for toxic gases. [1] One benefit of using MOFs is they can be tailored to specific adsorption applications by adding functional groups or metal sites which increase the MOF's affinity for the target molecules. [2]

We used density functional theory to study interactions between ammonia, phosphine, arsine, and 33 metal catecholates in order to examine periodic trends in the binding strength between the gases and metals. We find markedly different behavior in ammonia than in phosphine and arsine due to different processes governing the adsorption, despite the apparent similarity of these molecules. Phosphine and arsine adsorption is primarily driven by electron density being donated from the adsorbate molecule to the metal atom, while ammonia binding involves both some electron donation and a significant Coulombic attraction between the negatively-charged N atom and the positively-charged metal atom.

We will discuss how these different adsorption mechanisms are manifested through trends in binding strength across the periodic table, as well as the implications for adsorption of ammonia and phosphine in humid environments.

1. N. S. Bobbitt, M. L. Mendonca, A. J. Howarth, T. Islamoglu, J. T. Hupp, O. K. Farha, R. Q. Snurr, "Metal–organic frameworks for the removal of toxic industrial chemicals and chemical warfare agents." *Chemical Society Reviews, in press.*
2. D. Yu, P.Ghosh, and R. Q. Snurr, "Hierarchical modeling of ammonia adsorption in functionalized metal–organic frameworks." *Dalton Transactions*, *41*(14), 3962-3973 (2012)

# 493671 Mechanism of Sodium Adsorption on N-Doped Graphene Nanoribbons

**Abstract:**

We employed density functional theory (DFT) to investigate the mechanisms of sodium adsorption and diffusion on N-doped graphene nanoribbons (GNR). The calculated adsorption energy of edge region (-1.096 eV) on pristine GNR was stronger than that of center region (-0.836 eV). In addition, the adsorption energies on GNRs with nitrogen-doped defects (4ND, 3NV and pyrrolic: -3.794 ~ -1.837 eV) were stronger than that for the pristine-GNR edge (-1.096 eV). The adsorption of sodium on GNRs were strongly affected by the introduction of nitrogen defects. According to the band structure analysis, 3*s* valance orbital of sodium is raised when sodium is adsorbed on GNRs. Moreover, when sodium is adsorbed near nitrogen atoms, the *p* orbital of nitrogen is shifted down due to strong interaction between sodium and adjacent nitrogen. Although the adsorption of sodium on GNR with nitrogen-doped defects was improved, the diffusion energy barriers of sodium on the pyrrolic-GNR (0.009 ~ 0.021 eV) was lower than that for the pristine-GNR (0.019 ~ 0.079 eV).

**Acknowledgements:** This research was supported by Global Frontier Program through the Global Frontier Hybrid Interface Materials (GFHIM) of the National Research Foundation of Korea (NRF) funded by the Ministry of Science, ICT & Future Planning (No. 2013M3A6B1078882 and 2013M3A6B1078874). This research was supported by Basic Science Research Program through the National Research Foundation of Korea (NRF) funded by the Ministry of Science, ICT & Future Planning (2014R1A1A1004096).

# 496465 Influence of Solvent on the Thermodynamics of Molecular Adsorption on Metal Surfaces

**Abstract:**

Solid-liquid interfaces are central to the functional aspects of various applications, including the synthesis of nanomaterials and heterogeneous catalysis. In many applications that feature the binding of certain molecular species to a solid surface, the solvent can also play a key role. We investigate the influence of ethylene glycol solvent on the adsorption of polyvinylpyrrolidone (PVP) analog molecules on Ag surfaces by comparing systems in vacuum to solvated systems. Ethylene glycol and PVP are commonly present in the solution environment of Ag nanocrystal syntheses, in which the nanocrystal morphology is directly influenced by the choice of solvent and adsorbate. We calculate the potential of mean force between PVP analog molecules and surfaces of Ag using molecular dynamics simulations based on a many-body organic-metal force field. From our potential-of-mean-force calculations, we observe that the binding free energies of the PVP molecules are drastically smaller in the presence of solvent because these molecules are stabilized by solvent in the bulk solution environment. The observed entropy changes due to adsorption are negative in vacuum but they are zero or slightly positive in solvent. Our results demonstrate that the liberation of adsorbed solvent molecules when a molecule binds to a solid surface can contribute more to the system's entropy change than the configurational entropy loss due to molecular adsorption. We find the inclusion of solvent molecules to be a critical feature in correctly modelling molecular adsorption in solution.

# 497625 Computational Discovery of New Materials and Processes for Industrial Separations

**Abstract:**

Chemicals separations are responsible for nearly half of the US industrial energy consumption. Our goal is to contribute towards developing the next-generation technologies for cost-effective and energy-efficient separations. Two important industrial separations â€“ sour gas sweetening and ethane/ethylene separation â€“ are discussed here.

We use Monte Carlo (MC) simulations for predicting phase and adsorption equilibria. Accurate force fields are at the heart of any reliable computational prediction. These molecular models comprise of several parameters, that may be system-specific for matching to experimental data. However, this work takes a different approach, in that, a large amount of effort is first invested in a multi-dimensional, parallel, and physically guided parameter optimization for capturing a wide range of properties for the molecules of interest. The expense involved in parametrizing transferable force fields is recouped by allowing the use of these models for diverse systems, without the need for reoptimization each time a new system is investigated. This also provides truly predictive and reliable answers to problems where experimental data may not be available.

MC simulations are used to probe adsorption in zeolites â€“ nanoporous materials with highly specific pore sizes and shapes. Using tailored simulation protocols, a large database of zeolite structures is screened for adsorption performance and promising structures for efficient separations are discovered. Some of these zeolite frameworks are currently synthesized and assessed experimentally to validate the predictions. The new separation technologies that we are developing have the potential for far-reaching impacts on the global energy and chemical industries.

# 498011 Molecular Simulations of Fullerene Stabilization in Water By Fullerene-Oxides

**Abstract:**

Making materials out of buckminsterfullerene is challenging because it requires first dispersing the molecules in a solvent, and then getting the molecules to assemble in the desired arrangements. In this computational work we focus on the dispersion challenge: How can we conveniently solubilize buckminsterfullerene? Water is a desirable solvent because of its ubiquity and biocompatibility, but its polarity makes the dispersion of nonpolar fullerenes challenging. We perform molecular dynamics simulations of fullerenes in the presence of fullerene oxides in implicit water to elucidate the role of interactions (van der Waals and Coulombic) have on the self-assembly and structure of these aqueous mixtures. Two coarse-grained fullerene models are characterized over a range of temperatures and interaction strengths using HOOMD-Blue on the Blue Waters supercomputer. For both models, we find temperature-dependent structural transitions and observe phases where fullerene oxides form a shell around droplets of fullerenes, which is consistent with experimental hypotheses for how the oxides help stabilize aqueous dispersions of fullerenes. This work demonstrates how high-throughput simulations help test experimental hypotheses while providing additional insight into molecular structure that is not easy to determine experimentally.

# 498926 Interplay between Crystallization and Glass Transition in Bimetallic Nanoalloys

**Abstract:**

Metal nanocrystals and nanostructures are currently the focus of intense research because of their applications in a wide variety of fields including e.g. as DNA/ protein markers, drug carriers and in catalysis. In particular, bimetallic nanoparticles have been shown to exhibit enhanced catalytic properties, as a result of the synergistic effects that arise upon alloying different elements in a nanoparticle. In this work, using the quantum corrected Sutton-Chen potential to model the interactions in bimetallic systems, we carry out molecular simulations to shed light on the interplay between crystallization and glass transition during the nucleation process in bimetallic nanoalloys.

# 499028 Molecular Simulations of Bubble Formation in Metastable Liquids

**Abstract:**

The nucleation of gas bubbles in a metastable liquid is a phenomenon central to many applications, ranging from sonochemistry, cavitation and explosive boiling. However, despite its significance, the molecular mechanisms underlying this phenomenon have remained poorly understood. One of the reasons accounting for this is the difficulty in identifying and characterizing the incipient gas bubble, given a snapshot of the metastable system. In this work, we carry out molecular simulations of the bubble nucleation process, identify the free energy barrier of nucleation involved and discussed the interplay between void and bubble formation during this process.

# 499095 Molecular Simulation of Gas Adsorption in Metal-Organic Frameworks

**Abstract:**

In recent years, the development of new nanoporous materials, including e.g. metal-organic frameworks and covalent organic frameworks, for the adsorption and storage of gases has been the focus of intense research. Such nanoporous materials have potential applications in a wide range of field since they can be tailored to store efficiently environmental contaminants (CO2â€¦) or fuels (hydrogenâ€¦) and be used in separation applications (e.g. to separate branched alkanes with the purpose of increasing the octane number of gasoline). In this work, we discuss how we use molecular simulation methods to shed light on gas adsorption and storage in various types of metal organic frameworks.

# 493051 Leveraging Heterostructural Alloying to Design Metastable Nitrides with Improved Piezoelectric Properties

**Abstract:**

The tailoring of materials properties by alloying is routinely utilized to design materials for targeted technological applications. Despite the great successes of alloying in isostructural systems, heterostructural alloying remains a fundamentally unexplored area. In heterostructural alloys, the crossover between different crystal structures enables the control of the atomic structure by variation of the composition. The deliberate manipulation of local atomic coordination symmetry introduces an additional materials design parameter for tuning the piezoelectric response of a material. The piezoelectric effect underlies a wide range of industrial applications including quartz resonators in wristwatches, powerful hydrophones for naval sonar, medical ultrasound, and the microelectromechanical (MEMs) resonator-based bandpass filters in wireless handsets. However, many of the mature alloys used (eg. PZT) have been designed around large strain and electromechanical coupling coefficients for use in high power, low operating frequency applications while high frequency resonant, high temperature, and/or low power integrated devices typically use simple unary compounds such as quartz or aluminum nitride despite their relatively low piezoelectric response. In this talk, we present a complementary theoretical and experimental investigation of piezoelectrically active nitride alloys to develop design principles and approaches for utilizing heterostructural alloying as a materials design strategy. We use *ab initio* methods to predict the structural and electronic properties of $Al_{1-x}Sc_xN$ heterostructural alloys for the energetically competitive polymorphs and compute their corresponding equilibrium phase diagrams and materials properties. Combinatorial sputtering is employed as a non-equilibrium growth technique to overcome thermodynamic solubility limits and produce metastable thin-film samples spanning the alloy composition range. The experimentally validated predictions, theory guided combinatorial synthesis, and characterization of piezoelectric heterostructural alloys exemplify how our integrated research strategy is used to design and realize functional metastable materials which can be utilized in a variety of industrial applications. We apply these design principles to screen and computationally prototype new metastable nitride alloys with improved piezoelectric properties. Our approach establishes a new route for the control of structure-property and composition-structure relationships by accessing non-equilibrium phase space to develop new piezoelectric materials with uniquely tailored properties for specific applications.

# 502791 Discovery of High-Performing MOFs Via High-Throughput Computation and Machine Learning

**Abstract:**

We describe a data-driven approach to developing metal-organic frameworks (MOFs) with high hydrogen storage capacities. MOFs are crystalline nanoporous materials containing metal clusters bonded to organic linkers. High-throughput (HT) grand canonical Monte Carlo simulations and a Voronoi networks-based technique are used to compute the capacities and properties of nearly 500,000 compounds. The large properties-performance database resulting from these calculations presents a unique opportunity for machine learning (ML). Using ML, we predict the necessary structural properties of a MOF needed to achieve specific capacity targets. We refer to this approach as â €˜reverse crystal engineering.â€™ This method illustrates the possibility of developing purpose-built materials with specified functionalities.

# 487194 First-Principles Studies of the Interactions Between Chemical Species inside Vanadium Redox Flow Batteries

**Abstract:**

The operation of a large scale energy storage Redox Flow Battery (RFB)[i] is driven by reversible redox reactions of electro-active electrolyte species inside the two half-cells. The reversible redox reactions that happen during the charge – discharge cycles reuse all of the components of the system and thus eliminate all wastes. Therefore, RFBs should provide a solution to an environmentally clean energy alternative as all of the chemical wastes produced during the energy consumption process can be reversed by the charging process. One of the most studied system of RFBs is an all-Vanadium Redox Flow Battery (VRFB)[ii]. VRFB exploits the vast range of oxidation states that can be reached by vanadium. VRFB employs the redox couple of $V^{2+}/V^{3+}$ in the negative half-cell and $VO^{2+}/VO_2^{+}$in the positive half-cell. The introduction of Nafion as the membrane in the system of VRFB leads to the formation of all sorts of complexes between vanadium of all oxidation states and Nafion. Vanadium – Nafion complexes block the pores on the Nafion, which reduces the performance of the battery, and leads to VRFB eventual degradation. Molecular level understanding of these interactions would provide insights into the chemistry of RFBs at the real operation temperature to aid the further improvement of RFBs.

Our computational studies[iii] from the DFT based gas phase calculations of the enthalpy of formation for the vanadium - Nafion complexes indicate that all four vanadium species have driving force to covalently bind to sulfonic acid group of Nafion ionomer unit via the formation of a single V-O bond. From the Car-Parrinello molecular dynamics based metadynamics simulations of the aqueous solution at 298 K we find that the formation of covalently bonded vanadium-Nafion complexes is spontaneous for $V^{2+}$ and $V^{3+}$, while $V^{4+}$ and $V^{5+}$ have a sizable activation barrier to attach to Nafion and diffuse away from $SO_3^-$group.

In addition, we analyze the calculated IR spectra for the most energetically favorable vanadium complexes with Nafion.

---

[i] Prifti, Helen, et al. "Membranes for redox flow battery applications." Membranes 2.2 (2012): 275-306.

[ii] Skyllas-Kazacos, M., et al. "New all-vanadium redox flow cell." J. Electrochem. Soc.;(United States) 133 (1986).

[iii] Intan, N., et al. "Interactions between Vanadium Ions and Nafion Membrane: Insights from First - Principal Calculations" (In Preparation)

# 496068 Dehydrogenation Mechanism of Liquid Organic Hydrogen Carrier Materials: A Density Functional Theory Study

**Abstract:**

Storing hydrogen *via* liquid organic hydrogen carrier (LOHC) is one of well-known methods for chemical energy storage because it exhibits good performance in hydrogen storage by reversible reactions of hydrogenation and dehydrogenation. However, the mechanism of hydrogenation or dehydrogenation reaction in LOHC materials is still poorly understood. In this study, we analyzed the mechanism of dehydrogenation of a target material using density functional theory (DFT) calculations.

Benzyl toluene was selected as a representative material of the storage. We obtained a thermodynamic database of 196 intermediates and investigated plausible dehydrogenation pathways based on the information. The effect of doping N into benzyl toluene on the dehydrogenation mechanism was also assessed. Our findings will provide valuable insight for experimentalists to develop high-performance LOHC materials.

# 494872 The Crystal Structure and Surface Composition of Coalescing Ag-Au Nano-Alloys By Molecular Dynamics Simulations


**Abstract:**

Bimetallic nanoparticles exhibit catalytic, optical, electronic and magnetic synergy between their constituents. Typically, that synergy is traced to the structure and surface characteristics of such particles. Core-monolayer shell bimetallics exhibit often unique properties that are not an interpolation between those of their parent metals. This behavior is attributed to the lattice mismatch (strain effect) and charge transfer between layers (ligand effect) resulting in different electronic properties. Furthermore, the addition of an ad-metal may alter also the reactant adsorption geometries (ensemble effect) and the degree of ad-metal coverage and its diffusion into the bulk alter the adsorbate binding energies. Yet, there is only limited understanding of the structure and mixing of the two parent metals. Furthermore, understanding metal-adsorbate interactions is a key issue to controlling and improving the functionality of bimetallic nanoparticles in energy applications. Adsorption, however, depends on nanoparticle morphology.

Here, the effect of temperature and initial particle morphology on the sintering rate and crystalline structure is elucidated during sintering of such silver-gold nanoparticles by Molecular Dynamics (MD) simulations. The MD method is validated by the attainment of the melting temperature of pure Ag (Buesser and Pratsinis, 2015), Au (Goudeli and Pratsinis, 2016) and Ag-Au bimetallic nanoparticles as function of their size. The sintering rate of particle pairs with different initial morphology (Ag-Au alloys, segregated or Ag@Au and Au@Ag core-shell nanoparticles) is determined by tracking the evolution of the total surface area and the changes in surface composition of the formed nanoparticle are elucidated during coalescence at $T = 500$ â€“ 1000 K. The XRD pattern of Ag@Au and Au@Ag core-shell nanoparticles is calculated and compared with experiments (Sotiriou et al., 2014) as well as to those of pure Ag & Au and segregated or alloyed composites while their crystallite size is obtained by fitting these MD-obtained XRD patterns.

Pure Ag and segregated Ag-Au nanoparticles have larger XRD size than pure Au or alloyed Ag-Au ones. This can provide indication of the detailed structure of flame-made nanoparticles at the atomistic level as experimental measurements of Ag-Au particle size revealed smaller XRD diameter than pure Au or Ag particles (Sotiriou et al., 2014). Furthermore, sintering of Ag@Au core-shell nanoparticles leads to slight reduction (less than 10%) in Ag surface composition, in contrast to Au@Ag ones that increase their Ag surface composition more than 20%.


**References**


Buesser, B., and Pratsinis, S.E. (2015) *J. Phys. Chem. C*, 119, 10116-10122.

Goudeli, E., and Pratsinis, S.E. (2016) *AIChE J.*, 62, 589-598.

Sotiriou, G. A, Etterlin, G. D, Spyrogianni, A., Krumeich, F., Leroux, J-C., and Pratsinis, S. E. (2014) *Chem. Commun.* 50, 13559-13562.

# 490471 First-Principles Study of Atomistic Mechanisms in All-Vanadium Redox Flow Batteries

**Abstract:**

All-vanadium redox flow batteries (RFBs) is a promising large-scale grid energy storage which offers a number of advantages over their solid-state counterparts including the decoupling of power and energy, rapid response times, reversibility and long cycling lifespan. The operation of all-vanadium RFB is based on the electrochemical reactions of the $VO^{2+}/VO_2^+$ redox couple in the catholyte and the $V^{2+}/V^{3+}$ in the anolyte. Two critical parameters that dictate the overall efficiency of a RFB are energy density and power density. However, atomistic mechanisms of electrochemical reactions in the electrolyte and at the electrolyte/electrode interfaces responsible for enhanced energy and power density, respectively, are still poorly understood.

In this study we provide crucial insights into the mechanisms of vanadium self-discharge reactions in the electrolyte and interactions of graphite electrode surfaces with vanadium ions utilizing static DFT and Car-Parrinello molecular dynamics (CPMD) based metadynamics simulations.[1] Specifically, we find that vanadic acid formed in solution should play an important role in polymerization and self-discharge reactions, and we provide the energetics for various reaction pathways. For the electrode reactions, our results suggest that the edge surface is characterized by the formation of ketonic C=O functional groups due to complete water dissociation into H/O/H configuration with surface O atoms serving as active sites for adsorption of $V^{2+}/V^{3+}$ species.[2] The formation of V-O bonds at the surface should significantly improve the kinetics of electron transfer at the edge sites that is not the case for basal surface, in agreement with experimentally hypothesized mechanism.

1 Jiang Z., Klyukin K. and Alexandrov V. Structure, Hydrolysis and Diffusion of Aqueous Vanadium Ions from Car-Parrinello Molecular Dynamics. *J. Chem. Phys.* **2016**, *145*, 114303-114311

2 Jiang Z., Klyukin K. and Alexandrov V. First-Principles Study of Adsorption-Desorption Kinetics of Aqueous $V^{2+}/V^{3+}$ Redox Species on Graphite in a Vanadium Redox Flow Battery. *submitted.*

# 499502 Solvation Dynamics and Energetics of Single-Walled Carbon Nanotubes (SWCNTs) in Water/Alcohol Mixtures

**Abstract:**

The unique structural, optical, and electronic properties of single-walled carbon nanotubes (SWCNTs) lend themselves to a breadth of potential applications. However, a difficulty in realizing widespread use of these materials is the polydispersity of samples in terms of size, chirality, and handedness due to the uncontrolled nature of current synthesis techniques. Numerous separation techniques with related protocols have been developed for use on SWCNTs. Most protocols involve adding surfactants to suspend the tubes in an aqueous media before using techniques such as ion-exchange chromatography (IEX) or aqueous two phase extraction (ATPE) to separate the SWCNTs based on their physiochemical properties. The solution environment around these nanoparticles is extremely important and complex, multi-component solvents are often used in order to tune the colloidal solvation and improve the separation. In this study, we use molecular dynamics and free energy perturbation techniques to calculate the solvation energy of various nanotubes in mixed solvents. Additionally, the structure and dynamics of the solvent shell have been analyzed to better understand the behavior on a molecular level and to provide insight into how different solvent environments respond to the carbon nanoparticle. These investigations work towards our ultimate goal of providing better insight into how the local environment surrounding SWCNTs effects various separation techniques.

# 487499 Effect of Liquid-Liquid and Solid-Liquid Interfacial Resistance on Heat Transfer in Nanomaterials

**Abstract:**

We have used molecular dynamics simulations to demonstrate that significant role of solid-fluid and liquid-liquid interfaces on thermal transport in nano-devices. By manipulating solid-fluid and liquid-liquid interfaces between a thin solid wall and a liquid and two immiscible liquids we have also demonstrated the possibility of rectification as well as the direction of rectification. This thus opens the possibilities of more precise control of heat transfer rates. Our simulations also show that interface manipulations of the type introduced here can lead to over six fold increase in heat transfer rates in nano-devices. Dissipating heat from miniaturized semi-conductors of the future will become increasingly challenging since size limitations are often dictated by the heat dissipation rates.

# 491207 Molecular Dynamics of Inorganic and Polymer Interface with Force-FieldÂ parameter Based on DFT Simulation

**Abstract:**

Interface between inorganic and organic materials is a key factor in nanocomposites.

The structural and energetic properties are important points to understand physics of the composites. Molecular dynamics (MD) simulation is a powerful tool to investigate such system, however, force-field (FF) parameter set appropriate to the interfacial system is not always accessible. One approach to determine FF parameter is to optimize the parameters so that the potential surfaces calculated in FF and density functional theory (DFT) simulation coincide.

We have developed a tool to determine FF parameters for interfacial system, and all-atom (AA) or coarse-grained (CG) force-field parameter for inorganic and organic polymer systems are obtained. We have utilized SIESTA and J-0CTA for DFT and MD simulations respectively. Characteristics of polymer chain near the interface will be discussed.

# 490536 Multi Metric 3D Protein Descriptors: The Correlation Impact of Algebraic Forms and Its Analysis

**Abstract:**

Macro molecular descriptors serve to distinguish and represent proteins of different functional and structural profiles by exploring its characteristic features, components and distribution of the constituent amino acids, and its physicochemical properties. In the present work, we applied the calculation of protein 3D descriptors by using several algebraic forms in the Rn space, applying various generalization metrics matrices. For these descriptors `calculations, vectors conformed by numerical values representing each amino acid side chain properties were used as weighting schemes. Several normalization methods (simple stochastic, mutual probability) were applied on the inter amino acid distance matrices as a calculation standardization. The local amino acidic invariant (LAI) is introduced to characterize fragments of interest in proteins and study its properties. Also, In order to evaluate inter amino acid interactions, topological and spatial cut-offs were applied.

To assess the utility of global and local indices, a classification model for the prediction of the major four protein structural classes was built with the Linear Discriminant Analysis (LDA) technique. The developed model correctly 92.6% and 92.7% of the proteins on the training and test sets, respectively. The model yield high values of the generalized square correlation coefficient ($GC^2$) on both the training and test series. The statistical parameters derived from the validation procedures endorse the strength, stability and the high predictive power of the proposed model. The performance of the LDA-model demonstrates the competence of the proposed indices not only to codify relevant biochemical information related to the structural classes of proteins, but also to yield suitable interpretability. It is anticipated that the current method will benefit the prediction of other protein attributes or functions.

# 485870 Accurate Methods to Describe System-Specific Polarization and Dispersion Energies

**Abstract:**

We introduce several improvements to the description of polarization and dispersion energies in complex materials. First, we use DDEC6 atomic population analysis [1,2] to improve the computed r-cubed moments of atoms in diverse material types. These r-cubed moments are used as inputs to compute atomic polarizabilities and dispersion coefficients. Second, we calculate and include the conduction limit upper bound on the polarizabilities of buried atoms. This vastly improves the description of polarizabilities in solids and other dense materials. Third, we introduce improved van der Waals radii and multi-body screening radii for the frequency-dependent dipole-dipole interaction tensor used in self-consistent screening[3] to compute C6 dispersion coefficients and polarizabilities. Fourth, we separately compute and clarify the difference between polarizability due to an externally applied electric field, fluctuating polarizabilities used in the Casimir-Polder integral, and atomic polarizabilities optimized for polarizable force-fields. Fifth, we introduce a new form of the damping functional for DFT+dispersion calculations. This new damping functional implicitly includes C8 and C10 dispersion contributions. Sixth, we derive and program a more computationally efficient formulation that includes periodic boundary conditions exactly into the multi-body dispersion (MBD) equations for the range-separated coupled-fluctuating dipole model (rs-CFDM[3]). We present computational results for molecules and solids, including several benchmark sets.

[1] T. A. Manz and N. Gabaldon Limas, “Introducing DDEC6 atomic population analysis: part 1. Charge partitioning theory and methodology,” RSC Advances, 6 (2016) 47771-47801.

[2] N. Gabaldon Limas and T. A. Manz, “Introducing DDEC6 atomic population analysis: part 2. Computed results for a wide range of periodic and nonperiodic materials,” RSC Advances, 6 (2016) 45727-45747.

[3] A. Ambrosetti, A. M. Reilly, R. A. Distasio Jr., and A. Tkatchenko, “Long-range correlation energy calculated from coupled atomic response functions,” J. Chem. Phys. 140 (2014) 18A508.

# 499446 Applications of Atomistic Machine Learning for Estimating Adsorbate Free Energy and Entropy on Late-Transition Metal Surfaces

**Abstract:**

Free energies of adsorption are fundamental quantities in microkinetic modeling of chemical reactions on catalyst surfaces. Density functional theory (DFT) based approaches are well suited for calculating the electronic components of these free energies, but are too expensive to perform the sampling necessary to evaluate the finite temperature contributions. Consequently, these entropic contributions to the free energy are usually approximated, e.g. by treating the adsorbate as a harmonic oscillator or as a two-dimensional ideal gas. In this work, we test an alternative approach for computing the non-electronic contributions to the adsorption free energies by using neural network potentials [1] to sample the potential energy landscape. The neural network potentials are trained on a set of DFT calculations of simple adsorbates on transition metal surfaces. The adsorption free energy and entropy are then computed using the canonical partition function. We compare our free energy estimates to those obtained using standard approximations, and discuss tradeoffs in computational cost and accuracy.

**References**

[1] Khorshidi, A. and Peterson, A.A., 2016. Computer Physics Communications, 207, pp.310-324.

# 500937 Reconstructing Ancient Sequences to Understand the Structure and Function Relationships of Modern Proteins

**Abstract:**

**Reconstructing Ancient Sequences to Understand the Structure and Function Relationships of Modern Proteins**

Zahra Shamsi, Alexander Moffett and Diwakar Shukla*, Department of Chemical and Biomolecular Engineering, UIUC

**Abstract**

Understanding protein structure and function relationship lies at the heart of modern drug discovery efforts [1]. Molecular dynamics simulation is one of the most powerful tools to study proteins dynamics and functions. The major challenge in protein simulation is efficient sampling of pathways associated with rare conformational transitions [2]. Advances in the field of genomics over the last decade, have led to uncovering the substantial gene sequencing information, which can be utilized as valuable unused information to enhance the simulation samplings.

Proteins are not just simple sequences of amino acids; they can adopt thousands of different three-dimensional conformations, which affect their function. Evolution changes a protein's function by altering its conformational preferences. In this poster, we outline novel computational methods based on evolutionary information to guide the exploration of the long timescale behavior of the proteins [3]. We employ these methods to investigate the activation and drug-selectivity mechanisms of human kinases that are implicated in numerous diseases including Cancer. A "new" methods combined with "old" enzymes, the ancestors dating back up to about billion years, provides an unexpected outlook for future intelligent design of drugs.

**References**:

[1] Agafonov, *et al.* Front. Mol. Biosci., 2, 1-8, 2015; [2] Shukla, *et al.* Nat. Commun., 5, 3397, 2014; [3] Shamsi, *et al.* Sci. Rep. InPress;

# 500998 Improved Thermal Gradient Quasiharmonic Approximations for Thermodynamic Properties of Organic Crystals with the Inclusion of Anisotropy

**Abstract:**

We explore the utility of variations in the quasi-harmonic approximation (QHA) to calculate the thermodynamic properties of small organic polymorphs and compare each approach to results for full molecular dynamics (MD). Due to the fact that crystalline structures of small organic molecules can vary in properties, fields such as as pharmaceutical formulation, organic electronics processing and explosives preservation have interest in fast and efficient methods to predict favorable structures. Common computational methods fail to include the dependence of temperature and importance of entropic effects on determining favorable crystal structures, while MD simulations are computationally costly. Quasi-harmonic (lattice dynamics) approximations provide a useful compromise to efficiently approximate entropy.

In QHA, it is standard to describe thermal expansion by constructing a stepwise array of isotropically expanded structures from the lattice minimum crystals and at any given temperature determine what structure minimizes the Gibbs free energy. We have developed an alternative method that determines the local gradient of thermal expansion and uses a 4th order Runge-Kutta method to efficiently determine the minimum free energy structure for thermal expansion. We have found that the gradient method produces smoother volumetric versus temperature curves and, generalizes easily to anisotropic expansion, which a more common in organic crystals. Between the two isotropic and one anisotropic method the vibrational spectra can be calculated in two ways; by directly solving the mass-weighted Hessian or to approximate the changes in the vibrational spectra due to thermal expansion with the Gruneisen parameter.

For 12 systems, with up to 60 atoms/molecule and 4 degrees of torsional freedom, this approach deviates by by less than 0.01 kcal/mol in the case of isotropic expansion. Therefore, using a Gruneisen parameter with the gradient approach could cut the computational time by 95%. Previous research has shown that isotropic QHA computes free energy curves within error of MD for small rigid molecules, but fails for systems with molecular flexibility. For these more flexible molecules, we can efficiently apply anisotropic QHA with the gradient method and determine how much of this deviation is due to anisotropic expansion and how much is due to other anharmonic sources.

# 501335 Mosdef, a Python-Based Molecular Simulation and Design Framework


**Abstract:**

Here we present MoSDeF [1], a set of extensible Python tools designed to facilitate the initialization, atom-typing, and screening of soft matter systems using molecular dynamics simulations. MoSDeF is designed around two key pieces of software, mBuild [2,3] and Foyer [4,5]. mBuild is designed to facilitate the generation of starting configurations, designed around the concept of defining and connecting ports on individual building blocks. mBuild supports parameterized structures through generative modeling, allowing for declaratively expressing repetitive structures, e.g. polymer chains, crystal structures, planar or spatial tiling, as well as for parameterizing the affine transformations applied to subcomponents of a composite component. mBuild also includes a robust library for the generation of crystal structures. With knowledge of the crystallographic information of the system of interest, users can quickly build and implement a crystal structure, of both atomistic and molecular subunits, using mBuild's provided toolkit. mBuild's crystal builder currently supports multiple basis particles, 1D, 2D, or 3D unit cells, and generating and saving arbitrarily-sized crystal structures. For end users, this software design approach minimizes and often eliminates the need to explicitly rotate and translate components when assembling systems and allows parameters to be programmatically varied for a family of systems (e.g., polymer chain length, crystal size) or interchange individual components (e.g., polymer type, crystal basis), while still employing the same general script.

Foyer is a general tool for applying force fields to molecular systems (i.e., atom-typing) that does not rely upon rigid rule hierarchies to atom-type a system. Foyer is designed such that it separates the engine that evaluates the atom-typing rules from the rules themselves, making it easier to debug, extend, and disseminate force fields. Foyer provides a force field and simulator agnostic schema for defining parameter usage that relies upon SMARTS [6] based annotations of chemical context, providing both human and machine readable documentation of parameter usage along with the parameters themselves in a single XML document.

mBuild and Foyer are able to output fully parameterized structures to various file formats used by common simulation engines (E.g. GROMACS, LAMMPS, HOOMD-Blue etc.). These tools can be coupled with the Signac-flow workflow manager [7], allowing large scale screening over parameter space. As such, MoSDeF provides a flexible, yet clear approach to encapsulate the routines and parameters used when initializing and performing molecular simulations. Which increases transparency and reproducibility of simulation results.



[1] https://github.com/mosdef-hub

[2] C. Klein, J. Sallai, T. J. Jones, C. R. Iacovella, C. McCabe, and P. T. Cummings, "A Hierarchical, Component Based Approach to Screening Properties of Soft Matter," in Foundations of Molecular Modeling and Simulation, 2016, pp. 79–92.

[3] https://github.com/mosdef-hub/mbuild

[4] Iacovella, C. R.; Sallai, J.; Klein, C.; Ma, T. In Idea Paper: Development of a Software Framework for Formalizing Forcefield Atom-Typing for Molecular Simulation, 4th Workshop on Sustainable Software for Science: Practice and Experiences (WSSSPE4), 2016.

[5] https://github.com/mosdef-hub/foyer

[6] http://www.daylight.com/dayhtml/doc/theory/theory.smarts.html

[7] C. S. Adorf and P. Dodd, "Signac-Flow." [Online]. Available: https://bitbucket.org/glotzer/signac-flow.

# 501498 Screening Self-Assembled Monolayers for Lubrication Properties: Trends and Pitfalls

**Abstract:**

Self-assembled monolayers present an appealing platform for creating fine tuned lubrication schemes for micro- and nanoelectromechanical devices (MEMS and NEMS). In addition to the choice of substrate and its roughness, monomers can vary in length as well as functional group and mixtures of monomers further expand the chemical parameter space of possible materials. Using the mBuild[1] and Foyer[2] tools - developed to work together under the MoSDeF [3] project - we create a pipeline to rapidly screen self-assembled monolayers across alkane and PEG monomers with varying chain lengths and head groups on both crystalline and amorphous silica substrates.

Our results suggest that the added flexibility of PEG chains slightly decreases coefficient of friction on more realistic, amorphous substrates and even though it decreases monolayer ordering. On rougher substrates, the added flexibility provides more fluid pathways for frictional dissipation when surfaces slide past one another. Additionally, the automated procedure used to screen these monolayers revealed several quirks in the simulation setup that further strengthen the case for using realistic, amorphous silica substrates to model self-assembled monolayers. While crystalline silica substrates are frequently used in simulation studies for convenience, our results demonstrate that these idealized systems can result in anomalous behavior related to lateral vibrations during shearing.

[1] C. Klein, J. Sallai, T.J. Jones, C.R. Iacovella, C. McCabe, P.T. Cummings. "A hierarchical component based approach to screening properties of soft matter." accepted at Mol. Mod. and Sim.: Applications and Perspectives

[2] C. Klein, A. Z. Summers, T. Ma, C. R. Iacovella, and J. Sallai, "Foyer." [Online]. Available: https://github.com/mosdef-hub/foyer.

[3] https://github.com/mosdef-hub.

# 501741 Addressing Discrepancies in Hydrogen Abstraction By Ooh Radical Via Automatic Transition State Theory Calculations

**Abstract:**

Combustion of fuels is a ubiquitous process that is prevalent in many aspects of our lives, and understanding combustion allows us to smartly design new fuels and better engines. Detailed kinetic modeling is an imperative tool to thoroughly understand combustion, and the construction of these models requires the determination of thousands of kinetic and thermodynamic parameters through experimentation, electronic structure calculations, or estimations. For combustion models, reactions are initialized by the abstraction of a hydrogen by one of 6 radical molecules (HO·, H·, HOO·, O·, $H_3C$· and $O_2$) making hydrogen abstraction reactions key for combustion. For the scope of this project, we chose to focus on hydrogen abstraction via hydroperoxyl radical (HOO·), and address these discrepancies.

Given that a variety of methods exist to calculate or estimate reaction kinetics, discrepancies often arise and need to be addressed. Two common methods currently exist to determine reaction kinetics: rate rules, and transition state theory (TST) calculations. Rate rules can determine reaction kinetics quickly, but they can be misapplied, causing incorrectly calculated kinetics. On the other hand, TST calculations rely on quantum chemistry calculations to determine kinetics. These calculations are often more accurate than rate rules, but they require large amounts of computation time, a good initial guess for the transition state geometry, and human input.

To eliminate the need for human input, we have developed an Automated TST method (AutoTST) [1] a module included in the Reaction Mechanism Generator (RMG) software [2] which automatically estimates transition state geometries using a group contribution method. AutoTST, utilized in this work, optimizes the transition state using a quantum chemistry package such as Gaussian [3] and is integrated with a canonical transition state calculator, CanTherm [4].

From 72 detailed kinetic models collected from recent literature, 331 reactions were identified as possible hydrogen abstraction reactions via HOOÂ· radical. Using algorithms in RMG, the reactions that successfully matched the bimolecular hydrogen abstraction reaction template were fed through the AutoTST method. At the time of wring, 228 reaction kinetics have been calculated. Literature values for these reaction rates were then compared against AutoTST calculated rates to validate and observe discrepancies in existing models and the AutoTST algorithm. Future works include calculation of the existing 447 identified HOOÂ· reactions, continued validation of existing models, and refinement of the AutoTST method.

[1] P. L. Bhoorasingh and R. H. West, Transition state geometry prediction using molecular group contributions, *Phys. Chem. Chem. Phys.* 17 (2015) 32173â€“32182.

[2] C. W. Gao, J. W. Allen, W. H. Green, and R. H. West, Reaction Mechanism Generator: Automatic construction of chemical kinetic mechanisms, *Comput. Phys. Commun.* 203 (2016) 212 - 225. DOI: 10.1016/j.cpc.2016.02.013.

[3] M. J. Frisch et al., Gaussian 09, Gaussian, Inc., Wallingford, CT, 2009.

[4] J. W. Allen and W. H. Green, CanTherm: Open-source software for thermodynamics and kinetics, Included in: Reaction Mechanism Generator, v2.0.0. 2016, URL: http://reactionmechanismgenerator.github.io.

# 494293 Development of the Parallel Monte Carlo Simulation Engine Gomc

**Abstract:**

The use of molecular simulation to study complex physical phenomena at the atomic level has grown exponentially over the last decade with increasing CPU power and the development of parallel molecular dynamics codes that scale efficiently over thousands of processors[1-6]. Molecular dynamics codes that utilize parallel computation on CPUs and GPUs are relatively well developed, however, there are a number of problems that cannot be simulated with this methodology. Problems that require the simulation of an open system, such as adsorption, require an alternative methodology that allows for fluctuation in the number of molecules in the system. For these types for problems, Monte Carlo is the method of choice.

GPU Optimized Monte Carlo (GOMC) is an object-oriented Monte Carlo simulation engine, capable of performing simulations in canonical, isobaric-isothermal, grand canonical ensembles, as well as Gibbs ensemble Monte Carlo. GOMC is designed for the simulation of complex molecular topologies, and supports a variety of potential functions, such as Lennard-Jones and Mie potentials. Coulomb interactions are supported via the Ewald summation method[7].

In this talk, the optimization of GOMC on multicore CPUs via OpenMP, and Graphics Processing Units (GPUs) via NVIDIA CUDA is discussed. Performance comparisons are presented for simulations in a variety of ensembles for different molecule types to illustrate the strengths and weaknesses of each architecture. In addition, a number of new code features are introduced, including fixed atoms (for the simulation of adsorption), calculation of the pressure tensor, anisotropic volume moves and improved file I/O. Use cases are presented for automated force field optimization, prediction of vapor-liquid and liquid-solid equilibria, and adsorption in porous materials.

**References:**

1. Phillips, J.C., R. Braun, W. Wang, J. Gumbart, E. Tajkhorshid, E. Villa, C. Chipot, R.D. Skeel, L. Kale, and K. Schulten, *Scalable molecular dynamics with NAMD.* Journal of Computational Chemistry, 2005. **26**(16): p. 1781-1802.

2. Salomon-Ferrer, R., A.W. Gotz, D. Poole, S. Le Grand, and R.C. Walker, *Routine Microsecond Molecular Dynamics Simulations with AMBER on GPUs. 2. Explicit Solvent Particle Mesh Ewald.* Journal of Chemical Theory and Computation, 2013. **9**(9): p. 3878-3888.

3. Anderson, J.A. and A. Travesset, *Molecular Dynamics on Graphic Processing Units: Hoomd to the Rescue.* Computing in Science & Engineering, 2008. **10**(6): p. 8-.

4. Eastman, P., M.S. Friedrichs, J.D. Chodera, R.J. Radmer, C.M. Bruns, J.P. Ku, K.A. Beauchamp, T.J. Lane, L.P. Wang, D. Shukla, T. Tye, M. Houston, T. Stich, C. Klein, M.R. Shirts, and V.S. Pande, *OpenMM 4: A Reusable, Extensible, Hardware Independent Library for High Performance Molecular Simulation.* Journal of Chemical Theory and Computation, 2013. **9**(1): p. 461-469.

5. Brown, W.M., P. Wang, S.J. Plimpton, and A.N. Tharrington, *Implementing molecular dynamics on hybrid high performance computers - short range forces.* Computer Physics Communications, 2011. **182**(4): p. 898-911.

6. Plimpton, S., *Fast Parallel Algorithms for Short-Range Molecular-Dynamics.* Journal of Computational Physics, 1995. **117**(1): p. 1-19.

7. Ewald, P.P., *The calculation of optical and electrostatic grid potential.* Annalen Der Physik, 1921. **64**(3): p. 253-287.

# Theoretical study energetic ionic salts composed of nitrogen bridge 3,3'-dinitro-5,5'-bis-1,2,4-triazole-1,1'-diolate anion and various cations

**Guolin Xiong, Weihua Zhu, * Heming Xiao**

*Institute for Computation in Molecular and Materials Science and Department of Chemistry,*

*Nanjing University of Science and Technology, Nanjing 210094, China*

* E-mail: zhuwh@njust.edu.cn. Tel.: 86-25-84315947-805. Fax: 86-25-84303919

**ABSTRACT:** In this work, we have studied the densities and HOFs for a series of the energetical ionic salts by using density functional theory. Overall, our calculated results depend on at the MP2/6-311++G**//B3LYP/6-311++G** level. It is found that the title salts have high densities ranging from 1.65 to 1.86 g $cm^{-3}$. A1 and A2 is favorable for enhancing the density of ionic crystal. The salt containing the cation A1 has the highest density among the salt series. All the title salts possess high positive HOFs. Among the ionic salts, the salts containing triaminoguanidinium has the highest HOF. Meanwhile, the salt containing guanidinium also presents higher HOFs. Maybe it is because high nitrogen content contributes more to a positive HOF values. Many salts present comparable detonation performance with commonly used explosives RDX or HMX.

## 1. INTRODUCTION

To meet the continuing command for improved high-energy density materials(HEDMs), numerous studies have been focused on probing nitrogen-containing heterocyclic compounds with decreased sensitivity and excellent performance as well as environment compatibility.[1-5] Of the HEDMs, energetic nitrogen-rich ionic salts have received much attention owing to better properties(lower vapor pressures, higher densities, lower melting points, and higher cohesive energy densities) than their atomically similar nonionic compounds.

The triazole ring is a useful fragments for developing energetic compounds because of its inherently high nitrogen contents and good thermal stability. Their derivatives have played a vital role as energetic components in the production of high explosives, solid propellants, and pyrotechnic gas generators. It is reported that incorporating azo(-N=N-) linkage into the high-nitrogen compounds not only desensitizes but also increases the melting point dramatically. And numerous compounds with promising properties as energetic materials arose from the N-N connection of those triazole rings to 5,5'-bistriazoles.[1] A further way of azole functionalization is the oxidation of the heterocycle to its corresponding N-hydroxy compound. The introduction of N-oxides is a recently reintroduced method using oxidizing agents like trifluoroperacetic acid, potassium peroxomonosulfate(oxone), or hypofluorous acid. The additional oxygen atom generally leads to increased energetic properties due to a higher density and an even greater energy output. The oxidation of tetrazole compounds has been successfully accomplished recently, resulting in high performance explosives with low sensitivities. Our main purpose is to investigate the important role of nitrogen bridge conjugates with trizaole in the design of efficient high-energy ionic salts.

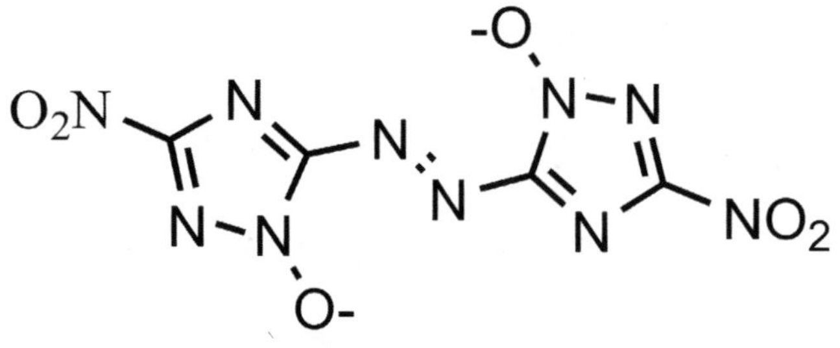


**Fig. 1** Framework of anion

## 2. COMPUTATIONAL METHODOLOGY

The calculations were performed with the Gaussian 03 suite of programs.[6] Here we used two methods (MP2/6-311++G**//B3LYP/6-311++G**) to calculate the properties of ionic salts. Figure 1 displays the frameworks of anions studied here.

Density is one of the most important physical properties of a solid energetic material. Thus, many approaches[7-11] have been developed to accurately predict the crystal density without a prior knowledge of the crystal structure. The traditional *M*/*V* procedure for predicting densities often leads to large errors, because it completely ignores intermolecular interactions within the crystal. The crystal density of ionic crystals can be improved by[12]

$$\rho = \alpha(M/V) + \beta(V_S^+ / A_S^+) + \gamma(V_S^- / A_S^-) + \delta \tag{1}$$

where *M* is the chemical formula mass of the compound and V is the volume of the isolated gas molecule. $A_S^+$ is the portion of a cation's surface that has a positive electrostatic potential. $V_S^+$ is the average value of that potential. $A_S^-$ and $V_S^-$ are the analogous quantities for an anion. The values of the coefficients $\alpha$, $\beta$, $\gamma$, and $\delta$ are 1.026, 0.0514, 0.0419, and 0.0227, respectively.

For an ionic crystal with formula unit $M_pX_q$, its volume is simply the sum of the volumes of the ions contained in the formula unit:[10]

$$V = pV_{M+} + qV_{X-} \tag{2}$$

where $V_{M+}$ is the volume of the cation $M^+$, $V_{X-}$ is the volume of the anion $X^-$, $p$ is the number of the cation $M^+$ per formula unit, and $q$ is the number of the anion $X^-$ per formula unit.

Since the volumes of individual ions can be obtained using the DFT procedure, we can us eq 2 to calculate formula unit volumes for ionic salts.

Based on a Born–Haber energy cycle, the HOF of a salt can be simplified by eq 3:

$$\Delta H^{\circ}_{f}(\text{salt,298K}) = \Delta H^{\circ}_{f}(\text{cation,298K}) + \Delta H^{\circ}_{f}(\text{anion,298K}) - \Delta H_{L} \quad (3)$$

Where $\Delta H_L$ is the lattice energy of the salts that can be predicted by the formula suggested by Jenkins et al. as[13]

$$\Delta H_L = U_{POT} + [p(n_M/2 - 2) + q(n_X/2 - 2)]RT \quad (4)$$

where $n_M$ and $n_X$ depend on the nature of the ions $M_{p+}$ and $X_{q-}$, respectively, and are equal to 3 for monatomic ions, 5 for linear polyatomic ions, and 6 for nonlinear polyatomic ions. The equation for lattice potential energy $U_{POT}$ ($kJ \cdot mol^{-1}$) is as follows:

$$U_{POT} = \gamma(\rho/M)^{1/3} + \delta \quad (5)$$

where $\rho$($g \cdot cm^{-3}$) is the density and M($g \cdot mol^{-1}$) is the chemical formula mass of the ionic material. For 1:2(charge ratio) salts, the coefficients $\gamma$ and $\delta$ are 8375.6 $kJ \cdot mol^{-1} \cdot cm$ and -178.8 $kJ \cdot mol^{-1}$, respectively.

The isodesmic reactions used to calculate the heats of formation (HOFs) of nitrogen bridge 3,3'-dinitro-5,5'-bis-1,2,4-triazole-1,1'-diolate anion and various cations at 298 K are illustrated in Scheme 2. For some small molecules, the experimental HOFs are unavailable; additional calculations were carried out for the atomization reaction: $C_aH_bO_cN_dF_e \rightarrow aC(g) + bH(g) + cO(g) + dN(g) + eF(g)$ using the G2 theory to get their HOFs.

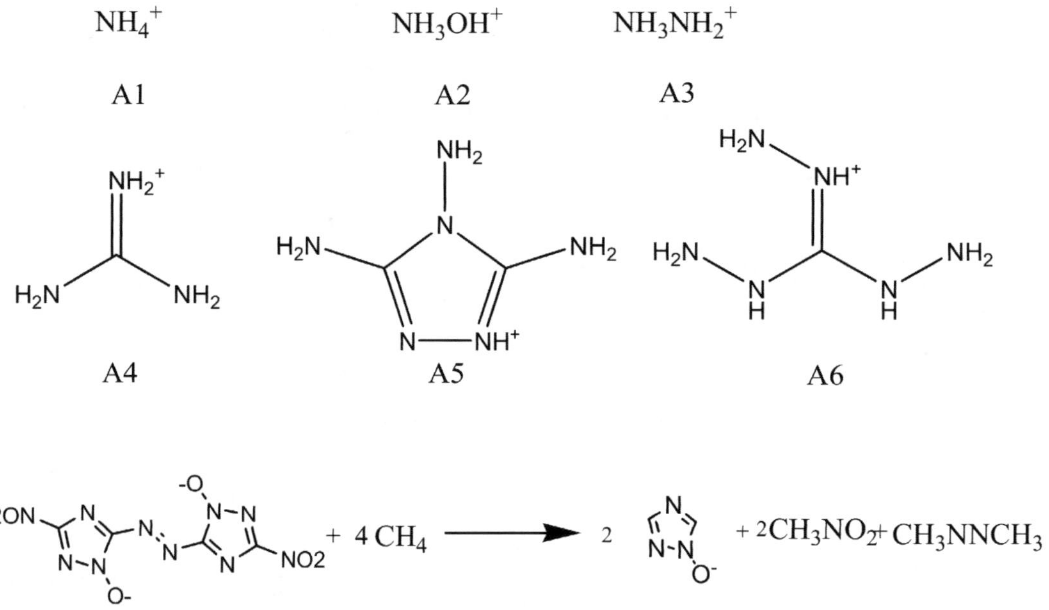


**Fig. 2** Framework of a series of cations and isodemic reactions for the anion.

## 3. Results and discussion

### 3.1 Crystal Density.

High density is a key parameter commonly required for energetic materials because more energy will be packed per unit volume. In this work, we investigate the effects of different cations on the densities of the energetic salts. Formula unit volumes to be used in predicting the crystal densities of ionic crystal were determined using eq 2. The volumes of the cations and anions contained in the salts were calculated using the optimized structures. The densities and other corresponding data of the salts are listed in Table 1. Figure 2 exhibits a comparison of the effects of different cations on the densities of ionic crystal.

**Table 1** Volumes(cm$^3$/mol) and Densities =(g/cm$^3$) of the salts

| Salt | | V | As- | Vs- | As+ | Vs+ | ρ |
|---|---|---|---|---|---|---|---|
| Anion | Cation | | | | | | |
| | A1 | 315.49 | | | 47.51 | 171.72 | 1.86 |
| | A2 | 338.39 | | | 60.76 | 153.27 | 1.84 |
| anion | A3 | 350.24 | 261.92 | -145.96 | 66.61 | 146.71 | 1.76 |
| | A4 | 409.15 | | | 95.02 | 122.72 | 1.69 |
| | A5 | 509.19 | | | 139.42 | 102.34 | 1.70 |
| | A6 | 505.95 | | | 139.54 | 100.90 | 1.65 |

As shown in Table 1, all the salts have relatively high densities ranging from 1.65 to 1.86 g cm$^{-3}$. When the cation is ammonium(A1) and A2, its energetic salt has higher density than the rest one. It is obviously that A1 and A2 is favorable for enhancing the density of ionic crystal. The salt containing the cation A1 has the highest density among the salt series. It is just the opposite for the cation A6. The evolution pattern of the HOFs for the salts containing the cations is different. Therefore, the effects of different cations on the density of their salts are variable.

Figure 3 illustrates a comparison of the densities of a series of energetic salts. It is obvious that when the cation is guanidinium and triaminoguanidinium, the density value decreases. It is mainly because of the steric hindrance inside the cation.

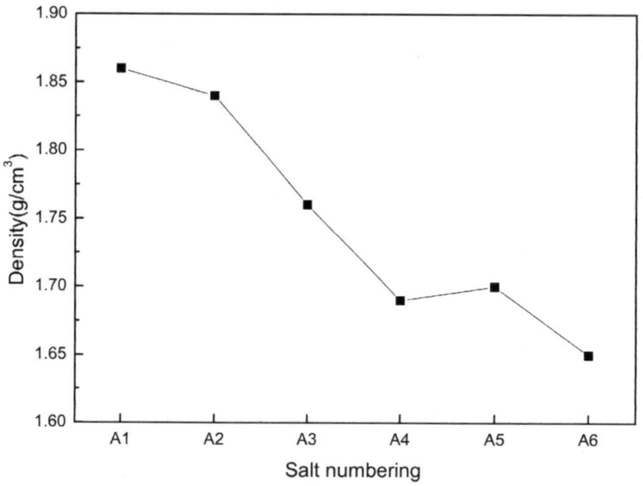


**Fig. 2** Comparison of the densities of the ionic salts.

### 3.2 Heats of formation.

As the HOF is frequently taken to be indicative of the "energy content" of an energetic compound, it is of great importance to predict its HOF. Table 2 lists the calculated HOFs of reference compounds in the isodesmic reactions along with available experimental and other calculated values. The calculated results indicate that the HOFs of small molecules are very close to corresponding experimental values. In addition, the calculated HOFs of the reference anions and cation are in agreement with other calculated results. Table 3 presents the HOFs of the anion, cations and their salts and lattice energies of these salts. The results demonstrate that the calculated HOFs are very close to other calculated values. Therefore, our computational method to predict the HOFs is expected to be reliable in this work.

**Table 2** Calculated and experimental heats of formation(kJ/mol) for small molecules and ions at 298 K

| Molecules | $\Delta H_f$(calc.)[a] | $\Delta H_f$(expt.)[b] | Molecules | $\Delta H_f$(calc.)[a] | $\Delta H_f$(expt.)[b] |
|---|---|---|---|---|---|
| $NH_3$ | -45.1 | -46.1 | $C_2H_6$ | -86.1 | -84.0 |
| $NH_4^+$ | 637.1 | | 1,2,4-trizole | 193.4 | 193.7 |
| $H^+$ | | 1536.2 | $NH_2NH_3^+$ | 762.7 | 759.7 |
| $CH_4$ | -77.8 | -74.4 | A0 | -316.9 | |
| $CH_3NO_2$ | -88.2 | -80.8 | Guanidinium | 589.3 | 575.9 |
| $NH_3OH^+$ | 683.8 | | Triaminoguanidinium | 893.4 | 890.3 |
| $NH_2NH_2$ | 110.5 | 95.4 | $CH_3NNCH_3$ | 151.9 | 151.5 |

As shown in Table 3, when the cation is A3 and A6, its salts have relatively larger HOFs than other salts. It means that containing the cation A3 or A6 is favorable for increasing the HOFs of the salts. The evolution pattern of the HOFs for the salts constituted of different cations is quite different. Except for the state containing triaminoguanidinium, it delivers the highest HOFs, which is triple or four times of the HOFs with the other cations. Consequently, the effects of different cations on the HOFs of their salts are quite different.

**Table 3** Heats of formation of different cations and their salts, and lattice energies of these salts

| | MP2/6-311++G**//B3LYP/6-311++G** | | | |
|---|---|---|---|---|
| cation | | | lattice energy | calc |
| A1 | 637.13 | | 1333.33 | 269.27 |
| A2 | 683.36 | | 1282.76 | 412.31 |
| A3 | 762.72 | 328.35 | 1263.09 | 590.70 |
| A4 | 588.82 | | 1178.67 | 327.32 |
| A5 | 568.42 | | 1076.41 | 388.78 |
| A6 | 893.40 | | 1079.23 | 1035.92 |

Figure 3 illustrates a comparison of the HOFs for a series of ionic salts. Among the ionic salts, the salts containing triaminoguanidinium has the highest HOF. Meanwhile, the salt containing guanidinium also presents higher HOFs. Maybe it is because high nitrogen content contributes more to a positive HOF values. However, it induces the dramatical decrease in the density of its salts. So it is contradictory between the density and HOFs. According to the computational study, it helps improving the balance of the density and HOFs and screening the outstanding properties of energetic materials. Many salts present comparable detonation performance with commonly used explosives RDX or HMX.

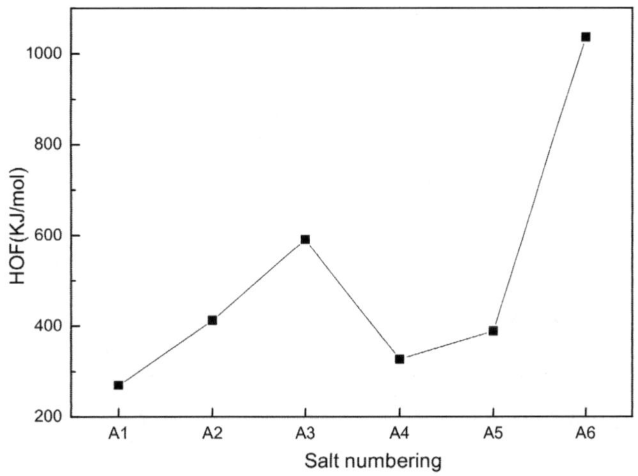


**Fig. 3** A comparison of the HOFs for a series of salts at the MP2/6-311++G**//B3LYP/6-311++G** level

## ■ Conclusion

In this work, we have studied the densities and HOFs for a series of the energetic ionic salts by using density functional theory. Overall, our calculated results depend on at the MP2/6-311++G**//B3LYP/6-311++G** level. It is found that the title salts have high densities ranging from 1.65 to 1.86 g $cm^{-3}$. A1 and A2 is favorable for enhancing the density of ionic crystal. The salt containing the cation A1 has the highest density among the salt series. All the title salts possess high positive HOFs. Among the ionic salts, the salts containing triaminoguanidinium has the highest HOF. Meanwhile, the salt containing guanidinium also presents higher HOFs. Maybe it is because high nitrogen content contributes more to a positive HOF values. Many salts present comparable detonation performance with commonly used explosives RDX or HMX.

**Reference**

(1) Dippold, A. A.; Klapotke, T. M., A study of dinitro-bis-1,2,4-triazole-1,1'-diol and derivatives: design of high-performance insensitive energetic materials by the introduction of N-oxides. *J Am Chem Soc* **2013,** *135* (26), 9931-8.
(2) Kamlet, M. J.; Jacobs, S. J., Chemistry of Detonations. I. A Simple Method for Calculating Detonation Properties of C–H–N–O Explosives. *J. Chem. Phys.* **1968,** *48* (1), 23-35.
(3) Keshavarz, M. H.; Motamedoshariati, H.; Moghayadnia, R.; Nazari, H. R.; Azarniamehraban, J., A new computer code to evaluate detonation performance of high explosives and their thermochemical properties, part I. *J Hazard Mater* **2009,** *172* (2-3), 1218-28.
(4) Xiang, F.; Wu, Q.; Zhu, W.; Xiao, H., Comparative Theoretical Studies on Energetic Ionic Salts Composed of Heterocycle-Functionalized Nitraminofurazanate-Based Anions and Triaminoguanidinium Cation. *J. Chem. Eng. Data* **2014,** *59* (2), 295-306.
(5) Xiang, F.; Zhu, W.; Xiao, H., Theoretical studies of energetic nitrogen-rich ionic salts composed of substituted 5-nitroiminotetrazolate anions and various cations. *J Mol Model* **2013,** *19* (8), 3103-18.
(6) Frisch, M. J. T., G. W.; Schlegel, H. B.; Scuseria, G. E.; Robb, M. A.; Cheeseman, J. R.; Montgomery, J. J. A.; Vreven, T.; Kudin, K. N.; Burant, J. C.; Millam, J. M.; Iyengar, S. S.; Tomasi, J.; Barone, V.; Mennucci, B.; Cossi, M.; Scalmani, G.; Rega, N.; Petersson, G. A.; Nakatsuji, H.; Hada, M.; Ehara, M.; Toyota, K.; Fukuda, R.; Hasegawa, J.; Ishida, M.; Nakajima, T.; Honda, Y.; Kitao, O.; Nakai, H.; Klene, M.; Li, X.; Knox, J. E.; Hratchian, H. P.; Cross, J. B.; Adamo, C.; Jaramillo, J.; Gomperts, R.; Stratmann, R. E.; Yazyev, O.; Austin, A. J.; Cammi, R.; Pomelli, C.; Ochterski, J. W.; Ayala, P. Y.; Morokuma, K.; Voth, G. A.; Salvador, P.; Dannenberg, J. J.; Zakrzewski, V. G.; Dapprich, S.; Daniels, A. D.; Strain, M. C.; Farkas, O.; Malick, D. K.; Rabuck, A. D.; Raghavachari, K.; Foresman, J. B.; Ortiz, J. V.; Cui, Q.; Baboul, A. G.; Clifford, S.; Cioslowski, J.; Stefanov, B. B.; Liu, G.; Liashenko, A.; Piskorz, P.; Komaromi, I.; Martin, R. L.; Fox, D. J.; Keith, T.; AlLaham, M. A.; Peng, C. Y.; Nanayakkara, A.; Challacombe, M.; Gill, P. M. W.; Johnson, B.; Chen, W.; Wong, M. W.; Gonzalez, C.; Pople, J. A., *Gaussian 03; Gaussian, Inc.: Pittsburgh, PA,* **2003**.
(7) Namazian, M., Density functional theory response to the calculation of electrode potentials of quinones in non-aqueous solution of acetonitrile. *Journal of Molecular Structure: THEOCHEM* **2003,** *664-665*, 273-278.
(8) Pan, J.-F.; Lee, Y.-W., Crystal density prediction for cyclic and cage compounds. *PCCP* **2004,** *6* (3), 471.
(9) Qiu, L.; Xiao, H.; Gong, X.; Ju, X.; Zhu, W., Crystal density predictions for nitramines based on quantum chemistry. *J Hazard Mater* **2007,** *141* (1), 280-8.
(10) Rice, B. M.; Hare, J. J.; Byrd, E. F. C., Accurate Predictions of Crystal Densities Using Quantum Mechanical Molecular Volumes. *The Journal of Physical Chemistry A* **2007,** *111* (42), 10874-10879.
(11) Bouhmaida, N.; Ghermani, N. E., Elusive contribution of the experimental surface molecular electrostatic potential and promolecule approximation in the empirical estimate of the crystal density. *J Chem Phys* **2005,** *122* (11), 114101.

(12)Politzer, P.; Martinez, J.; Murray, J. S.; Concha, M. C., An electrostatic correction for improved crystal density predictions of energetic ionic compounds. *Mol. Phys.* **2010,** *108* (10), 1391-1396.
(13)Jenkins, H. D. B.; Tudela, D.; Glasser, L., Lattice Potential Energy Estimation for Complex Ionic Salts from Density Measurements. *Inorg. Chem.* **2002,** *41* (9), 2364-2367.

# Structural Transformations and Absorption Properties of Crystalline 4,10-dinitro-2,6,8,12-tetraoxa-4,10-diazaisowurtzitane under High Pressures

Dong Xiang, Weihua Zhu*

*Institute for Computation in Molecular and Materials Science and Department of Chemistry, Nanjing University of Science and Technology, Nanjing 210094, China*

**Abstract:** The results indicate that the PW91-OBS is the best functional for studying 4,10-dinitro-2,6,8,12-tetraoxa-4,10-diazaisowurtzitane (TEX). As the pressure increases, the lattice of parameters, band gap, density of states and absorption spectra change regularly except for 203 GPa, where TEX begins to decompose and form the N1-N1' new bond. An analysis of the band gap and density of states indicates that TEX becomes more and more sensitive under compression. The absorption spectra show that TEX has relatively high optical activity with increasing pressure.

**Keywords.** DFT with dispersion-correction (DFT-D); 4,10-dinitro-2,6,8,12-tetraoxa-4,10-diazaisowurtzitane; hydrostatic pressure; structures; optical absorption

---

*To whom correspondence should be addressed. E-mail: zhuwh@njust.edu.cn.

1. INTRODUCTION

Cage crystal molecules containing the nitro groups are important and widely used as energetic materials [1–3] due to their high energy, density, detonation pressure, and low sensitivity. isowurtzitane cage structure. isowurtzitane cage structure containing nitramine functionalities as the most prominent and relatively powerful explosives have attracted much interest. [4,5] 4,10-dinitro-2,6,8,12-tetraoxa-4,10-diazaisowurtzitane (Figure 1a) is better known as CL-20 (Figure 1b)which is the most dense stable CHNO compound with an ambient density of the -polymorph of $\rho$ =2.044 g/cm$^3$. [6] TEX possesses the same isowurzitane cage structure as CL-20 (2,4,6,8,10,12-hexanitro-2,4,6,8,10,12-hexaazaisowurtzitane or HNIW) (Figure 1), which is one of the most powerful high explosives that have been synthesized to date, and it also has a relatively high ambient temperature density of 1.99 g/cm$^3$. The absence of sterically demanding nitramine groups greatly reduces its sensitivity greatly and makes it an interesting insensitive energetic material ($h_{50}$ >177 cm).[7] Its heat of formation, detonation velocity, and detonation pressure were calculated to be -541 kJ/mol, 8665 m/s, and 41.03 GPa, respectively. [8] To understand the complex behavior of explosives, to control risk during their usage and storage, and to develop new high-energy explosives, it is necessary to study the structure and properties of TEX as it has a special structure and exhibits excellent performance under high pressure.

*< Fig. 1 about here>*

Although several studies have reported its synthesis [9] and thermal decomposition mechanism, [10] many fundamental and practical problems of TEX are still not well understood because it possesses a complex chemical behavior. For instance, to the best of our knowledge, its behavior under different pressures has not been investigated. It is important to study and obtain the structure information over a range of pressures, especially under high pressures. It is acknowledged that the performance of an explosive depends on several main factors like crystal density and morphology, thermal stability, detonation velocity and pressure, and chemical reactivity which are dominated by their solid-state structure. In addition, the information on explosive under extreme conditions is helpful for its safe use and effective and cost-effective disposal. Therefore, in order to understand the properties of TEX clearly, it is necessary to study the behavior of TEX under high pressure.

The investigation of the microscopic properties of energetic materials under high pressures still remains a challenging task for experiment measurements. Theoretical calculations may be an effective way to model the physical and chemical properties of complex solids under high pressure at the atomic level as a complement to experimental work. [11,12] Nowadays, density functional theory (DFT) with pseudopotentials and a plane-wave basis set has been well established and has been applied successfully to study the structures and properties of energetic solids under hydrostatic compression. [13]

In this study we performed periodic DFT with vdW correction (DFT-D) calculations to study the structural, electronic, and absorption properties of crystalline TEX. Our main purpose here is to examine the differences in their electronic structure and properties.

The remainder of this paper is organized as follows. A brief description of our computational method is given and the results and discussion are then presented followed by a summary of our conclusions.

**Computational method**

The calculations performed in this study were done based on DFT [14] as implemented in the Cambridge Serial Total Energy Package (CASTEP) code. [15] The calculations were done using Vanderbilt ultrasoft pseudopotentials for all atoms [16] and generalized gradient approximation [17] with the PBE (Perdew Burke Ernzerhof) [18] and PW91 (Perdew and Wang) [19] functionals. To correct DFT for missing vdW interaction, we used three types of corrections such as the Grimme (G06) [20] and Tkatchenko and Scheffler (TS) [21] corrections to the PBE and the Ortmann, Bechstedt, and Schmidt (OBS) [22] correction to the PW91. The self-consistent ground state of the system was determined by using a band-by-band conjugate gradient technique to minimize the total energy of the system with respect to the plane-wave coefficients. The electronic wave functions were obtained by a density-mixing minimization method [23] and the structures were relaxed by using the Broyden, Fletcher, Goldfard, and Shannon (BFGS) method [24] Geometry optimization is based on reducing the magnitude of calculated forces and stresses until they become smaller than defined convergence tolerances. Therefore, it is possible to specify an external stress tensor to model the behavior of the system under tension, compression, shear, etc. In these cases, the internal stress tensor is iterated until it becomes equal to the applied external stress. The cutoff energy of plane waves was set to 600 eV. Brillouin zone sampling was performed by using the Monkhost Pack scheme with a *k*-point grid of 2 × 1 × 1. The values of the kinetic energy cutoff and the *k*-point grid were determined to ensure the convergence of total energy.

TEX crystallizes in an orthorhombic lattice with *P-1* space group and contains two C6N6O4H6 molecules per unit cell. Figure 2 displays the crystal and atomic structures of TEX. The experimental crystal structure of TEX was first relaxed to allow the atomic configurations, cell shape, and volume to change at zero pressure. Then from the relaxed structure, we applied hydrostatic compression of 0–220 GPa to relax the crystal structure without any symmetry constraints. All calculations are based on the same crystal structure of TEX. In the geometry relaxation, the total energy of the system was converged less than $2.0 \times 10^{-5}$ eV, the residual force less than 0.05 eV/Å, the displacement of atoms less than 0.002 Å, and the residual bulk stress less than 0.1 GPa for all of the equilibrium structures.

*< Fig. 2 about here>*

**Results and discussion**

Here we applied the G06 and TS corrections to the PBE (labeled as PBE-G06 and PBE-TS) functional and the OBS correction to the PW91 (labeled as PW91-OBS) functional to fully relax TEX crystal at ambient pressure without any constraint, respectively. Table 1 lists the experimental and relaxed cell parameters of TEX crystal. It is found that the lattice parameters and cell volume estimated by the PW91-OBS here are much closer to the experimental value than those by the PBE-G06 and

PBE-TS. In addition, the results calculated by the PW91-OBS and PBE-TS are obviously different from the experimental values. This indicates that the PW91-OBS is suitable for study of the three crystals. Although the cell angle values estimated by the PW91-OBS here are much larger to the experimental value than those by the PBE-G06 and PBE-TS. The errors estimated by the PW91-OBS are acceptable. Considering the errors of cell angle, lattice parameters and cell volume, we adopted the PW91 with the OBS correction to study their structure and properties under high pressure here.

*< Table 1 about here>*

**Crystal and molecular structure**

Figure 3 displays the relaxed lattice constant (*a*, *b*, *c)* of TEX in the pressure range of 0-220 GPa. Overall, *a, b*, and *c* decrease with the increasing pressure. But the pressure-induced change trends of the lattice constants are different. It is seen in Figure 3 that the *c* axis is the most compressible. This indicates that the compressibility of TEX crystal is anisotropic. Figure 4 displays the relaxed lattice constant (*a*, *b*, *c)* compression ratio of TEX. In the pressure range of 0-220 GPa, the total compression ratio increases in the order of $b < a < c$, indicating that the structure is much softer in the *c* direction than in the *a* and *b* axis. Pressure induced changes of the unit-cell volumes are presented in Figure 5. The unit cell volume decreases monotonically. At about 50 GPa, the volume compression is up to 50 %.

*< Fig. 3, Fig. 4 and Fig. 5 about here>*

The applied pressure causes the changes not only in the unit cell of TEX but also in its molecular geometry like the bond lengths and bond angles. Figure 6 presents the selected three bond angles (C1-N1-N2, C1-N1-C4, and O7-C3-N3 angles) of TEX at different pressures. In the range of 0-203 GPa, the angles of C1-N1-N2 and O7-C3-N3 decrease monotonically, while the C1-N1-C4 increase monotonically. At 203 GPa, all the angles decrease suddenly. In the range of 203-220 GPa, the C1-N1-N2, and C1-N1-C4 angles increase slightly, whereas the O7-C3-N3 angle decreases a little. This indicates that the structure is changed at the pressure of 203 GPa.

*< Fig. 6 about here>*

Figure 7 presents the selected two twist angles (O5-C1-N1-C4, and C2-C1-N1-N2 angles) of TEX at different pressures. In the range of 0-203 GPa, the angles of O5-C1-N1-C4 increases monotonically. The C1-N1-C4 increase monotonically during 0-50 GPa, and it decreases monotonically in the range of 50-220 GPa. However, both the two twist angles decrease abruptly at 203 GPa. This indicates that the structure is changed at the pressure of 203 GPa.

*< Fig. 7 about here>*

Figure 8 presents the atom distance between N1 and N1' in TEX at different pressures. When the pressure increases from 0 to 200 GPa, the atomic distance between N1 and N1' decreases monotonically. At 203 GPa, the distance between N1 and N1' atoms decreases abruptly from 1.738 to 1.307 Å, which shows that N1 and N1' atoms form a new bond at this pressure; moreover, the N1 and N1' atoms are sensitive to compression. It may thus be inferred that the formation of the N1 and N1' bond may possibly trigger the TEX decomposition.

*< Fig. 8 about here>*

**Electronic Structure**

Band gap is an important parameter to characterize the electronic structure of solids. Figure 9 displays the band gaps of NADNP at different pressures. The band gap of TEX gradually decreases with the increase of pressure in 0–200 GPa range, it is a constant zero in 203–220 GPa. At 203 GPa, the abruptly reduction to zero of the band gap is caused by the structural change and it has metal property. The smaller the band gap, easier is the electron transfer from the valence band to the conduction band, and higher the probability that the energetic materials become decomposed and explode. [25-27] Therefore, it may be inferred that the impact sensitivity for TEX becomes more and more sensitive with the increase of pressure.

*< Fig. 9 about here>*

In order to understand the band nature and electronic structure of the crystal TEX at different pressures, we investigated its total density of states (DOS) and partial DOS (PDOS). Figure 10 displays the total and partial DOS of the C, N states, and H states of TEX at 0 GPa. The PDOS of the O states and N states are larger than those of the H states. It may be expected that the former makes more important contributions to the valence bands than the latter, indicating that the O states and N states act as active centers.

*< Fig. 10 about here>*

Figure 11 presents the total density of states (DOS) of NADNP at 10, 100, 200,

203, 205 and 210 GPa. It is seen that the peaks become more and more wide and dispersed and have a tendency of shifting to lower energy with increasing pressure. This indicates that the band formation and band dispersion increase is accompanied by a broadening of DOS because the intermolecular distances are reduced and thus the intermolecular interactions are enhanced with the increasing pressure. Upon compression the DOS peaks in the conduction bands have a tendency of shifting to lower energy. Therefore, the electronic delocalization in bulk TEX increases. At 203 GPa, the DOS connects together and appears a peak at the Fermi energy level, indicating that the electrons have become delocalized completely and can transfer freely between the valence band and conductions band.

*< Fig. 11 about here>*

**Absorption spectra**

In this section, we turn to investigate the optical absorption coefficients of NADNP at different pressures. The interaction of a photon with the electrons in the system can result in transitions between occupied and unoccupied states. The spectra resulting from these excitations can be described as a joint density of states between the valence and conduction bands. The imaginary part $\varepsilon_2(\omega)$ of the dielectric function can be obtained from the momentum matrix elements between the occupied and unoccupied wave functions within the selection rules, and the real part $\varepsilon 1(\omega)$ of dielectric function can be calculated from imaginary part $\varepsilon_2(\omega)$ by Kramer–Kronig relationship. Absorption coefficient $\alpha(\omega)$ can be evaluated from $\varepsilon_1(\omega)$ and $\varepsilon_2(\omega)$:[28]

$$\alpha(\omega) = \sqrt{2}\omega(\sqrt{\varepsilon_1^2(\omega) + \varepsilon_2^2(\omega)} - \varepsilon_1(\omega))^{1/2}$$

The absorption coefficients $\alpha(\omega)$ of TEX at different pressures are shown in Figure 12. The absorption spectra are active over various regions corresponding to the molecular or lattice structures of the individual material. The evolution pattern of absorption spectra for TEX at different pressures is qualitatively similar.

They have an absorption band covering from 0 to 35.0 eV and more strong optical absorption from 12.5 to 17.5 eV. The magnitude of the absorption coefficients of these peaks allows an optical transition due to excitons. At ambient pressure, TEX exhibits a relatively high absorption coefficient over a relatively few, closely spaced bands. The strongest peak occurs at 14.0 eV, and other three weak peaks occur at 5.0, 9.0, and 12.9 eV. In the energy below 2.5 eV or above 22.5 eV, the absorption coefficients are close to zero, showing that TEX is transparent in the wavelength range above 885.7 nm or below 72.9 nm. In the pressure range of 0-200 GPa, the absorption coefficient of the strongest peak enhances consistently, showing that TEX has relatively high optical activity at high pressure, and the structural transformation that occurred at 203 GPa obviously improves the optical activity of TEX. When the pressure increases from 100 to 200 GPa, the strongest peak undergoes a obvious shift toward higher energy, and its absorption coefficient decreases significantly. This is because of the sudden reduction in the band gap and abnormal changing in the structure at 203 GPa. It is seen that the absorption width at 203 GPa are much lager than that at 100 GPa in the energy range of 0-2 eV, and the more peaks are appeared, showing that the structural transformation at 203 GPa broadens the absorption region of TEX in the low energy range. The absorption spectra at 205 GPa, and 210 GPa are similar to that at 203 GPa, suggesting that the structures at the three pressures are much alike.

< *Fig. 12 about here*>

## 4. CONCLUSIONS

In this work, DFT calculations have been performed to study the structural, electronic, and absorption properties of crystalline TEX under hydrostatic compression of 0-110 GPa. Our results show that structural transformations by forming N1-N1' occurred at the pressure of 203 GPa. An analysis of the band gap and density of states of TEX indicates that its electronic character transforms into a metallic system at 203 GPa, and it becomes more sensitive under compression. The absorption spectra show that TEX has relatively high optical activity with the increasing pressure, and the structural transformations that occurred at 203 GPa can broaden its absorption region in the energy range of below 2 eV, respectively. This work may provide useful information in understanding how TEX behaves under high pressures. Our results may be helpful for discovering some new structures that are unobserved at the ambient conditions.

[1] Eaton, P. E. and T. W. Cole. **1964**. Cubane. *J. Am. Chem. Soc.*, 86: 3157–3158.

[2] Eaton, P. E., Y. S. Or, and S. J. Branca. **1981**. Pentaprismane. *J. Am. Chem. Soc.*, 103: 2134–2136.

[3] Dauben, W. G. and A. F. Cunningham. **1983**. Formal synthesis of pentaprismane. *J. Org. Chem.*, 48: 2842–2847.

[4] E. C. Koch, *Propellants, Explos., Pyrotech.*, **2015**, 40, 374–387

[5] E. D. M. Antoine, Heijden van der and H. B. Richard, *Crystal Growth Design*, **2004**, **5**, 999–1007.

[6] R. L. Simpson, P. A. Urtiew, D. L. Ornellas, G. L. Moody, K. J. Scribner, D. M. Hoffman, CL-20 Performance Exceeds That of HMX and its Sensitivity is Moderate, *Propellants Explos. Pyrotech*. **1997**, 22, 249–255.

[7] J. A. Vagenknecht, P. Marecek and W. A. Trzcinski, *J. Energ. Mater.*, **2002**, 20, 245–253.

[8] M. B. Talwar, J. K. Nair, R. S. Palaiah, T. Mukundan and S. P. Gejji, *Def. Sci. J.*, **2002**, 52, 157–163

[9] Y. Bayat, V. Azizkhani, Synthesis of 4,10-Dinitro-2,6,8,12-tetraoxa-4,10-diazaiso-wurtzitane (TEX) Using Heteropolyacids as Efficient and Recyclable Heterogeneous Catalysts. *J. Energ. Mater.*, **2012**, 30, 209–219.

[10] D. Xiang, W. H. Zhu, Thermal decomposition of isolated and crystal 4,10-dinitro-2,6,8,12-tetraoxa-4,10-diazaisowurtzitane according to ab initio molecular dynamics simulations. *RSC Adv.*, **2017**, 7, 8347–8356.

[11] Zhu, W.; Wei, T.; Zhang, X.; Xiao, H. *J. Mol. Struct. Theochem* **2009**, *895*, 131.

[12] Chen, X.; Hou, C.; Han, K. *J. Phys. Chem. A* **2010**, *114*, 1169.

[13] Zhu, W.; Xiao, H. *J. Solid State Chem.* **2007**, *180*, 3521.

[14] Segall, M.D.; Lindan, P.J.D.; Probert, M.J.T.; Pickard, C.J.; Hasnip, P.J.; Clark, S.J.; Payne, M.C. *J. Phys. Condens. Matter* **2002**, *14*, 2717.

[15] Payne, M.C.; Teter, M.P.; Allan, D.C.; Arias, T.A.; Joannopoulos, J.D. *Rev. Mod. Phys.* **1992**, *64*, 1045.

[16] Vanderbilt, D. *Phys. Rev. B* **1990**, *41*, 7892.

[17] Perdew, J.P.; Zunger, A. *Phys. Rev. B* **1981**, *23*, 5048.

[18] Perdew, J.P.; Burke, K.; Ernzerhof, M. *Phys. Rev. Lett.* **1996**, *77*, 3865.

[19] Perdew, J.P.; Wang, Y. *Phys. Rev. B* **1992**, *45*, 13244.

[20] Grimme, S. *J. Comput. Chem.* **2006**, *27*, 1787.

[21] Tkatchenko, A.; Scheffler, M. *Phys. Rev. Lett.* **2009**, *102*, 073005.

[22] Ortmann, F.; Bechstedt, F.; Schmidt, W.G. *Phys. Rev. B* **2006**, *73*, 205101.

[23] Kresse, G.; Furthmüller, J. *Phys. Rev. B* **1996**, *54*, 11169.

[24] Fletcher, R. *Practical Methods of Optimization*; Vol. 1, Wiley: New York, 1980

[25] Kurbatov, S.; Goumont, R.; Lakhdar, S.; Marrot, J.; Terrier, F. 4-Nitrobenzodifuroxan: a Highly Reactive Nitroolefin in Diels-Alder Reactions. Tetrahedron 2005, 61, 8167-8176.

[26] Christian, S. L.; Chafin, A. P.; Nielsen, A. T.; Atkins, R. L.; Chafin, A. P. Synthesis of Aminonitrobenzodifuroxan. US 5149818, 1992.

[27] Fabbiani, F. P. A.; Pulham, C. R. High-Pressure Studies of Pharmaceutical Compounds and Energetic Materials. Chem. Soc. Rev. 2006, 35, 932-942.

[28] Saha, S.; Sinha, T. P.; Mookerjee, A. Electronic Structure,Chemical Bonding, and Optical Properties of Paraelectric BaTiO3. Phys. Rev. B 2000, 62, 8828-8834.

**Table 1.** Comparison of relaxed lattice parameters of TEX with experimenta data at ambient conditions

| | $\alpha$(°) | $\beta$(°) | $\gamma$(°) | $a$ (Å) | $b$ (Å) | $c$ (Å) | V (Å$^3$) |
|---|---|---|---|---|---|---|---|
| Experimental | 82.37 | 75.05 | 79.46 | 6.84 | 7.64 | 8.78 | 433.64 |
| PW91-OBS | 83.05 | 75.31 | 79.26 | 6.84 | 7.63 | 8.82 | 436.26 |
| | (0.83) | (0.35) | (-0.25) | (-0.01) | (-0.08) | (0.53) | (0.61) |
| PBE-TS | 82.69 | 75.16 | 78.90 | 6.95 | 7.72 | 8.97 | 454.81 |
| | (0.38) | (0.15) | (-0.70) | (1.62) | (0.99) | (2.26) | (4.88) |
| PBE-G06 | 82.46 | 75.56 | 78.99 | 6.90 | 7.68 | 8.79 | 440.77 |
| | (0.10) | (0.68) | (-0.58) | (0.94) | (0.50) | (0.10) | (1.65) |

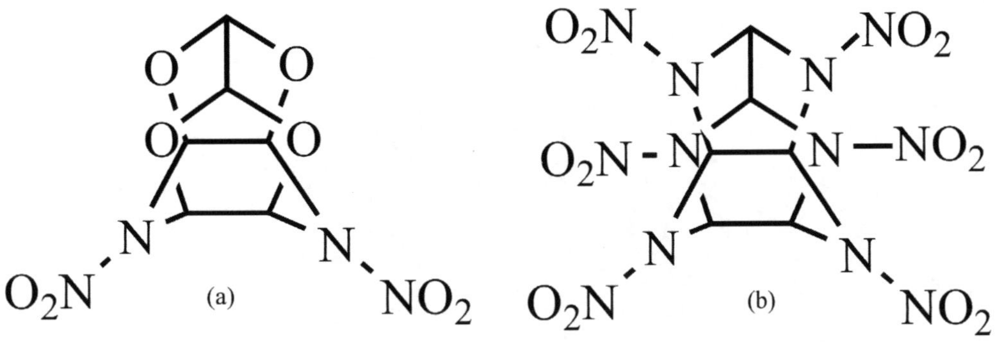


**Figure 1.** (a) Molecular structure of TEX. (b) Molecular structure of CL-20.

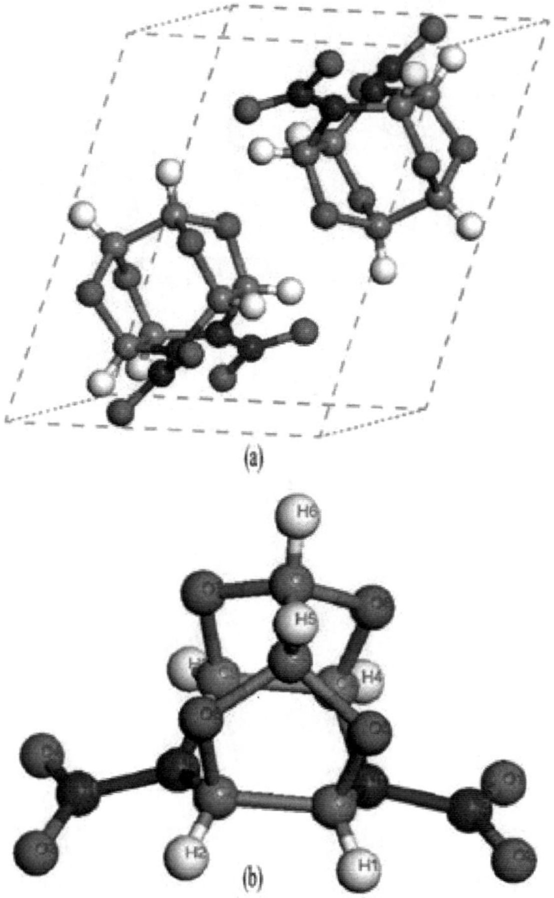


**Figure 2.** (a) Unit cell of TEX cystals; (b) Conformations and atomic numberings of TEX molecules.

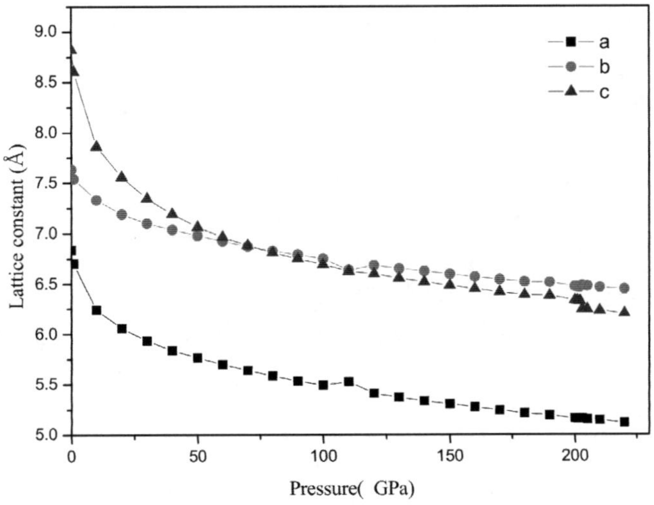


**Figure 3.** Lattice constants *(a, b, c)* of TEX in the pressure range of 0-220 GPa.

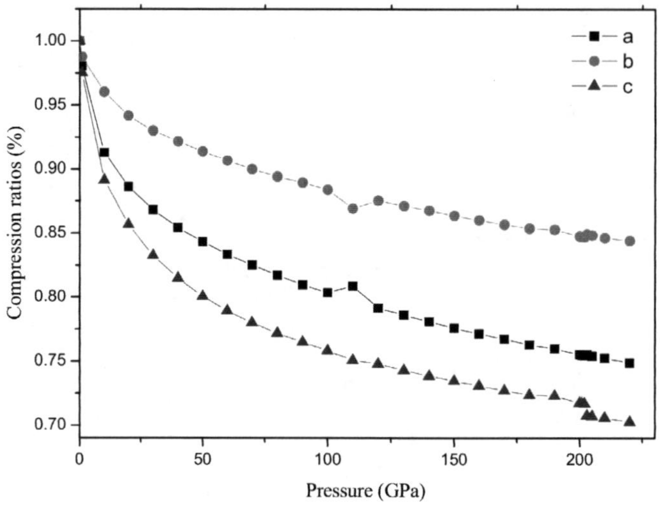


**Figure 4.** Lattice constant (*a*, *b*, c) compression ratio of TEX in the pressure range of 0-220 GPa.

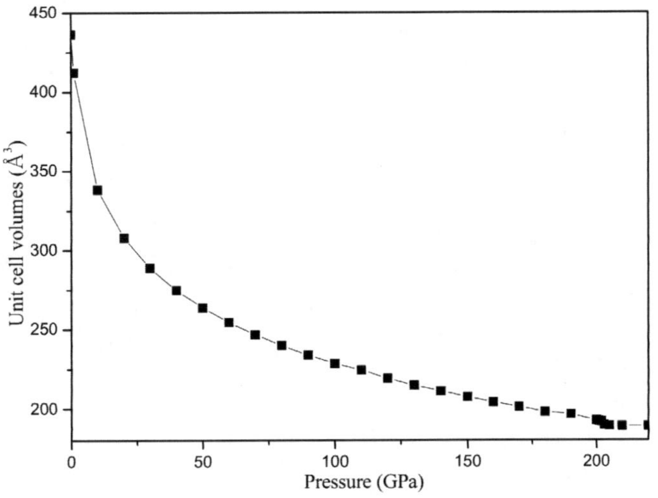


**Figure 5.** Unit cell volumes of TEX in the pressure range of 0-220 GPa.

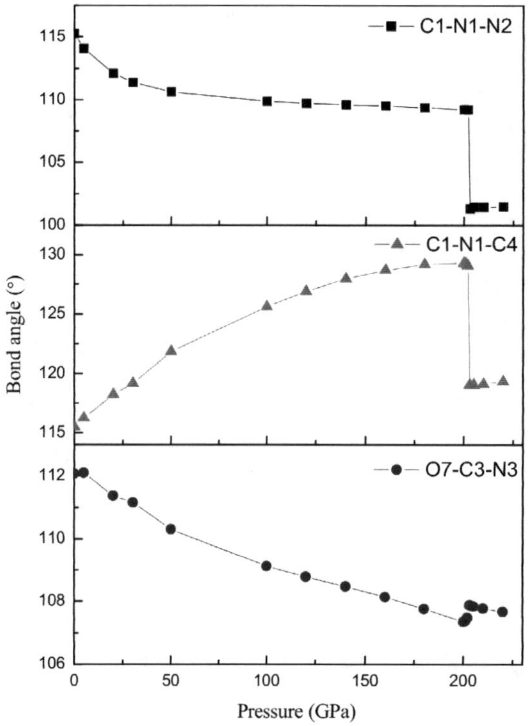


**Figure 6.** Bond angles of NADNP in the pressure range of 0-220 GPa.

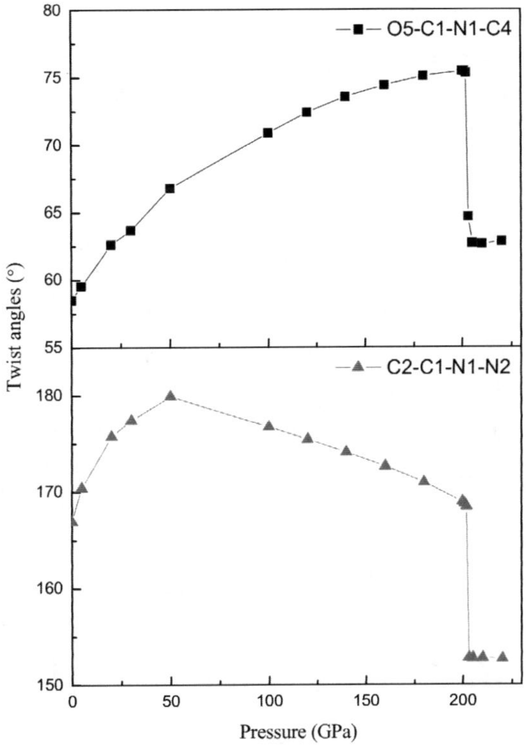


**Figure 7.** Twist angles of TEX in the pressure range of 0-220 GPa.

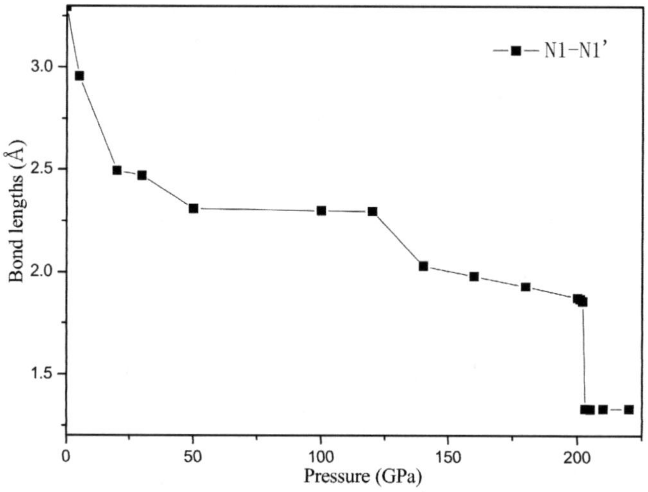


**Figure 8.** N1-N1' bond length in TEX at different pressures.

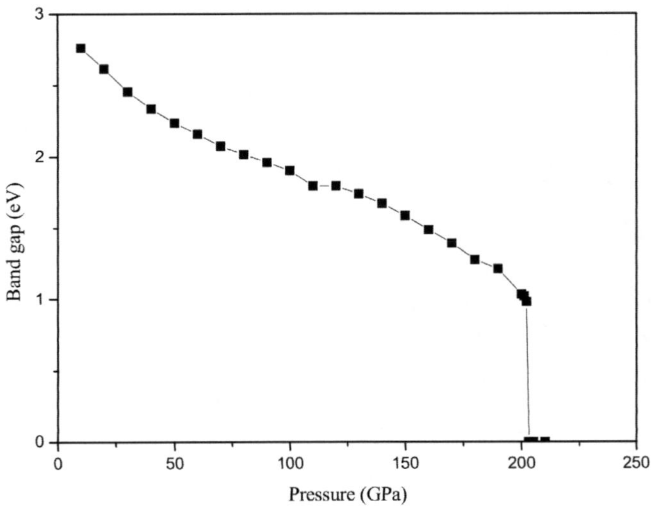


**Figure 9.** Band gap of TEX as a function of hydrostatic pressure.

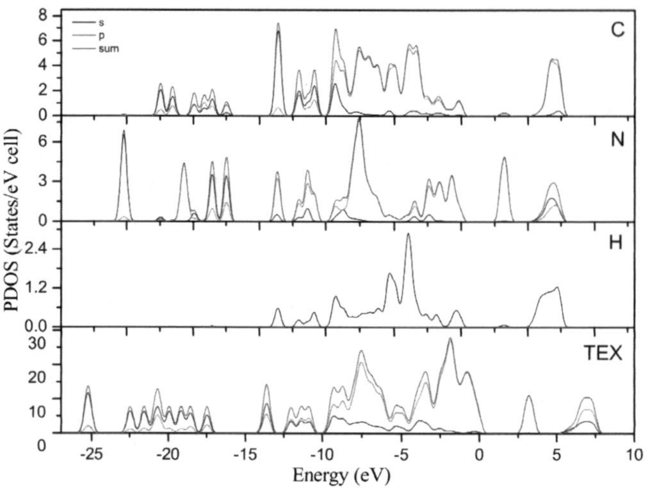


**Figure 10.** Total and partial density of states of the C states, N states in NO2, N states in NH2, and N states in furoxan ring of ANBDF at pressure of 0 GPa. The Fermi energy is shown as a vertical dashed line.

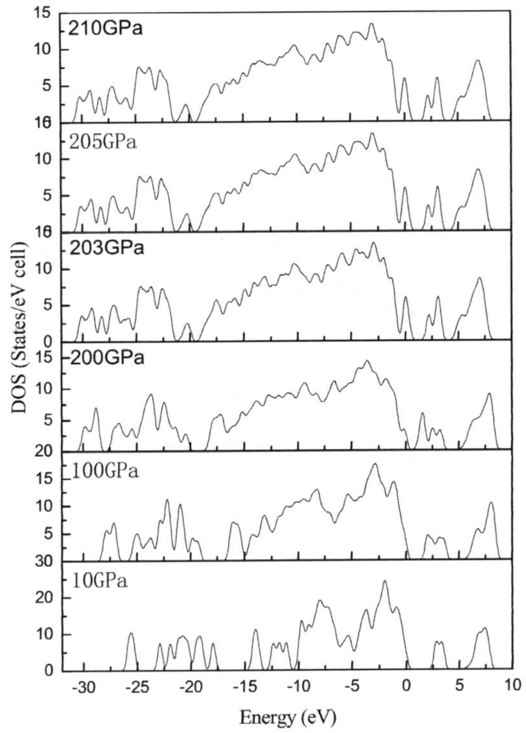


**Figure 11.** DOS of TEX at pressures of 10, 100, 200, 70, 203, 205, and 210 GPa.

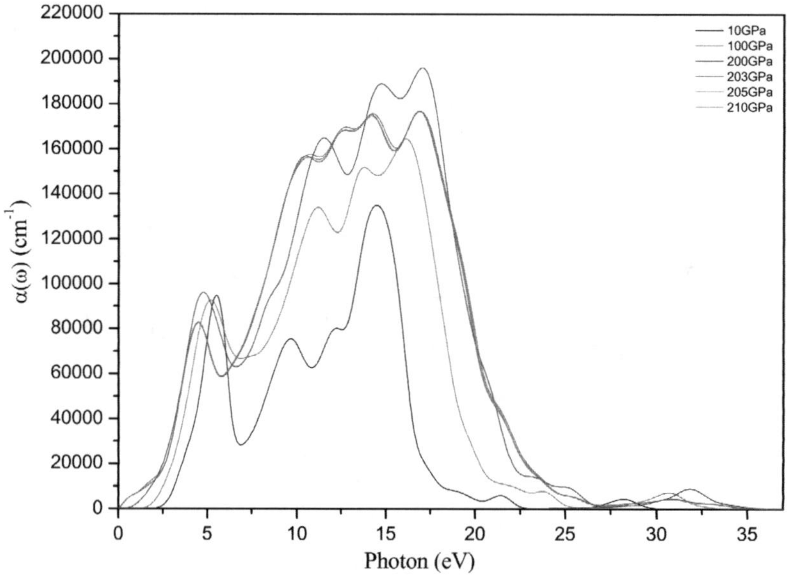


**Figure 12.** Absorption coefficients $\alpha(\omega)$ of TEX at different pressures.

# 492394 Comprehensive Generation of Libraries of Lignin Structures As an Exploration of Lignin Space


**Abstract:**

The structure of lignin has yet to be entirely understood owing to its polydispersity, complexity, and hyperbranched topology. In previous work, the Broadbelt group has developed an algorithm capable of generating libraries of structural representations of wheatstraw lignin whose average properties – monomer distribution, bond distribution, molecular weight distribution, and branching coefficient – statistically agree with the experimental values. In this work, we have extended this stochastic method to accommodate more complexity, in addition to any type of biomass – softwood, hardwood, or herbaceous. The individual lignin molecules were constructed by random walks along decision trees, where the structure of the trees and the values of the edge weights were informed by unique structural and mechanistic details. With the additional complexity, two methods were investigated and ultimately utilized to reduce the degrees of freedom of the optimization of the edge weights. The first method was pruning the decision trees so as to have fewer parameters, and the second method was biasing the edge weights to within a certain range of their experimental values so as to decrease the size of the parameter space. Utilizing these two methods, we were able to obtain edge weights that produced libraries for miscanthus (herbaceous), spruce (softwood), and birch (hardwood). Notably, wheatstraw and beech proved problematic, revealing that drawing experimental values from disparate sources may lead to a combination of properties that cannot be physically realized. Building on these results, we put forth the concept of feasible "lignin space", i.e., ranges of experimental characteristics quantified by the four main properties above that can be realized simultaneously. Thus, the advantages of the approach are twofold: generation of lignin libraries for any biomass source to be used in kinetic modeling studies, and guidance for experimentalists as they seek to design and characterize lignin.

# 500854 Understanding the Nanoscopic Structure of Lyotropic Liquid Crystal Membranes Using Molecular Dynamics Simulations

**Abstract:**

For the first time, to our knowledge, a lyotropic liquid crystal (LLC) system is studied atomistically using molecular dynamics simulations. LLC's are a class of nanostructured materials that can be modified, assembled and crosslinked into stable membranes with uniform straight pores that can perform solute specific separations and offer the ability to control pore architecture at the atomic lengthscale. Pore diameters on the order of 1 nm make LLC membranes well-suited for aqueous separations such as desalination and biorefinement. A molecular model which is consistent with experimentally measured structural features and material properties can provide a link between monomer structure and macroscopic separation performance. We have used simulations with experimental data to create such a model.

In our model, we have developed a method to simulate the crosslinking mechanism, a key step during synthesis. We have also adopted and compared multiple methods for calculating ionic conductivity from atomistic simulations. Simulated X-ray diffraction patterns based on atomic coordinates generate 1D and 2D diffraction patterns containing all major features present experimentally. Using these methods to characterize the molecular model, we have discovered an additional metastable state which may form under certain experimental conditions. Each configuration gives different structural properties but persists for 100's of simulated nanoseconds.

We have also studied the transport of a range of solutes with various charges and hydrodynamic radii. With a clear understanding of mechanisms of transport in these complex self-assembled systems, one can choose monomers to achieve specific separation goals. We can use this information to draw correlations between pore structure and selective preferences. These studies will help guide monomer choice for separation-specific objectives.

# 492737 Thermodynamics of Block Polymers – Monte Carlo Simulations and Self-Consistent Field Theory Study


**Abstract:**

Block polymers self-assemble at mesoscopic length scales to form a wide variety of ordered structures that are instrumental for several emerging technologies such as nanostructured membranes and photonic crystals. Their equilibrium structure depends on numerous molecular parameters: chemistry, length, sequence, and the number of blocks bonded together to form a polymer molecule. Usually, the existence of these ordered structures is not known a priori and many past discoveries are the result of immense experimental effort. However, synthesizing every possible block polymer in such a high-dimensional parameter space is expensive and time-consuming. Therefore, efficient computational methods are key to discovering new materials.

In this poster presentation, we discuss our results for two different class of block polymers: AB diblocks and ABA′C tetrablocks. Our first study focusses on "short" diblock copolymers, motivated by the ever growing demand in microelectronics industry for ordered structures with characteristic dimensions as small as possible. We implemented an efficient lattice-based Monte Carlo simulation tool to predict nanostuctures and their associated properties in linear AB diblocks. In this study, we resolved long-standing artifacts in lattice simulations, namely, commensurability and finite-size effects. This enabled us to accurately determine the characteristics dimensions of a lamellar structure as well as the fluctuation-driven weakly first-order transition in symmetric diblocks.

In order to study multiblock polymers, i.e., ABA′C tetrablocks, we use a field-theoretic simulation method called self-consistent field theory (SCFT). The SCFT framework involves highly non-linear partial differential equations, which are solved using a pseudospectral algorithm employing efficient Fast Fourier Transform, and Anderson-mixing iteration scheme. The SCFT solver begins with a guess structure, solves the governing equations iteratively, and outputs a defect-free ordered structure. The convergence of the iteration procedure depends strongly on the guess structure. We developed a robust, physically-informed approach to generate guess structures that is required to initiate the SCFT iterator, and a efficient method to converge the solutions to stress-free crystals. We used this approach to perform SCFT simulations for different sphere-forming structures in ABA′C tetrablock terpolymers and predicted the most promising range of experimental parameters for discovering new phases. Our experimental colleagues synthesized these polymers, and together we discovered nine nanostructures, including the complex Frank-Kasper σ and A15 phases. In order to accelerate the discovery of block polymer materials, we have shared our SCFT tool online (http://pscf.cems.umn.edu) for public download.

# 500873 Refinement of Techniques in Molecular Modeling of Multicompartment Nanoreactors

**Abstract:**

Chemical reaction schemes often require multiple mutually orthogonal catalysts that preclude the use of one-pot systems. Single compartment micelle nanoreactors offer the benefits of chemical reaction separability of desired products. It follows that a system which immobilizes catalysts within separate compartments of a micelle for multistep reactions is highly desirable in order to gain the benefits of reaction separability and enhanced selectivity. An estimate of the thermodynamic interaction between blocks of copolymers is required to assess the ability of micelles to form distinct compartments. Our analysis utilizes the Flory-Huggins theory of polymer miscibility, which introduces a miscibility parameter  as a measure of the favorability of the interaction between polymers in blends. In our current work, we compare the miscibility of binary polymer systems, using a series of computational methods to calculate the  parameter for each pair of molecules as they might be found in a micelle. In order to explore the effect of molecular size on  parameter, we extended the number of monomeric units from one to three. By refining our parameters and methods, our results have converged to the experimental observations of miscibility for these candidates. More specifically, we will investigate the relationship between number of lowest energy samples, number of cluster samples, reference temperature and accuracy of calculated  values. An increase in the number of lowest energy samples has been shown to increase precision for averaged  values, especially for trimeric systems. Similarly, we can improve the accuracy of the obtained  values by increasing the number of molecular cluster samples probed to estimate the coordination between a given pair of molecules. Finally, we will explore whether the choice of reference temperature greatly affects the final results, or if post-calculation Boltzmann averaging performed at the desired temperature is sufficient. Our primary goal is to investigate the existence of a quantitative model between molecular size and the parameters we have established here. Through further research, we will show that computational modeling can confidently and consistently predict miscibility behavior.

# 495663 Rapid and Accurate Property Prediction for Polymer Systems Using Atomistic-Scale Simulation


**Abstract:**

Polymeric materials are critical to the performance of many advanced material systems such as thermoset composites. The chemistry of the polymer, along with processing parameters, determines the behavior of the final polymer product. Tools to rapidly screen the effect of chemical changes on polymer performance can enable faster development of improved polymer systems. Simulations are a powerful tools for polymer development but they must provide accurate predictions in an efficient manner in order to be widely adopted.

A series of properties and behaviors from reaction kinetics to mechanical properties were explored through chemically informed simulations to illustrate the capabilities and accuracy of atomic level simulation for glassy polymer systems. Quantum mechanical simulations were used to develop reaction pathways and kinetic rates for epoxy-amine curing reactions yielding the relative reaction rates for various intermediates.

For thermophysical properties, 3D polymer structures were constructed by creation of amorphous combinations of multiple polymeric chains or by cross linking simulations utilizing a reaction chemistry informed algorithm to develop a network topology. With the representative 3D structure, Tg and mechanical properties (stress vs. strain behavior, elastic constants) were efficiently calculated by using GPU-enabled molecular dynamics simulations. GPU technology allows for improved accuracy in prediction of polymer properties through extended time-scale simulation (in excess of 1 μs) and increased simulation throughput. The rapid property predictions were further leveraged to explore the effect of monomer ratios by simulating numerous stoichiometries for epoxy-amine systems, yielding the dependence of Tg on epoxy molarity.

This poster with illustrate these combinations of intuitive simulation construction and efficient quantum mechanic and molecular dynamic simulation engines. These capabilities allow for rapid knowledge capture for new and existing polymer systems and are a powerful tool to be leveraged in developing polymers to meet desired properties characteristics and advancing a design for materials properties framework.

# 499201 Flow Properties of Model Alkanes in Nanopores

**Abstract:**

The elucidation of the transport mechanisms of chain molecules in nanoconfined systems is of great interest for a wide range of applications such as e.g. for separation processes. In this work, we use molecular simulation to determine the transport coefficients for model alkanes confined in nanopores. For this purpose, we develop an approach based on the transient-time correlation function and carry out nonequilibrium molecular dynamics simulations to determine the flow properties and viscosity of the confined fluid. The results highlight the interplay between the relaxation time of the molecules and the characteristics of the flow.

# 499337 Pure and Mixed Gas Absorption in Nonideal Binary Ionic Liquid Mixtures, a Molecular Simulation Study


**Abstract:**

Ionic liquids (ILs) are novel chemical substances composed entirely of ions and are attractive candidates for replacement of volatile organic compounds used in chemical industry. One of the widely studied research topics in the field of ILs is their ability to absorb polar substances such as carbon dioxide ($CO_2$) and sulfur dioxide ($SO_2$) owing to their ionic nature. At the same time, conventional ILs exhibit rather poor solubility for nonpolar substances such as alkanes. A large repertoire of ionic liquids has been designed given that the solubility is governed by the anion identity and ionic liquid free volume. In this study, we take a different approach of ionic liquid mixtures in tuning the gas solubility. Our work is guided by the hypothesis that positive deviation in the excess molar volume will translate into nonideality in gas solubility. This study focuses on bulk fluid solubility of $CO_2$, $SO_2$ and $CH_4$ and their mixtures in two sets of binary IL systems: (a) 1-n-butyl-3-methylimidazolium $[C_4mim]^+$, chloride $Cl^-$ and bis(trifluoromethanesulfonyl)imide $[NTf_2]^-$ and (b) $[C_4mim]^+$ , $Cl^-$ and trifluoroacetate $[CF_3COO]^-$ as a function of temperature. The choice of ionic components was influenced from our prior work suggesting that these ionic liquid systems are characterized by small yet positive excess molar volumes independent of the temperature. We will present the results of the Henryâ€™s law constants along with single and binary component gas absorption isotherms.

# 501746 Construction of a Hydrogel System for Bioadsorption and Bioseparations By Molecular Modeling and Simulation


**Abstract:**

Hydrogels are formed by three-dimensionally cross-linked hydrophilic polymers. They have highly porous structures, good affinity for water, and consequently extraordinary capability to swell in water and retain a large amount of water. Hydrogels can also possess desirable biocompatibility either by starting from biocompatible polymers or through functionalization with biocompatible ligands. More importantly, they can exhibit dramatic volume phase transition (VPT), i.e., quasi-reversible swelling and collapse of hydrogel structures, in response to external physical or chemical stimuli. A great deal of effort has thus been devoted to developing hydrogels into material systems that fulfill important needs ranging from super absorbers to innovative biomedical applications including drug storage and delivery. In order to fully realize their potential, a thorough understanding of their various structural and dynamic properties is of essential importance and necessitates molecular-level characterization of their porous structures and volume phase transitions. However, our understanding in these aspects are still quite limited to date because (i) current hydrogel studies are mostly *experimental and experiential with limited resolution* and (ii) the few relevant theoretical studies are either based on statistical mechanics of *idealized network* (*lattice*) *models* or focused on specific fine-scale factors (e.g., hydrogen bond) using *atomistic modeling and simulations of individual polymer chains or segments*, which are not sufficient for representing the properties and structures of hydrogels. Additionally, dynamic properties, for the most part, have been overlooked. In this work, dextran hydrogel is selected to be studied because it has significant and desirable hydrophilicity and biocompatibility, and consequently greater potential for industrial and biotechnological applications. In order to examine practically important aspects of hydrogel properties and applications that are mostly unexplored and poorly understood, model systems are constructed with multiple dextran chains cross-linked together by $H_2N(-CH_2)_n-NH_2$. This cross-linking agent has two terminal $-NH_2$ function groups that can react with two -OH groups in different dextran chains to form -C-O-NH- cross-linkages. The number of methylene groups, *n*, take on a few different values in order to study the effects of the cross-linker length. We have successfully modeled and simulated such dextran hydrogel using a coarse-grain force field derived from the so-called M3B force field. The model hydrogel system is immersed in water which is also coarse grained into proper spherical beads. The current focus is on the volume phase transition (VPT) phenomenon, i.e., quasi-reversible swelling and collapse of hydrogel structures, which is triggered by temperature change and accompanied by significant swings of water content in the hydrogel. Water molecules are found to exist in diverse energetic states and hence different mobility in the hydrogel, which underlies the VPT phenomenon. The results and mechanisms will be presented and discussed in detail in this presentation.

# 501784 Modeling Alkane Partitioning and Phase Behavior in Non-Permeable and Permeable Slit Pores

**Abstract:**

In unconventional shale reservoirs, hydrocarbon adsorption and absorption both affect reservoir fluid distribution and phase behavior, which makes the prediction of fluid thermodynamic properties (pressure/volume/temperature) increasingly challenging. It has been recognized that the nano-confinement lowers the bubble point and critical point of fluid mixtures, and solvation of hydrocarbon in kerogen is also significant. For a hydrocarbon mixture, the pore fluid composition has a large deviation from extracted bulk fluid due to different surface affinities of different components. Although sorption can be described by some macroscopic models, density functional theory (DFT) can provide detailed information of the fluid structure in the pore-wall system.

In this presentation, we apply interfacial Statistical Associating Fluid Theory (iSAFT) to investigate the fluid adsorption in a non-permeable graphite pore and both adsorption and absorption in a permeable polymer pore. We show the shifted pure and binary phase diagrams in a non-permeable 3 nm pore and the pore size dependency of pure component critical properties, where the results all have a reasonable agreement with molecular simulation. The competitive adsorption of mixtures is predicted and suggests different dominating factors, enthalpic or entropic, determine the pore selectivity depending on the bulk conditions. For a permeable pore wall, we show the adsorbed density in the pore and the penetrated density in polymer wall, and meanwhile the sorption isotherms obtained from density distribution. We compare the enhanced sorption of fluid in a pore-wall system with bulk phase equilibrium solvation and discuss the phase equilibrium in different pore width and wall thickness.

# 489448 Enhancing the Oxidation of Toluene with External Electric Fields: A Reactive Molecular Dynamics Study

**Abstract:**

**Enhancing the Oxidation of Toluene with External Electric Fields: a Reactive Molecular Dynamics Study**

Shen Tan, Tao Xia, Yao Shi, and Yi He*

*College of Chemical and Biological Engineering, Zhejiang University, Hangzhou, 310027, P. R. China*

*Corresponding author: yihezj@zju.edu.cn

Tuning chemical reactions with external electric field (Efield) has long been a desirable goal for researchers. Molecular simulation methods based on reactive force fields (ReaxFF) is recently developed for the study of chemical reactions in large systems at a micro-scale. To this end, the effects of Efields on chemical reactions were studied with the ReaxFF MD simulations by using the oxidation of toluene as a model system. We observed that Efields may greatly enhance the oxidation rate of toluene. The initial reaction time of toluene is also reduced remarkably in Efields. A stronger Efield leads to a faster oxidation rate of toluene. Further studies reveal that the applying of Efields may results in the oxidation of toluene at 2100 K which is otherwise not able to happen when Efield is not at present. The oxidation rate of toluene at 2100 K in Efield is comparable with the oxidation rate of toluene at 2900 K when Efield is not applied. In addition, Efield was observed to significantly enhance the occurrence of the initial radical generation for different pathways of toluene oxidation but it does not seem to favor any of the pathways. Finally, Efields do not seem to enhance the polarization of toluene during its transition state, which suggests that a polarizable charge equilibration method (QEq) method might be needed to take the effects of Efields into consideration.

**Figure. 1** Evolution of the toluene and oxygen molecules at density ¦Ñ = 0.15, 0.35 g/cm$^3$, Efield = 0, 0.2 V/Å (ReaxFF NVT-MD simulation, T = 2500 K, toluene/oxygen = 1:100).

# 498709 Molecular Simulation of Ionic Polyimides and Ionic Liquid Composites for Gas Separation

**Abstract:**

Molecular Simulation of Ionic Polyimides and Ionic Liquid Composites for Gas Separation

Asghar Abedini, Ellis Crabtree, Jason E. Bara, and C. Heath Turner*

Abstract:

Industrial gas separation and storage are primary challenges in energy and power related processes. For instance, the existence of acid gases such as H2S and CO2 in natural gas increases the equipment maintenance and separation cost for the energy industry.1 To make energy supply lines more economically feasible and minimize corrosion, these impurities must be removed through natural gas sweetening with precombustion CO2 capturing units prior to routing natural gas to the supply line.2 Furthermore, CO2 is a main component of power plant exhaust gas and is a side product of many processes of the petrochemical industry, such as ammonia production. Palliating carbon dioxide from emission sources is a critical need in industry, and it is necessary in order to meet current and future environmental regulations. Aqueous alkanolamine solvents are the most common process liquids for CO2 absorption units.3 However, high energy demands for recovery4, volatility, corrosivity, and degeneracy to toxic products are the main drawbacks of these solvents in gas sweetening process.5

Adsorption in porous materials is an alternative energy-efficient CO2 separation method, and the development of high carbon dioxide selectivity and adsorption capacity are key factors. There are some traditional adsorbents that have been widely studied such as activated carbon6, zeolites7, metal organic frameworks (MOFs),8 and silica gel.9 The low cost, high surface area, and large sorption capacity of activated carbons make them widely used in industry for similar applications.10 Additionally, polymers of intrinsic microporosity (PIMs) are another potential material that can be used as a membrane for gas separation, and they can provide increased stability, as compared to other adsorbents.11 The porosity of PIMs comes from their unique molecular structure,12 which provides a significant free volume and microstructure resulting from adjacent polymer chains.13, 14, 15 In order to tailor the permeability of these materials, we are exploring the behavior of composite structures composed of a PIM support plus an ionic liquid (IL) coating. The intent is to leverage the microporous structure of the PIM support with the tunability of the IL selectivity. Otherwise, a pure IL solvent suffers from high transport resistance and low free volume.

In the past, ionic liquids have been shown high selectivity for CO216, 17 versus CH4,18 and this makes ILs promising for applications in natural gas sweetening and precombustion CO2 capturing. The energy required for solvent recovery in ILs can be reduced due to the physical absorption mechanism.

In our study, 1-Butyl-3-methylimidazolium bis(trifluoromethylsulfonyl)imide ([C4mim+][Tf2N-]) molecules have been chosen to infuse PIMs because of the structural consistency between PIMs and [C4mim+][Tf2N-] and because this particular IL has a strong tendency to absorb CO2 gas.1 Using a combination of molecular dynamics and Monte Carlo simulations, the solubility of CO2 and CH4 is modeled over a wide range of different pressures within the PIM and PIM + IL materials. Our results are benchmarked against experimental results, and modeling work shows excellent agreement with the experimental solubility for CO2 and CH4. Moreover, the modeling results indicate that the composite (PIM + IL) has the ability to surpass the predicted adsorption behavior that is typically correlated to the FFV of the adsorbent.

References:

1. Budhathoki, S.; Shah, J. K.; Maginn, E. J. Molecular Simulation Study of the Solubility, Diffusivity and Permselectivity of Pure and Binary Mixtures of CO2 and CH4 in the Ionic Liquid 1-n-Butyl-3-methylimidazolium bis(trifluoromethylsulfonyl)imide. Industrial & Engineering Chemistry Research 2015, 54 (35), 8821-8828.
2. Aparicio, S.; Atilhan, M. Computational Study of Hexamethylguanidinium Lactate Ionic Liquid: A Candidate for Natural Gas Sweetening. Energy & Fuels 2010, 24 (9), 4989-5001.
3. Chang, H.; Shih, C. M. Simulation and optimization for power plant flue gas CO2 absorption-stripping systems. Separ Sci Technol 2005, 40 (4), 877-909.
4. Turner, C. H.; Johnson, J. K.; Gubbins, K. E. Effect of confinement on chemical reaction equilibria: The reactions 2NO <->(NO)(2) and N-2+3H(2)<-> 2NH(3) in carbon micropores. Journal of Chemical Physics 2001, 114 (4), 1851-1859.
5. Dupart, M. S.; Bacon, T. R.; Edwards, D. J. Understanding Corrosion in Alkanolamine Gas Treating Plants. Hydrocarb Process 1993, 72 (5), 89-&.
6. Peng, X.; Wang, W.; Xue, R.; Shen, Z. Adsorption separation of CH4/CO2 on mesocarbon microbeads: Experiment and modeling. AIChE Journal 2006, 52 (3), 994-1003.
7. Heymans, N.; Alban, B.; Moreau, S.; De Weireld, G. Experimental and theoretical study of the adsorption of pure molecules and binary systems containing methane, carbon monoxide, carbon dioxide and nitrogen. Application to the syngas generation. Chemical Engineering Science 2011, 66 (17), 3850-3858.
8. García, E. J.; Mowat, J. P. S.; Wright, P. A.; Pérez-Pellitero, J.; Jallut, C.; Pirngruber, G. D. Role of Structure and Chemistry in Controlling Separations of CO2/CH4 and CO2/CH4/CO Mixtures over Honeycomb MOFs with Coordinatively Unsaturated Metal Sites. The Journal of Physical Chemistry C 2012, 116 (50), 26636-26648.
9. Morishige, K. Adsorption and Separation of CO2/CH4 on Amorphous Silica Molecular Sieve. The Journal of Physical Chemistry C 2011, 115 (19), 9713-9718.
10. Larsen, G. S.; Lin, P.; Hart, K. E.; Colina, C. M. Molecular Simulations of PIM-1-like Polymers of Intrinsic Microporosity. Macromolecules 2011, 44 (17), 6944-6951.
11. Slater, A. G.; Cooper, A. I. Porous materials. Function-led design of new porous materials. Science 2015, 348 (6238), aaa8075.
12. McKeown, N. B.; Budd, P. M. Polymers of intrinsic microporosity (PIMs): organic materials for membrane separations, heterogeneous catalysis and hydrogen storage. Chem Soc Rev 2006, 35 (8), 675-83.
13. Hart, K. E.; Colina, C. M. Estimating gas permeability and permselectivity of microporous polymers. Journal of Membrane Science 2014, 468, 259-268.
14. Tocci, E.; De Lorenzo, L.; Bernardo, P.; Clarizia, G.; Bazzarelli, F.; McKeown, N. B.; Carta, M.; Malpass-Evans, R.; Friess, K.; Pilnáček, K.; Lanč, M.; Yampolskii, Y. P.; Strarannikova, L.; Shantarovich, V.; Mauri, M.; Jansen, J. C. Molecular Modeling and Gas Permeation Properties of a Polymer of Intrinsic Microporosity Composed of Ethanoanthracene and Tröger's Base Units. Macromolecules 2014, 47 (22), 7900-7916.
15. Swaidan, R.; Ghanem, B.; Litwiller, E.; Pinnau, I. Physical Aging, Plasticization and Their Effects on Gas Permeation in "Rigid" Polymers of Intrinsic Microporosity. Macromolecules 2015, 48 (18), 6553-6561.
16. Finotello, A.; Bara, J. E.; Camper, D.; Noble, R. D. Room-temperature ionic liquids: Temperature dependence of gas solubility selectivity. Industrial & Engineering Chemistry Research 2008, 47 (10), 3453-3459.
17. Turner, C. H.; Cooper, A.; Zhang, Z.; Shannon, M. S.; Bara, J. E. Molecular simulation of the thermophysical properties of N-functionalized alkylimidazoles. J Phys Chem B 2012, 116 (22), 6529-35.
18. Ramdin, M.; Balaji, S. P.; Vicent-Luna, J. M.; Gutiérrez-Sevillano, J. J.; Calero, S.; de Loos, T. W.; Vlugt, T. J. H. Solubility of the Precombustion Gases CO2, CH4, CO, H2, N2, and H2S in the Ionic Liquid [bmim][Tf2N] from Monte Carlo Simulations. The Journal of Physical Chemistry C 2014, 118 (41), 23599-23604.

# 501749 Molecular Simulation of Ionic Liquid Mixtures: Applications to Capacitive Energy Storage

**Abstract:**

Room temperature ionic liquids (RTILs) are a class of molten organic salts with remarkable physical and electrochemical properties that make them appealing candidates for electrolytes in high-performance supercapacitors. Already boasting an enormous library of binary neat ionic liquids from the billions of possible combinations of chemical moieties, properties of ionic liquid electrolytes can further be tuned by forming mixtures. Here, we present molecular dynamics (MD) contributions to combined experimental and simulation studies on ionic liquid mixtures. We begin with a relatively simple case of mixtures of the ionic liquids EMIM-Tf2N and EMIM-BF4 and examine how composition impacts electric double layer (EDL) structure and capacitance on planar graphene. Then, we study mixtures of EMIM-Tf2N and HMIM-Tf2N in a bimodally nanoporous carbon and examine how ions selectively partition themselves into different sized pores and connect this to macroscopic capacitance data from electrochemistry experiments. Together, these studies demonstrate concrete cases of unique electrochemical behavior with a mixed ionic liquid electrolytes.

# 501776 Molecular Simulation of Ionic Liquid Systems: Effects of Solvation and Humidification

**Abstract:**

Room temperature ionic liquids (RTILs) are a class of molten organic salts with high electrochemical stability that drive their utility in energy-relevant applications, such as electrical energy storage, gas separation, and cellulose processing. Two potential pitfalls in many applications are water uptake and slow dynamics. Despite typically containing hydrophobic constituents, common ionic liquids are extremely hygroscopic, even taking up water vapor found in ambient air. Water content can be beneficial or detrimental to a given application, thereby motivating its study in applied systems. The slow transport properties in neat RTILs are due to the strong electrostatic interaction between counter-ions, which is often mitigated by solvation in organic solvents; however, this parameter space is not fully explored. To this end, we use molecular dynamics (MD) simulations in combination with quasi-elastic neutron scattering (QENS) experiments to examine two specific systems to better understand the effects of water and solvation. First, we examine how bulk RTIL transport properties are impacted by solvent concentration and solvation in five organic solvents of different polarity and found that increasing solvent concentration and solvent polarity both enhance the diffusivity of the cation. Second, we study the same RTIL confined in a microporous carbon and exposed to deuterated water vapor. We find humidity-dependent changes in the structure of the confined RTIL, which correlate with increased diffusivity reported by QENS and increased capacitive performance reported by electrochemistry experiments. Together, these two studies can aid in the design of RTIL-based supercapacitors.

# 501328 Aerosol Formation in Post Combustion $CO_2$ Capture Columns – Molecular Dynamic Simulation

**Abstract:**

The significant amount of solvent is lost due to the aerosol formation in the post combustion CO2 capture columns. The present work investigates the intermolecular interactions between $CO_2$ and Mono Ethanol Amine (MEA) and cluster formation in the gaseous phase, that contribute to the aerosol fromation. Molecular Dynamics (MD) simulation in GROMACS was conducted to compute cluster number and size. The Radial Distribution Function (RDF) analysis was done to observe the distribution of the carbon atoms inside the system. In addition, the force field parameters of pure $CO_2$ and MEA systems were examined by verifying densities and self-diffusion coefficients. Three cases were considered: 1, 10 and 25 molecules of MEA with 1081 molecules of CO2. The cluster analysis shows the formation of the one big cluster with MEA and CO2 inside and few CO2 molecules moving around the cluster. It is observed that addition of more amine molecules to the system will significantly influence the cluster formation. In the RDF analysis, increased intensity of C..C (of MEA) correlation in the mixture system indicates the presence of the cluster. The importance of this research is that it opens the roots for modeling aerosol formation in the absorption columns. This work is being continued to include air, water and other compounds in the system.

# 500022 Molecular Simulation of Transport of DNA Grafted Nanoparticles

**Abstract:**

Gold nanoparticles grafted with DNA chains (DNA-AuNPs) have been the focus of intense research in recent years due to their potential applications in nanomedicine. However, the exact role played by hybridization in the self-assembly of these systems, as well as its impact on their flow properties have remained poorly understood. In this work, we carry out non-equilibrium molecular dynamics (NEMD) simulation to study transport processes of DNA-AuNPs in solution and to send light on the extent of hybridization under shear.

# 492232 Protein Adsorption on Surfaces: The Role of Forcefield and Surface Ions

**Abstract:**

Understanding the interaction of proteins with surfaces is central to biomineralization. Molecular dynamics simulations can provide comprehensive details about the conformation, surface coverage, binding energy, and adsorption mechanism of proteins on surfaces. But the efficacy of these predictions is conditional on the accurate description of the system by both the forcefield and the simulation setup. In this study, we investigate the effect of parameters like surface ions, surface charge and forcefields to determine how they contribute to the binding of proteins on silica. For this, we compare two commonly used forcefields - CLAYFF[1] and INTERFACE[2] - to describe the silica surface. We see how different parameterization schemes change the hydrophobicity of the surface and consequently its interaction with proteins and electrolytes. We also investigate the role of surface ions, namely sodium. We see that the presence or absence of ions on the surface changes the energetic penalty to protein adsorption and alters the distance at which the protein is stabilized near the surface. For these studies, we simulate model peptides, both small (GGKGG) and large helical (LKa14). This allows us to compare the behavior of peptides with single and multiple binding side-chains. Ultimately, we use lessons from these studies to resolve the behavior of experimentally-relevant peptides on silica. Car9, a tag protein, was recently shown to bind to silica[3]. Using the optimal forcefield and ion setup from above, we compare the binding of car9 its mutants to the results of SPR experiments. Our simulations make extensive use of metadynamics[4,5], an efficient enhanced sampling method, to explore the high dimensional free energy landscapes germane to our problem.

**References:**

(1) Cygan, R. T.; Liang, J.-J.; Kalinichev, A. G. *J. Phys. Chem. B* 2004, *108*(4), 1255–1266.

(2) Heinz, H.; Lin, T.-J.; Kishore Mishra, R.; Emami, F. S. *Langmuir* 2013, *29*(6), 1754–1765.

(3) Coyle, B. L.; Baneyx, F. *Biotechnol. Bioeng.* 2014, *111*(10), 2019–2026.

(4) Pfaendtner, J.; Bonomi, M. *J. Chem. Theory Comput.* 2015, *11*(11), 5062–5067.

(5) Deighan, M.; Bonomi, M.; Pfaendtner, J. *J. Chem. Theory Comput.* 2012, *8* (7), 2189–2192.

# 500562 Scaling of Peptide Sequence-Dependent Hydrophobic Interactions from Experiment and Simulation

**Abstract:**

Hydrophobic interactions (HIs) drive structural transitions and self-assembly processes in many natural and synthetic systems. Past experimental investigations into the HI have probed macroscopic interfaces or indirectly investigated molecular-level details. In contrast, simulations have historically focused on HIs for microscopic, idealized systems. Here, we use both atomic force microscopy experiments and molecular dynamics simulations to directly probe the HI between peptides of controlled hydrophobic content and an extended hydrophobic self-assembled monolayer (SAM) surface. Specifically, a hydrophilic glycine-serine repeat scaffold is systematically modified to contain one to four leucine residues, which are allowed to interact with the SAM surface. Simulated systems closely mirror the experimental set-up, even employing the same non-equilibrium technique, specifically Jarzynskiâ€™s equality, to evaluate free energy differences. This constitutes a unique convergence of simulation and experiment in directly probing the HI on a molecular level for a realistic, soft-matter system. In addition to qualitatively agreeing with experiment, our simulations also identify key molecular structural factors driving the increase in the strength of the observed HI. While all peptides are relatively disordered in both solution and absorbed to surfaces, we observe differences in the tetrahedral structure of nearby water. We find that the three-body angle distributions of waters nearby solutes clearly distinguish between large and small scale hydrophobes, as well as hydrophobic and hydrophilic residues. As the hydrophobic content of a peptide increases, three-body angle distributions shift towards that of an isolated, small hydrophobe. Using these shifts as a metric, we construct an empirical model that well-predicts the free energy change of removing these peptides from the SAM surface. This work may be extended to probe the response of three-body angle distributions of water to various solutes and the ability of models based on this quantity to estimate solvation or interaction free energies.

# 492561 Mapping of Gas Diffusion Pathways in [FeFe]-Hydrogenase

**Abstract:**

The $H_2$ production potential of [FeFe]-Hydrogenase, a hydrogen-producing enzyme from green algae, is reported to be promising for economical and large-scale production of $H_2$ as an alternative source of renewable energy. The production of hydrogen takes place at the catalytic center buried in the enzyme core. Unfortunately, oxygen binding to the catalytic center of the enzyme irreversibly inactivates it, essentially blocking hydrogen production. Therefore, a better understanding of the mechanism of enzyme inhibition by $O_2$ is necessary to develop strategies for designing oxygen-tolerant enzymes. In this work, we are investigating the pathways and diffusion channels of $O_2$ gas in this hydrogenase. Through exhaustive mapping of oxygen diffusion channels, we have computed a full thermodynamic map of preferred binding locations of $O_2$ gas within the enzyme interior. Our results suggest that $O_2$ can enter and exit the enzyme through multiple pathways along which are key residues that are known to perturb rates of $O_2$ binding. We find that the global minimum in the free-energy landscape is located near the H-cluster, a metallic center within the enzyme. Along oxygen diffusion channels, we further identify residues that could be potential candidates for mutations to increase the oxygen tolerance of these enzymes.

# 496665 Studying the Structure and Dynamics of Amyloidβ(21−30) with Experiments and Simulations

**Abstract:**

**Studying the Structure and Dynamics of Amyloid$\beta_{21-30}$ with Experiments and Simulations**

Amyloid peptide, Aβ42, is a 42 amino acids long biomolecule that is found in amyloid fibrils and plaques in the brain tissue of patients suffering from Alzheimer's disease or cerebrovascular disease cerebral amyloid angiopathy [1]. Aβ42 has a 10 residue long, protease and temperature denaturation resistant region called A$\beta_{21-10}$, which is also known to be a nucleation site for aggregation of Aβ42 to mature plaques [2]. Experiment Directed Simulations (EDS) was used to bias Molecular Dynamics simulations with experimental NMR backbone chemical shifts [3]. EDS biases the simulation by match- ing the ensemble average of a collective variable (a variable that depends on atomic positions) with experimentally observed values of the same variable [3]. EDS provides a constant minimal bias without the use of replicas so that dynamic properties can be studied [3]. During the simulation, Nuclear Overhauser Effect Spectroscopy spectrum was also generated. The poten- tial of mean force was studied for a few collective variables. The dynamics and transitions between different low energy conformations of A$\beta_{21-10}$ were thoroughly studied with EDS. EDS can be applied to small dynamic biomolecules such as peptides, to study their conformational space as well as their dynamics which are hard to pinpoint experimentally. EDS gives an opportunity to use experimental data with simulations to study microscopic and transient processes that are hard to grasp with just experiments alone.

References

[1] C. R. Watts. et al, "Structural Properties of Amyloid β(1-40) Dimer Explored by Replica Exchange Molecular Dynamics Simulations," *Proteins*, 2017.

[2] M. M. Murray. et al, "Amyloid beta-protein: experiment and theory on the 21-30 fragment.," *J. Phys. Chem.*B, 2009.

[3] A. D. White. et al, "Efficient and Minimal Method to Bias Molecular Simulations with Experimental Data," *J. Chem. Theory Comput.*, 2014.

# 498992 Solute Transport across Blood-Brain Barrier Tight Junction Pores

**Abstract:**

The blood-brain barrier tight junctions present a formidable challenge in the development of therapeutics for the treatment of neurodegenerative diseases. Claudin-5 membrane proteins that are the functional component of the tight junctions assemble in the paracellular space to constitute the tight junction strands. Here using a multiscale molecular dynamics simulation approach, we have elucidated the nature of claudin-5 assembly at the tight junction interface. Significantly, claudin-5 protomers assemble in four preferred dimeric conformations, which subsequently interact across the membranes of two adjacent cells to constitute tight junction pores. Two distinct tight junction pores were observed that corroborate with the experimental findings reported in the literature. Starting from the pore assemblies that were observed in the simulations, we have characterized the potential of mean force (PMF) of different biologically relevant solutes across these pores. It is evident from these calculations that claudin-5 pores form a formidable barrier for small molecule transport while still being permissive of water. The main conclusions from this study will serve to further the understanding of the blood-brain barrier.

# 499318 Capturing Differences in Dynamics of Structurally Similar Signaling Proteins

**Abstract:**

Although many homologous proteins have highly conserved structural motifs as well as overall structural folds, the functional characteristics of such proteins are strikingly different in solution, indicating a role of dynamic motions. The differences in dynamics of closely-related proteins present significant hurdles for structure-based design of small molecules targeting such proteins. In this work, we have studied the interplay of protein dynamics and small-molecule inhibition in a set of signaling proteins, known as Regulators of G protein signaling (RGS) proteins, that modulate signaling in G-protein coupled receptors by binding to GÎ±-subunits of heterotrimeric G proteins and accelerating hydrolysis of GTP. While structures of several known RGS proteins largely contain Î±-helical motifs, some *thiadiazolidinone* (TDZD) compounds that target cysteine residues have shown different levels of specificities and potencies for closely related RGS proteins, thereby indicating intrinsic differences in their dynamics. We have characterized such differences in three different RGS proteins (apo-RGS4, apo-RGS8, and apo-RGS19) using microsecond-scale classical molecular dynamics (MD) simulations with CHARMM and AMBER force-fields. Analyses of these trajectories reveal high fluctuations in Î±5 and Î±6 helices and in the loops connecting them that facilitate exposure of deeply buried cysteine residues to allosteric inhibitors. We further validate our findings using hydrogen-deuterium exchange mass spectrometry (HDX-MS) analysis of all three proteins and provide details on exchange rates for hydrogen atoms at the resolution of single residues.

# Liang Zhang


Email: zhangbright1986@gmail.com
Tel: (512)-658-1239
Link: Google Scholar
Address: Department of Chemical and Biomolecular Engineering
University of Pennsylvania, Philadelphia, PA


## EDUCATION

2009~2015 University of Texas at Austin, Austin, USA
**Doctor of Philosophy** in Physical Chemistry (GPA: 4.0/4.0)
Advisor: **Prof. Graeme Henkelman**
**Thesis:** "Theoretical study of correlation between structure and function for nanoparticle catalysts"

2005~2009 University of Science and Technology of China (USTC), Hefei, China
**Bachelor of Science** in Chemical Physics (GPA: 3.73/4.3)
Advisor: **Prof. Jinlong Yang**
**Thesis:** "First principles study of size effect on the electronic structure of g-$C_3N_4$ nanoribbons"

## RESEARCH EXPERIENCE

2016~present **Post-doc Researcher,** University of Pennsylvania, Philadelphia, PA
Advisor: **Prof. Aleksandra Vojvodic**

- Developed the theoretical framework of designing hetero-structured perovskites catalysts for water splitting.
- Tailored electronic structures and catalytic activities of MXenes with functionalization, strain, and support.
- Simulated STM and water splitting process on Fe-doped $CoO_x$ bilayer.

2015~2016 **Post-doc Researcher,** Stanford University, CA
Advisor: **Prof. Jens. K. Nørskov** and **Dr. Aleksandra Vojvodic**

- Illuminated the coupling between surface structure and reaction mechanism for $O_2$ incorporation on $(La_{0.6}Sr_{0.4})FeO_{3-\delta}$ electrode.
- Revealed the role of cover defect in CO activation under confinement.

2009~2015 **Graduate Research Assistant,** University of Texas, Austin, TX
Advisor: **Prof. Graeme Henkelman**

- Built structure-function correlations for nanoparticle (NP) catalysts, including core@shell, random alloy, and alloy-core@shell NPs and guided experimental design for ORR, HER, CO oxidation and alkene hydrogenation.
- Developed accelerated molecular dynamics methods for long-timescale simulation.
- Elucidated the involvement of surface intermediates in controlling the activity and selectivity of ally alcohol partial oxidation.

## TEACHING EXPERIENCE

2016-2017 **Teaching Assistant:** Computational Science of Energy and Chemical Transformations
University of Pennsylvania, Philadelphia, PA

- Hosted recitation sessions and delivered guest lectures.
- Prepared computational platform and website for projects.

2011-2013 **Research Educator:** Freshman Research Initiative Program, Computational Nanoparticle Stream
University of Texas at Austin, Austin, TX

- Introduced freshman students into theoretical chemistry research, including DFT, molecular dynamics, and Monte Carlo simulations.
- Prepared syllabus, computational platform, course website, and research projects.
- Delivered laboratory lectures and individual mentoring.

2012 **Teaching Assistant:** Physical Chemistry Lab
University of Texas at Austin, Austin, TX

- Instructed undergraduate students to perform physical chemistry experiments, including thermochemistry, kinetics, electrochemistry and so on.

## HONORS AND AWARDS

| | |
|---|---|
| 2013 | Graduate Student Professional Development Awards, University of Texas at Austin |
| 2008 | China National Awards for Inspiration, USTC |
| 2005-2007 | USTC Outstanding Student Scholarship, USTC |
| 2005-2008 | Cyrus Tang Scholarship for Personal Development and Community Service, USTC |

## PUBLICATIONS


1 Akbashev, A.R.; **Zhang, L.**; Chueh, W.C.; Vojvodic, A. "Activation of $SrTiO_3$ with Subsurface $SrRuO_3$ for Oxygen Evolution Reaction", **2017**, *submitted* (**co-first author**).

2 **Zhang, L.**; Ng, M. L.; Vojvodic, A. "CO Dissociation Under the Confinement between *h*-BN and Rh", **2017**, *submitted*.

3 Fredrickson, K. D.; **Zhang, L.**; Vojvodic, A. "Large Gapped Dirac Cones in Functionalized Two-dimensional Carbides (MXenes)", **2017**, *submitted*.

4 **Zhang, L.**; Raman, S. A.; Vojvodic, A. "Subsurface Engineering of $SrTiO_3$ for Electrochemical Water Splitting", **2017**, *manuscript in preparation*.

5 Guan, Z.; **Zhang, L.**; Chueh, W.C.; Vojvodic, A. "Mechanistic Study of Oxygen Incorporation Reaction on $(La_{0.6}Sr_{0.4})FeO_{3-\delta}$ Thin Film Electrodes", **2017**, *manuscript in preparation* (**co-first author**).

6 **Zhang, L.**; Fredrickson, K. D.; Vojvodic, A. "Tuning MXenes for Hydrogen Evolution: Strain and Support Effect", **2017**, *manuscript in preparation*.

7 Rodríguez-Fernández, J.; Sun Z.; **Zhang, L.**; Tan, T.; Fester J.; Vojvodic, A.; Lauritsen, J. V. "xxxxx", **2017**, *manuscript in preparation*.

8 Shin, K.; **Zhang, L.**; An, H.; Ha, H.; Yoo, M.; Lee, H. M.; Henkelman, G.; Kim, H. Y. "Interface Engineering for A Rational Design of Poison-free Bimetallic CO Oxidation Catalysts", *Nanoscale*, **2017**, 9, 5244-5253.

9 Corona, B.; Howard, M.; **Zhang, L.**; Henkelman, G. "Computational Screening of Core-Shell Nanoparticles for the Hydrogen Evolution and Oxygen Reduction Reactions", *J. Chem. Phys.*, **2016**, 145, 244708.

10 Luo, L.; **Zhang, L.**; Duan, Z.; Lapp, A. S.; Henkelman, G.; Crooks R. M. "Efficient CO Oxidation Using Dendrimer-Encapsulated Pt Nanoparticles Activated with <2% Cu Surface Atoms", *ACS Nano.*, **2016**, 10, 8760-8769.

11 Kunal, P.; Li, H.; Dewing, B. L.; **Zhang, L.**; Jarvis, K.; Henkelman, G.; Humphrey S. M. "Microwave-Assisted Synthesis of PdAu Alloy Nanoparticles: A Combined Experimental and Theoretical Assessment of Synthetic and Compositional Effects upon Catalytic Reactivity", *ACS Catal.*, **2016**, 6, 4882-4893.

12 Xiao, P.; Duncan, J.; **Zhang, L.**; Henkelman, G. "Ridge-Based Bias Potentials to Accelerate Molecular Dynamics", *J. Chem. Phys.*, **2015**, 143, 244104.

13 Anderson, R. M.; **Zhang, L.**; Wu, D.; Brankovic, S. R.; Henkelman, G.; Crooks R. M. "A Theoretical and Experimental In-Situ Electrochemical Infrared Spectroscopy Study of Adsorbed CO on Pt Dendrimer-Encapsulated Nanoparticles", *J. Electrochem. Soc.*, **2015**, 163, H3061.

14 **Zhang, L.**; Chill, S.; Henkelman, G. "Distributed Replica Dynamics", *J. Chem. Phys.*, **2015**, 143, 174112.

15 Luo, L.; **Zhang, L.**; Henkelman, G.; Crooks R. M. "Unusual Activity Trend for CO Oxidation on $Pd_xAu_{140-x}$ @Pt Core@shell Nanoparticle Electrocatalysts", *J. Phys. Chem. Lett.*, **2015**, 6, 2562-2568. (**co-first author**)

16 Yu, W.-Y.; **Zhang, L.**, Mullen, G.; Henkelman G.; Mullins, C. B. "Effect of Annealing in Oxygen on Alloy Structures of Pd-Au Bimetallic Model Catalysts", *Phys. Chem. Chem. Phys.*, **2015**, 17, 20588-20596.

17 Yu, W.-Y.; **Zhang, L.**, Mullen, G.; Henkelman G.; Mullins, C. B. "Oxygen Activation and Reaction on Pd-Au Bimetallic Surfaces", *J. Phys. Chem. C*, **2015**, 119, 11754-11762.

18 Anderson, R. M.; Yancey, D. F; **Zhang, L.**; Chill, S. T.; Henkelman, G.; Crooks R. M. "A Theoretical and Experimental Approach for Correlating Nanoparticle Structure and Electrocatalytic Activity", *Acc. Chem. Res.*, **2015**, 48, 1351-1357.

19 **Zhang, L.**; Anderson, R. M; Crooks R. M; Henkelman, G. "Correlating Structure and Function of Metal Nanoparticles for Catalysis", *Surf. Sci.*, **2015**, 640, 65-72.

20 Mullen, G. M.; **Zhang, L.**; Evans, E. J. Jr; Yan, T.; Henkelman, G.; Mullins, C. B. "Control of Selectivity in Allylic Alcohol Oxidation on Gold Surfaces: The Role of Oxygen Adatoms and Hydroxyl Species", *Phys. Chem. Chem. Phys.*, **2015**, 17, 4730-4738.

21 **Zhang, L.**; Henkelman, G. "Computational Design of Alloy-Core@Shell Metal Nanoparticle Catalysts", *ACS Catal.*, **2015**, 5, 655-660.

22 Garcia, S.; **Zhang, L.;** Piburn, G. W.; Henkelman, G.; Humphrey, S. M. "Microwave Synthesis of Classically Immiscible Rhodium-Silver and Rhodium-Gold Alloy Nanoparticles: Highly Active Hydrogenation Catalysts", *ACS Nano,* **2014**, 8, 11512-11521.

23 Chill, S. T.; Welborn, M.; Terrell, R.; **Zhang, L.**; Berthet, J-C.; Pedersen, A; Jónsson, H.; Henkelman, G. "*EON*: Software for Long Time Simulations of Atomic Scale Systems", *Model. Simul. Mater. Sci. Eng.*, **2014**, 22, 055002.

24 Mullen, G. M.; **Zhang, L.**; Evans, E. J. Jr; Yan, T; Henkelman, G.; Mullins, C. B. "Oxygen and Hydroxyl Species Induce Multiple Reaction Pathways for the Partial Oxidation of Allyl Alcohol on Gold", *J. Am. Chem. Soc.*, **2014**, 136(17) 6489-6498.

25 Anderson, R. M.; **Zhang, L.**; Frenkel, A. I.; Henkelman, G.; Crooks R. M. "Structural Study of the Inversion of Pd@Pt Core@Shell Dendrimer-Encapsulated Nanoparticles and Comparison to Theory", *ACS Nano,* **2013**, 7, 9345-9353.

26 **Zhang, L.** and Henkelman, G. "Design of Pt-shell Nanoparticles with Alloy Cores for the Oxygen Reduction Reaction", *ACS Nano,* **2013**, 7, 9168-9172.

27 **Zhang, L.**; Kim, H. Y.; Henkelman, G. "CO Oxidation at the Au-Cu Interface of Bimetallic Nanoclusters Supported on $CeO_2$ (111)", *J. Phys. Chem. Lett.*, **2013**, 4, 2943-2947.

28 Yancey, D. F.; Chill, S. T.; **Zhang, L.**; Frenkel, A. I.; Henkelman, G.; Crooks, R. M. "A Theoretical and Experimental Examination of Systematic Ligand-Induced Disorder in Au Dendrimer Encapsulated Nanoparticles", *Chem. Sci.*, **2013**, 4, 2912-2921.

29 Iyyamperumal, R.; **Zhang, L.;** Henkelman, G.; Crooks, R. M. "Efficient Electrocatalytic Oxidation of Formic Acid Using Au@Pt Dendrimer Encapsulated Nanoparticles", *J. Am. Chem. Soc.*, **2013**, 135, 5521-5524.

30 **Zhang, L.**; Henkelman, G. "Tuning the Oxygen Reduction Activity of Pd Shell Nanoparticles with Random Alloy Cores", *J. Phys. Chem. C*, **2012**, 111, 120860-120865.

31 Yancey, D. F.; **Zhang, L.**; Crooks, R. M.; Henkelman, G. "Au@Pt Dendrimer Encapsulated Nanoparticles as Model Electrocatalysts for Comparison of Experiment and Theory", *Chem. Sci.*, **2012**, 4, 1033.

32 Tang, W.; **Zhang, L.**; Henkelman, G. "Catalytic Activity of Pd/Cu Random Alloy Nanoparticles for Oxygen Reduction", *J. Phys. Chem. Lett.*, **2011**, 11, 1328-1331.


## Conference Presentations

### Talks:

Nov 2016 **Zhang, L.**; Vojvodic, A. "Mechanistic Study of Oxygen Incorporation of $LaSrFeO_{3-\delta}$ (001) Surface", *2016 AIChE Annual Meeting*, San Francisco, CA.

Nov 2016 **Zhang, L.**; Henkelman, G. "Computational Design of Novel Core@shell Nanoparticles for Catalysis", *2016 AIChE Annual Meeting*, San Francisco, CA.

Jun 2016 **Zhang, L.**; Vojvodic, A. "Under and Above the Cover: Towards Understanding the Role of the 2D Cover", *Institute for Pure and Applied Math (IPAM): Materials for a Sustainable Energy Future Reunion Conference*, Lake Arrowhead, CA.

Oct 2013 Henkelman, G.; **Zhang, L.** "Modeling the Structure and Activity of Nanoparticle Catalysts", *IPAM: Materials for a Sustainable Energy Future*, Los Angeles, CA. (split-talk with Prof. Henkelman)

Apr 2013 **Zhang, L.**; Henkelman, G. "Alloy-Core@Shell Nanoparticles for Oxygen Reduction Reaction(ORR)", *245th ACS National Meeting*, New Orleans, LA.

**Posters:**

Nov 2016 **Zhang, L.**; Fredrickson, K. D.; Vojvodic, A. *2016 AIChE Annual Meeting*, San Francisco, CA.

Aug 2015 **Zhang, L.**; Henkelman, G. *SUNCAT Summer School 2015: Heterogeneous Catalysis for Energy and Society*, Menlo Park, CA.

Feb 2015 **Zhang, L.**; Henkelman, G. *Center for Electrochemistry (CEC) Annual Workshop on Electrochemistry*, Austin, TX.

Jul 2014 **Zhang, L.**; Henkelman, G. *American Conference on Theoretical Chemistry (ACTC)*, Telluride, CO.

Jun 2014 **Zhang, L.**; Henkelman, G. *Gordon Research Conference: Catalysis*, New London, NH.

May 2013 **Zhang, L.**; Henkelman, G. *Freshman Research Initiative (FRI): 2013 Inaugural Annual Conference*, Austin, TX.

Nov 2012 **Zhang, L.**; Henkelman, G. *2012 Materials Research Society(MRS) Fall Meeting*, Boston, MA.

## TECHNICAL SKILLS

Ab Initio simulation (DFT, DFT+U, hybrid functional), Molecular Dynamics (MD) and Accelerated MD, Mote Carlo (MC) and Kinetic MC, Micro-kinetic modeling, STM simulation, Vibrational Spectroscopy (IR and Raman) simulation, Cyclic Voltammogram simulation, XRD simulation.
Packages: VASP, Quantum Espresso, Lammps, GPAW, dMol3, ASE
Programming Languages: Python, C++, Fortran, Perl

## REVIEW

Regularly review articles for *J. Phys. Chem C, Phys. Chem. Chem. Phys., J. Mater. Chem. A*

## REFERENCES

1. **Prof. Graeme A. Henkelman**
   Department of Chemistry
   The University of Texas at Austin
   1 University Station, A5300,
   Austin, TX 78712-0165
   Phone: (512) 471-4179
   Email: henkelman@cm.utexas.edu

2. **Prof. Richard M. Crooks**
   Department of Chemistry
   The University of Texas at Austin
   1 University Station, A5300,
   Austin, TX 78712-0165
   Phone: (512) 475-8674
   Email: crooks@cm.utexas.edu

3. **Prof. Aleksandra Vojvodic**
   Department of Chemical & Biomolecular Engineering
   University of Pennsylvania
   220 South 33rd Street, 371 Towne Building
   Philadelphia, PA 19104
   Phone: (215)-746-2264 (work)
   Email: alevoj@seas.upenn.edu

4. **Prof. William Chueh**
   Department of Materials Science & Engineering
   Stanford University,
   496 Lomita Mall, 137 Durand Building,
   Stanford, CA 94305- 4034
   Phone: (650)-725-7515
   Email: wchueh@stanford.edu

# 499574 Climbing the Volcano: Active-Site Engineering at the Atomic Scale

**Abstract:**

In recent years, molecular simulations of catalyst surfaces have yielded considerable insights into how the electronic structure of catalyst materials influences activity. One of the most notable examples is that of the scaling relations, or correlations between adsorption energies of molecules onto surfaces. Using these correlations, models for turnover rates, coverages, and theoretical overpotentials (in the case of electrocatalysis) may be constructed, often yielding "volcano" relationships which may define fundamental limits on the ideal performance of catalysts. Some of the clearest examples of these limits are present in electrocatalysis of earth-ambient molecules into fuels and chemicals, which include reactions like water oxidation and CO2 electroreduction, in which even the most efficient catalysts require significant overpotentials to achieve moderate turnover rates.

In this work, I identify materials for which assumptions of scaling-relation-based models might be strategically broken in order to circumvent fundamental limits on activity. These include bifunctional materials, in which adsorbates interact with multiple sites on a surface and may be selectivity stabilized by distinct moities in alloys and other bulk binary compounds. Electrolytes, interfaces, ligand-functionalized surfaces and confined nanostructures may also achieve a similar selective stabilization or destabilization, being perturbed off of the traditional scaling relation in order to achieve higher activity. Lastly, I describe a high-throughput methodology developed using software infrastructure from the Materials Project which may aid in the more rapid identification of materials which meet these criteria.

# 496619 Developing Ab Initio Methodology for Advancing Catalytic Reactions

**Abstract:**

Ab initio investigations have advanced the molecular-scale understanding of catalytic reactions. Current methods allow for predictions of catalytic reaction behavior, although reliability of calculations is a barrier to widespread adoption. The growing string method automates molecular-scale reaction path discovery and is a rapid, unbiased approach. It is enhanced for electrochemical systems which experience electrical potential bias. Although calculations without potential bias treatment are capable to make meaningful predictions for electrochemical systems, a more reliable approach considers the charge of the electrocatalyst surface and solvent effects on adsorbates. This improved growing string method is demonstrated on copper and silver for elementary reactions of electrochemical carbon dioxide reduction. In a second example of ab initio method strengthening, a study of platinum nanoparticles supported by titanium-dioxide catalyzing the water-gas shift reaction is presented. This study incorporates uncertainty quantification and the Bayesian inference from experiments to complement calculations. A microkinetic model connects the atomic-scale calculations with the reactor-scale kinetic experiments. Through the Bayesian inference process, the uncertainties of the ab initio calculations are refined, and a conclusion about the catalytically active site is achieved.

# 493253 First-Principles-Derived Structure-Energy Relationship for Surface Oxides

**Abstract:**

Metal-supported, 2-dimensional surface oxide monolayers represent an emerging class of catalysts for a variety of applications, ranging from biomass conversion to electrocatalysis. A tremendous challenge with modeling such materials is their atomic structure, which is typically unknown. Commonly employed structure prediction methods (simulated annealing, evolutionary algorithms) require accurate force fields, which are either expensive (density functional theory), or too complex and have transferability issues (ReaxFF).

Herein, we show that the systematic reduction of Kohn-Sham equations via the Harris energy functional leads to a simple structure-energy relationship involving only interatomic distances and coordination numbers for systems, in which angular dependence of energy is not important. The model reproduces the functional dependence of the metal cohesive energy on geometric characteristics, and its connection to molecular orbital theory allows generalizations to surface oxides. We assess the model performance for $Co_xO_y$/Pt(111) monolayer structures and investigate the transferability of its parameters.

# 491413 First-Principles Based Design of Reaction Conditions for the Catalytic Conversion of Methane to Methanol over Cu-Exchanged SSZ-13

**Abstract:**

Due to its perfect symmetry, methane is the most stable hydrocarbon and its selective conversion to methanol is one of the long-standing challenges in catalysis. Cu containing zeolites are able to catalyze this reaction under relatively mild conditions but a variety of $Cu_xO_yH_z$ clusters have been suggested as active sites. Results in the literature indicate that the type of active site does not only vary with the zeolite framework, but different types of active sites have been observed for the same zeolite structure prepared by different groups or under varying conditions and clear rules for the formation of these sites have not been established so far.

In this contribution, we construct a thermodynamic model for Cu in zeolite SSZ-13 based on RPA Density Functional Theory, and show that under almost all conditions the formation of a distribution of $Cu_xO_yH_z$ clusters is favorable. We identify the chemical potential of O and H (Î¼$^{O}$ and Î¼$^{H}$), the local Al distribution, the reference state and the temperature as important parameters in predicting the exact stoichiometry of these sites. We furthermore model the $Cu_xO_yH_z$ distribution of several systems studied in the literature; comparing these results to experimental studies for the catalytic conversion of methane to methanol significantly narrows down candidates for the most active sites. In a subsequent step we then use our model to predict conditions, which lead to equally as good or better catalytic performance.

The systematic work presented here shows that the Cu speciation in zeolites is condition-dependent and that there is not one single Cu species present. Furthermore, we establish basic rules for choosing the proper conditions to optimize the performance of this catalytic system.

# 498697 Reaction Ensemble Monte Carlo Simulations of Xylene Isomerization Under Confinement

**Abstract:**

Xylene exists as three isomers: meta- (m-), ortho- (o-), and para- (p-). m-Xylene is the most abundant isomer in bulk fluids at equilibrium, but p-xylene is a valuable feedstock in the production of terephthalic acid. p-Xylene can be selectively produced using ZSM-5, a zeolite with narrow channels that prevent m- and o-xylene from escaping interior pores. Using reaction ensemble Monte Carlo (RxMC), we demonstrate that confining a xylene fluid to the interior of a carbon nanotube shifts the equilibrium composition as a function of nanotube diameter.

We adapt the original RxMC reaction move to align both the position and orientation of inserted product molecules and deleted reactant molecules. The accuracy and efficiency of this move is demonstrated for xylene isomerization in vapor, liquid, and supercritical phases. Classical RxMC requires the ideal gas free energy of reaction as an input. We compare three methods for computing this free energy: using tabulated enthalpies and entropies of formation, using the harmonic oscillator and rigid rotor approximation, and using QM/MM alchemical transformation combined with the multistate Bennett acceptance ratio (MBAR). We find that the tabulated free energies of reaction give the best agreement with experimental equilibrium compositions in bulk fluids. RxMC simulations in a carbon nanotube with an

consistent with the production of p-xylene in ZSM-5. We also show that o-xylene becomes the dominant isomer in

nanotubes of this diameter, effectively allowing o-xylene to fit into cavities inaccessible to the other isomers.

# 501357 Understanding Separation and Catalysis in Nanoporous Materials

**Abstract:**

The sustainable production of energy and chemicals requires efficient separation and catalytic systems, and an improved understanding of the principles driving the various physical and chemical processes is clearly critical to our success in this endeavor. This talk will highlight applications using crystalline nanoporous materials, primarily zeolites, and show how molecular modeling can be used both to gain a microscopic understanding into phenomena difficult to probe experimentally and to predict accurate data at a scale beyond the reach of the traditional trial-and-error approach. Examples discussed include the aqueous-phase adsorption of oxygenated compounds, the unique adsorption/diffusion behaviors in hierarchical materials and ultrathin membranes, and the molecular structure and framework confinement effects on hydrocarbon cracking. New insights and methodological developments from these studies provide the basis for an effort in which a large number of materials were screened with the purpose of finding better candidates for two energy-related applications, ethanol/water separation and hydrocarbon iso-dewaxing, and some of our predictions were subsequently verified by experiments. Together they demonstrate the possibility of using theory and computation in a predictive mode to expedite the pace of materials discovery for complicated, realistic, and industrially relevant applications.

# 493009 Modeling Self-Assembly of Metal-Organic Frameworks with Enhanced Sampling Techniques


**Abstract:**

Self-assembly is a process by which building blocks form functional materials. In principle, instructions can be imparted onto the building blocks to engineer and design the final structure. One class of materials which takes advantage of this building block approach is metal-organic frameworks (MOFs), nanoporous crystalline materials formed in solution through coordination-driven self-assembly of inorganic nodes and organic linkers. Experimental observations of MOF self-assembly reveal complex and interesting phenomena where synthetic conditions (solvent, temperature, concentration of building blocks) alter the final MOF structure as well as intermediate metastable states. However, computational efforts to elucidate MOF self-assembly are scarce. For this study, we focus on the self-assembly of MOF-5.

We employ a combination of enhanced sampling techniques to study MOF self-assembly for different scenarios and system sizes. We model MOF-5 with an atomistic model as either a discrete unit cell or four unit cells in explicit solvent. To validate the model, we perform replica exchange with solute scaling (REST2) simulations and calculate the free energy profile using the average distance and average angle between nodes as collective variables. The free energy minima observed in these systems are close to the values in the experimental unit cell, therefore validating that the chosen model represents at least a metastable state.

To model self-assembly, we perform finite temperature string (FTS) method calculations using the average distance and angle between nodes and the total coordination between nodes and linkers. The process was modeled for both single and four unit cells systems from two starting points: fully disassembled and amorphous. The self-assembly starting from the disassembled state was found to be downhill in free energy for both the single unit cell and four unit cells systems. In contrast, the self-assembly starting from the amorphous state reveals energetic barriers on the way to the final MOF structure. Finally, we discuss the mechanism of self-assembly of MOF-5 for the scenarios and system sizes considered.

# 492643 Computationally-Efficient High-Throughput Screening of Metal-Organic Frameworks for Hydrogen Storage

**Abstract:**

Metal-organic frameworks (MOFs) are porous materials consisting of metal nodes connected with organic linkers. These materials can have high surface area and porosity, which make them interesting candidates for applications in gas separation and storage. Since MOFs can be made from many different nodes, linkers, and functional groups, there are a huge number of possible MOF structures. High-throughput computational screening can efficiently characterize large numbers of materials to identify the best candidates for applications such as gas separation and storage. [1,2]

We have used high-throughput screening techniques paired with grand canonical Monte Carlo simulations to calculate the hydrogen storage capacity for tens of thousands of MOFs. We will discuss commonalities in the structural and textual properties among the top-performing candidates from this screening and further explore how the information gleaned from high-throughput screening can inform the design of new materials that can be synthesized for hydrogen storage applications.

Due to the significant computational cost of performing GCMC simulations for a large number of MOFs, we will introduce an efficient screening method that can estimate the storage capacity for thousands of structures in a small fraction of the time required for full GCMC simulations. This allows us to quickly identify the most promising candidates for more detailed study, and this technique can be applied to other gases, such as methane, as well as a wide range of temperatures and pressures.

1. Y.J. Colon, R.Q. Snurr, “High-throughput computational screening of metal-organic frameworks.” *Chem. Soc. Rev.* 43, 5735-5749 (2014)
2. N. S. Bobbitt, J. Chen, R. Q. Snurr, "High-Throughput Screening of Metal–Organic Frameworks for Hydrogen Storage at Cryogenic Temperature." *The Journal of Physical Chemistry C* 120(48), 27328-27341 (2016)

# 502903 Effects of Ion Self-Energy on the Double Layer Structure and Properties at the Dielectric Interface

**Abstract:**

A systematic, unified and predictive theoretical framework that is able to capture all the essential physics of the interfacial behaviors of ions, such as the Hofmeister series effect, Jones-Ray effect and the salt effect on the bubble coalescence remains an outstanding challenge. The most common approach is the Poisson-Boltzmann (PB) theory; however, there are many systems for which the PB theory fails to offer even a qualitative explanation, especially for ions distributed in the vicinity of an interface with dielectric contrast between the two medium. A key factor missing in the PB theory is the self-energy of the ion. We develop a self-consistent theory that treats the electrostatic self-energy (including both the short-range Born solvation energy and the long-range image charge interactions), the nonelectrostatic contribution of the self-energy, the ion-ion correlation and the screening effect systematically in a single framework. The theory gives a continuous self-energy across the interface, which allows ions on the water side and the vapor/oil side of the interface to be treated in a unified framework. Using the theory, we demonstrate three essential effects, the image charge effect, the inhomogeneous screening effect and the specific ion effect, for the ions near the dielectric interfaces. First, we show that the image charge repulsion creates a depletion boundary layer that cannot be captured by a regular perturbation approach. The image force qualitatively alters the double layer structure and properties, and gives rise to many non-PB effects, such as like-charge attraction and charge inversion. Then, we show that the double layer structure and interfacial properties is drastically affected by the inhomogeneous screening due to the nonuniform ion distribution particularly at high salt concentrations when the bulk Debye screening length is comparable to the Bjerrum length. The characteristic length of the depletion layer scales with the Bjerrum length instead of the Debye screening length, resulting in a linear increase of the negative adsorption of ions with concentration, in agreement with experiments. Finally, we study the self-energy of a single ion across the dielectric interface. Using intrinsic parameters of the ions, such as the valency, radius, and polarizability, we predict the specific ion effect on the interfacial affinity of halogen anions at the water/air interface, and the strong adsorption of hydrophobic ions at the water/oil interface, in agreement with experiments and atomistic simulations.

# 497540 Molecular Investigation into the Transport across the Blood-Brain Barrier Interface [Invited Talk]

**Abstract:**

The blood-brain barrier is an interfacial barrier constituted by claudin-5 membrane proteins. The selective permeability of chemical across this interfacial barrier is responsible for the lack of treatment options for neurodegenerative diseases such as Alzheimerâ€™s and Parkinsonâ€™s. In an effort to overcome this hurdle, we use multiscale molecular simulations to study the intricate self-assembly of claudin-5 transmembrane proteins in explicit membrane and solvent environment. Based on the simulation we were able to identify the stable dimer conformations that result from membrane driven claudin-5Â *cisÂ* interactions and the corresponding pore structures. Our results reveal the nature of the molecular interactions that drive claudin-5 aggregation in forming the tight junctions, and are in good agreement with the previous experimental predictions. The insights from this study will help in designing the next generation of smart therapeutics that can permeate the blood-brain barrier interface.

# 493826 Adsorption and Self-Assembly of Surfactants on Metallic Surfaces Studied Using Molecular Simulations


**Abstract:**

Surfactant molecules, such as hydrocarbons with imidazoline-based polar groups, are employed in the oil and gas industry for inhibiting corrosion inside transportation pipelines. The adsorption of these molecules at the metal-water interface is understood to be important for their efficacy as corrosion inhibitors. Experimental and theoretical investigations suggest that these molecules self-assemble at the metal-water interface in organized layers. The mechanism of adsorption and self-assembly, while considered critical for their inhibition properties, are only poorly understood because of the difficulties in probing the interface at molecular length-scales. Using classical molecular simulations and a coarse-grained representation of surfactant molecules, we study the adsorption of these molecules on metallic surfaces, focusing on the effect of hydrophobic character of the alkyl tail, the tail length and the size of the polar head group. Our simulations reveal that the hydrophobic character of the alkyl tail facilitates adsorption by promoting formation of a self-assembled monolayer on the metal surface. Surfactant molecules with longer alkyl tails show a decrease in the adsorbed amount, because of larger entropic loss of these molecules upon adsorption. Upon increasing the size of the polar group, we observe that the nature of the adsorbed layer changes, with a significant fraction of molecules adsorbed with their polar groups towards the aqueous phase. This arrangement of adsorbed molecules achieves better packing efficiency and hence lower free energy because of enthalpic effects.

# 501676 Modeling Alkane Partitioning and Phase Behavior on Graphite Pores: A Discussion on Dispersion Free Energy Formalism

**Abstract:**

In reservoir simulations, thermodynamic properties of fluid such as pressure /volume /temperature, bubble/dew point and critical point are required to achieve a good estimation of reservoir production performance. Lots of studies have been done in unconventional shale showing the fluid phase behavior is shifted from what is predicted by a bulk equation of state. This is due to the significant adsorption of fluids in the organic rich shale. It has been recognized that the solid composition, pore size and geometry can all affect the fluid phase boundaries, which are shown to be much narrower than for bulk phase behavior. Therefore, it is highly important to establish a phase diagram for confined fluids in a way that is both accurate and efficient. Molecular density functional theory is preferred in solving this problem since it relates intermolecular forces to fluid structure and phase behavior similar to molecular simulation.

We model the mixed alkane fluid adsorption on graphite slit pores with interfacial Statistical Associating Fluid Theory (iSAFT) and study the phase behavior of pure and mixed alkanes in different pore sizes. The excess free energy is a summation of intermolecular repulsion and attraction contributions where the repulsion is accounted for by hard sphere volume exclusion and chain connectivity that is consistent with the SAFT equation of state for a bulk fluid. However, the contribution due to long range attraction is more challenging since the pair correlation function is unavailable for inhomogeneous fluids. There are two types of approximation mainly used by people, which are the mean field approximation (MFA) and a weighted density approximation (WDA).In this study, we compare the performance (accuracy and efficiency) of three schemes, which are MFA, WDA and a combination of these two, in predicting the fluid structure and adsorption isotherms. While MFA is the simplest and most widely used method, the accuracy of bulk fluid property prediction is limited, which makes the confined fluid property also shifted systematically. Thus we explore the performance of different types of WDA or a combination of MFA and WDA in approximating the local dispersion free energy and predicting fluid phase behavior in a wide temperature, pressure, pore size and species range. The accuracy of different methods will be evaluated by comparing to simulation data and computational time of each method will also be shown.

# 490686 Effect of Solvent on the Binding Energies of MoleculesÂ on Metal Surfaces

**Abstract:**

The adsorption of molecules on surfaces in a liquid environment is an important phenomena in many cutting-edge applications such as nanomaterials, biosensors, and catalysis. Although the interaction between molecules and surfaces has been extensively studied in the past, the solvent's influence on the adsorption thermodynamics has not yet been fully understood. Here, we examine the adsorption of polyvinylpyrrolidone (PVP) on Ag surfaces in both ethylene glycol solvent and in vacuum, for comparison. This interface is commonly present in the shape-selective synthesis of Ag nanocrystals, in which the choice of solvent and adsorbate directly influences nanocrystal morphology. We use molecular dynamics simulations based on a many-body organic-metal force field to calculate the potential of mean force between PVP analog molecules and surfaces of Ag. In the resulting binding free energies, we see that solvent drastically decreases the binding free energy of the molecules because the solvent stabilizes the molecule in the bulk solution environment. We find that the system's entropy changes due to adsorption are generally negative in vacuum but zero or slightly positive in solvent. These results show that the liberation of adsorbed solvent molecules when a ligand binds to a solid surface can contribute more to the system's entropy change than the configurational entropy loss of the ligand due to adsorption. Our study underscores the importance of including solvent in theoretical models for molecular adsorption.

# 495799 Tuning Proximal Water Diffusion Via Silanol Patterning on Quartz Surfaces

**Abstract:**

Controlling the diffusivities of liquid species proximal to solid surfaces is a critical design variable in many processes involving heterogeneous catalysis and separations, for which surface mobility determines the optimal operating conditions. Silica is commonly used as a support structure in many such processes or, as in the case of zeolites, is itself used as a molecular sieve for purification. We investigate how variations in the patterning of surface chemistry, specifically the arrangement of surface hydroxyls on quartz, determines the diffusion coefficient of nearby water. We develop a novel genetic algorithm optimization, coupled to iterative molecular dynamic simulations, that designs the geometric arrangement of surface hydroxyl groups so as to minimize or maximize the diffusion coefficient of nearby water molecules. Applied to a range of silanol densities, our results reveal the possibility of a wide spread of surface diffusivities at low, fixed surface densities of hydroxyls. Such information may be used to map observed experimental diffusivities to specific, heterogeneous patterns of silanols and vice versa. Additionally, our findings explain recent experimental measurements of sharp increases in the diffusivity of surface water as silanol density reaches low values. We attribute these jumps to changes in the thermodynamically favorable arrangement of surface hydroxyls as silanol density is decreased, which in turn leads to large shifts in the diffusion coefficient.

# 483707 Molecular Dynamics Studies of the Effects of Ionic Liquid Molecular Properties and Particle Concentration on the Behavior of Nanoparticles at the Ionic Liquid/Water Interface

**Abstract:**

Ionic liquid (IL)/liquid interfaces have exhibited unique phenomena when interacting with hard particles--including particle "bridging" between emulsion droplets and, perhaps most interestingly, the spontaneous transport of microparticles across the liquid/liquid interface, something which acts against the accepted thermodynamics of such systems. The current theory behind this phenomena is only thinly supported. Here, extensive molecular dynamics simulations of hydrophobic nanoparticles at the IL/water interface are described. These simulations are the first of their kind to vary both the concentration of nanoparticles and the ionic liquid involved, allowing more general insights into the molecular-scale behavior of IL/water/particle systems. This study compares the interfaces of water with four ILs--1-ethyl-3-methylimidazolium hexafluorophosphate ([EMIM][$PF_6$]), 1-butyl-3-methylimidazolium hexafluorophosphate ([BMIM][$PF_6$]), 1-hexyl-3-methylimidazolium hexafluorophosphate ([HMIM][$PF_6$]), and 1-butyl-3-methylimidazolium bis(trifluoromethylsulfonate) ([BMIM][$Tf_2N$]). In addition, three levels of particle concentration are simulated for a total of twelve simulation systems. The three different imidazolium cations increase in hydrophobicity with the increasing length of the carbon chain, and so give insight into the effects of IL hydrophobicity on the studied systems and behaviors. Meanwhile, the two anions have markedly different shapes. Tools including density profiles and interaction-energy comparisons are used to analyze these simulations. It is found that the presence of hydrophobic nanoparticles profoundly affects the interface itself by moderating the repulsive forces between hydrophobic ILs and water. This leads to the creation of a "pseudo-phase" of intermingled water, IL, and particles. This effect is more prominent in systems with less hydrophobic ILs, smaller and more symmetrical ILs, and higher particle concentrations. This It is postulated that this pseudophase explains some of the unusual behavior of particles observed experimentally at the IL/liquid interface.

# 501755 Evaporation Induced Nucleation of NaCl in Clay Minerals: Mechanism and Potential Sites

**Abstract:**

Salt crystallization and nucleation in confined media occur in Earth's crust, damaging physical properties of porous media. It also occurs in other solid materials of high practical importance, such as asphalt and concrete. Clays are a main component of sedimentary geological formations and, thus, salt crystallization and precipitation in clays is the prototype of the more general phenomenon. Despite numerous efforts to understand the mechanisms of nucleation of salt crystals, the effect of the chemical composition of nanopores of geological formations and their fluid content on the growth, distribution and properties of salt crystals has not been studied. To address this shortcoming, we have carried out extensive molecular dynamics simulation using models of two types of clay minerals: Na-montmorillonite (Na-MMT) and Kaolinite. The simulations provide new and deep insight into the mechanism of nucleation and crystal growth in clay minerals. We have investigated the effect of the evaporation rate of brine and the clay mineral type on the spatial distribution of the precipitated NaCl crystals from supersaturated brine.

17AIChE
Annual Meeting, Minneapolis, MN

# Computational Investigation of the Role of Topology and Functionalization on ZIF Stability

**Rebecca Han**, Souryadeep Bhattacharyya, Sankar Nair, David Sholl

School of Chemical & Biomolecular Engineering, Georgia Institute of Technology

10.31.2017

Georgia Tech Energy Frontier Research Center

UNCAGE ME

# Roadmap

Motivation: bulk stability of zeolitic imidazolate frameworks (ZIFs)

Experiment: exposure conditions and observations

Simulation: methodology and modeling

Results: stability dependence on pKa, thermodynamics, kinetics

Conclusion

Georgia Tech Energy Frontier Research Center

UNCAGE ME

# Motivation: Why ZIFs?

Zeolitic imidazolate frameworks (ZIFs) are a subset of metal organic frameworks

$Zn^{2+}$ centers, imidazolate (IM) linkers

Topologically isomorphic to zeolites

Industrially relevant for separations

Tunable porosity/chemistry, high thermal stability

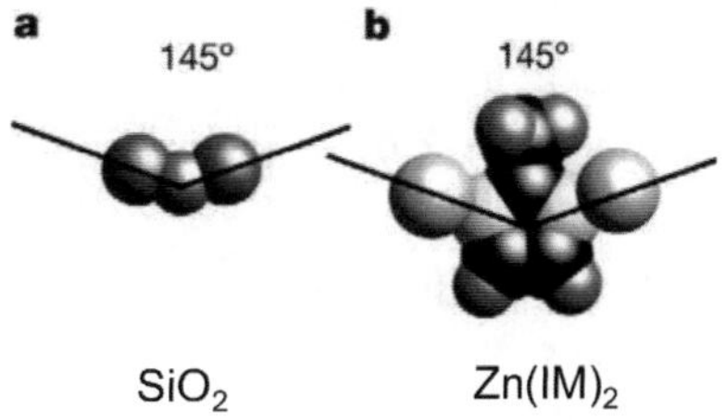


Georgia Tech | Energy Frontier Research Center

Banerjee et al. *J Am Chem Soc* **2009**, 131 (11).
Phan et al. *Acc Chem Res* **2010**, 43 (1).
Pimental et al. *Chem Sus Chem* **2014**, 7 (12).
Sholl and Lively. *Nature* **2016**, 533 (7603).

3

# Motivation: Chemical Stability

Acid gases ($CO_2$, $SO_x$, $NO_x$, etc) are common in industrial applications

Severely limit material performance and lifetime

Limited knowledge of degradation mechanisms

Bulk stability depends on topology and linker functionalization

Systematic investigation of ZIF chemical stability at different process conditions

Georgia Tech | Energy Frontier Research Center

Burtch et al. *Chem Rev* **2014**, 114 (20).
Sholl and Lively. *J Phys Chem Lett* **2015**, 6 (17).
Yu et al. *J Phys Chem C* **2012**, 116 (38).

4

# Experiment: Exposure Conditions

10 ZIFs – 6 linkers, 3 topologies
- 2 mixed-linker hybrids

Exposure up to 5 days
- $H_2O$ – vapor (.5 to .95 rh), aqueous
- $SO_2$ – dry, humid (.85 rh), aq. (20 ppm)
- $CO_2$ – dry, humid (.9 rh)

Measuring degradation
- Loss of crystallinity (XRD)
- Loss of pore volume, BET surface area

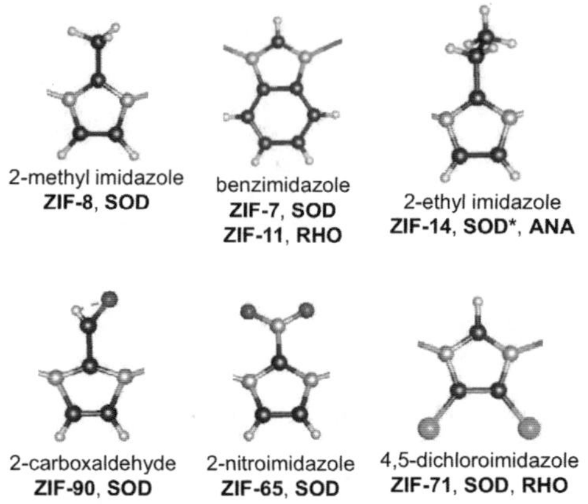


Georgia Tech | Energy Frontier Research Center

Park et al. *Proc Natl Acad Sci USA* **2006**, 103 (27).
Schweinefuss et al. *Dalton Trans* **2014**, 43 (9).
Zhang et al. *Chem Commun* **2013**, 49 (31).
Thompson et al. *Chem Mater* **2012**, 24 (10).



# Experiment: Exposure Observations

| Adsorbent | Topology | Dry $SO_2$ | Dry $CO_2$ | Humid air | Liquid $H_2O$ | Aq. $SO_2$ | Humid $SO_2$ | Humid $CO_2$ |
|---|---|---|---|---|---|---|---|---|
| ZIF-8 | SOD | Stable | Stable | Stable | Stable | Stable | Degrades | Degrades |
| ZIF-14 | ANA | Stable | Stable | Stable | Degrades | Degrades | Degrades | Degrades |
| ZIF-$8_{15}14_{85}$ | SOD | Stable | Stable | Stable | Degrades | Degrades | Degrades | Degrades |
| ZIF-7 | SOD | Stable | Stable | Stable | Phase change | Phase change | Degrades | Degrades |
| ZIF-11 | RHO | Stable | Stable | Stable | Phase change | Phase change | Degrades | Degrades |
| ZIF-90 | SOD | Stable | Stable | Stable | Stable | Stable | Degrades | Degrades |
| ZIF-$8_{50}90_{50}$ | SOD | Stable | Stable | Stable | Stable | Stable | Degrades | Degrades |
| ZIF-71 | RHO | Stable | Stable | Stable | Phase change | Phase change | Stable | Stable |
| ZIF-71 | SOD | Stable | Stable | Phase change | Phase change | Phase change | Phase change | Phase change |
| ZIF-65 | SOD | Stable | Stable | Phase change | Phase change | Phase change | Phase change | Phase change |

Legend: Stable | Phase change | Degrades

Georgia Tech | Energy Frontier Research Center

Zhao et al. *Chem Mater* **2014**, 26 (5).
Schweinefuss et al. *Dalton Trans* **2014**, 43 (9).

# Simulation: Determining Accurate Models

Goal: achieve mechanistic understanding and explanation for observations

Accurate degradation model
- local metal-linker geometry
- protonating species
- defect configuration

Identify predictive "stability indicators" – computed properties that correlate with experimental stability

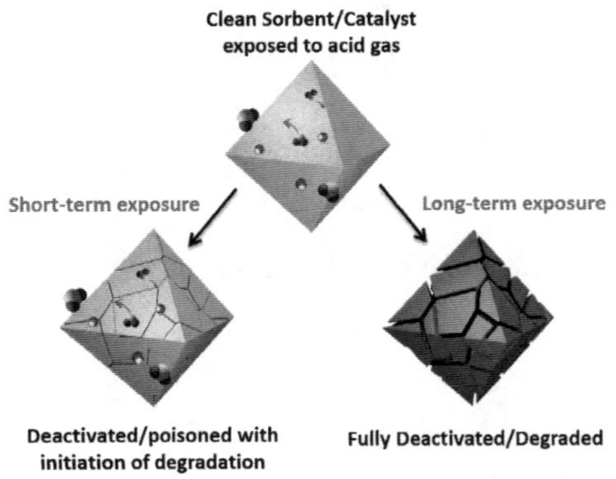


Zhu et al. *Chem Eur J* **2016**, 22 (31).

Georgia Tech Energy Frontier Research Center

Low et al. *J Am Chem Soc* **2009**, 131 (43).
Lu et al. *J Mater Chem A* **2014**, 2 (38).

7

# Stability Indicators

Ligand pKa and atomic charges

Calculated gas phase free energy with DFT(B3LYP)/6-311++G(d,p)

Calculated solvation free energy with Solvation Model based on Density (SMD)

Density derived electrostatic and chemical (DDEC) atomic charges

Defect formation energy, thermodynamic stability

Optimized with DFT(GGA-PBE)+D3

Hydrophobicity, kinetic stability

Grand canonical Monte Carlo (GCMC)

Georgia Tech Energy Frontier Research Center

Liptak et al. *J Am Chem Soc* **2001**, 123 (30).
Manz and Sholl. *J Chem Theory Comput* **2010**, 6 (8).
Wierzchowski and Kofke. *J Phys Chem B* **2003**, 107 (46).

8

# DFT: Degradation Pathway

Degradation begins at defect sites

Dangling linker "point defects" – adsorbate inserts into Zn-IM bond

Optimize periodic **SOD** structures (<1000 atoms) with DFT

What is the actual adsorbent, adsorbate?

Defect formation energy:

$$\Delta E = E_{defect} - (E_{pristine} + E_{adsorbate})$$

Density functional theory (DFT)

- Vienna ab initio simulation package
- PBE-GGA functional, Grimme DFT-D3 dispersion corrections
- Projector-augmented wave pseudopotentials, $E_{cutoff} = 600$ eV
- Conjugate gradient optimizer, $F_{tol} = 0.05$ eV/Å
- Γ-point k-space sampling

Georgia Tech Energy Frontier Research Center

Blöchl. *Phys Rev B* **1994**, 50.
Perdew et al. *Phys Rev Lett* **1996**, 77.

9

# DFT: What is the Absorbent?

Determining an unambiguous ZIF structure

SOD isomorphs

Mixed-linker hybrids

Bulk is only as stable as the weakest bond

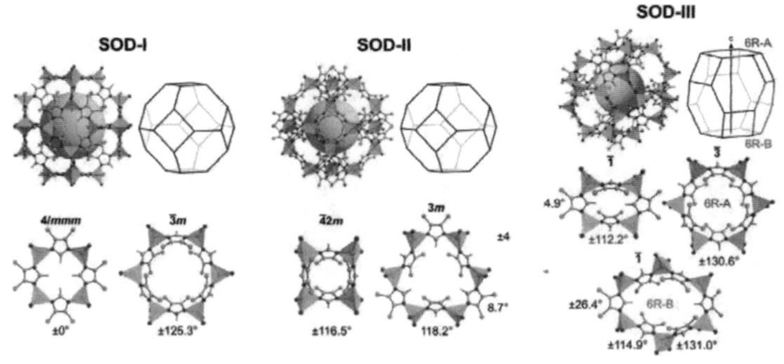


Georgia Tech Energy Frontier Research Center

Springer et al. *Cryst Eng Comm* **2016**, 18.
Schweinefuss et al. *Dalton Trans* **2014**, 43 (9).
Zhang et al. *Phys Chem Chem Phys* **2014**, 16.

10

# DFT: What is the Adsorbate?

Determining the acidic protonating species

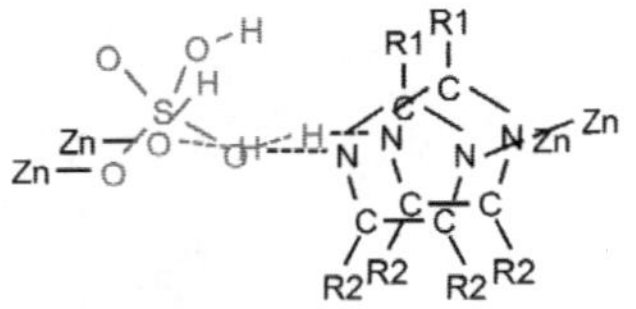

$SO_2$ oxidizes to $H_2SO_4$ which then inserts

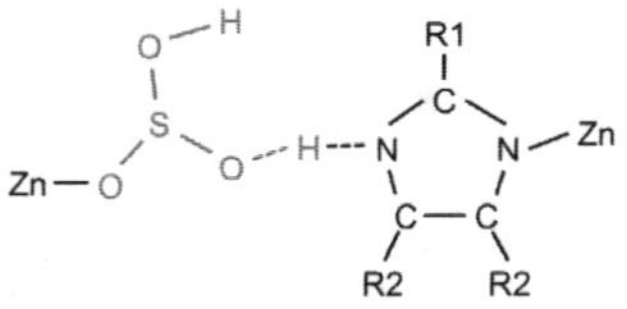

$H_2O$ and $SO_2$ pre-react to form $H_2SO_3$ which then inserts

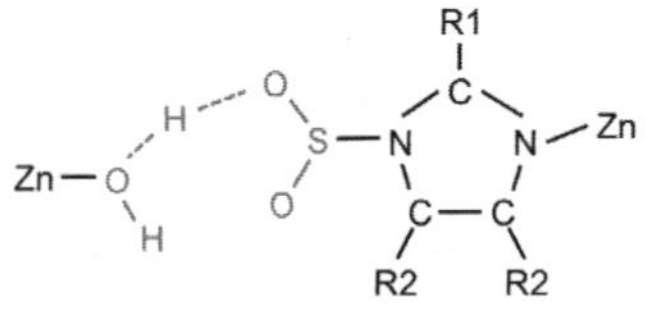

$H_2O$ first inserts into Zn-N bond followed by $SO_2$ adsorbing to N

Georgia Tech | Energy Frontier Research Center

Flamm et al. *J Electrochem Soc* **1981**, 128 (3).
Li et al. *Environ Pollut* **2018**, 232.
Temelso et al. *J Phys Chem A* **2012**, 116 (9).

11

# DFT: What is the Final State?

Three-dimensional configuration space

Local vs global minima for defect
~ 0.1 eV variance on heavy atom placement
~ 0.01 eV variance on H placement

| dE (eV) | ZIF-8 | ZIF-7 | ZIF-71 (I) | ZIF-71 (II) |
|---|---|---|---|---|
| "inline" | 0.61 | 0.48 | 0.10 | 0.79 |
| "nonplanar" | | 0.53 | 0.16 | 0.49 |

Defines an upper bound on (in)stability

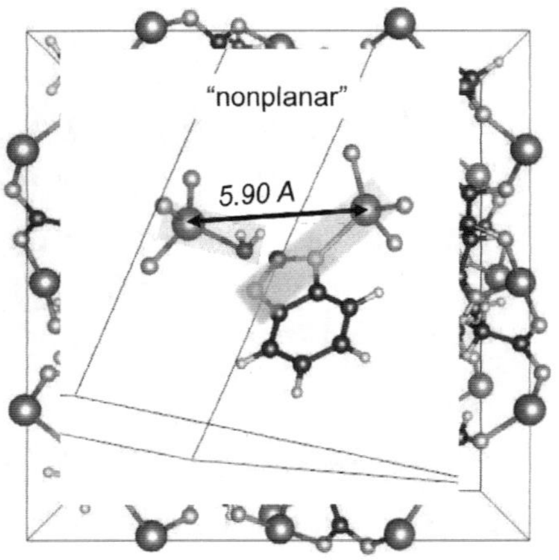


Georgia Tech | Energy Frontier Research Center

UNCAGE-ME

12

# Results: Ligand pKa and Atomic Charges

Is stability predicted by linker species?

Lewis acid-base theory suggests higher basicity results in stronger Zn-N bond

Higher ligand pKa correlates with higher stability in carboxylate MOFs

Calculated pKa and atomic charges do not predict ZIF stability to acid gas degradation

| Adsorbent | Ligand pKa | Aq. $SO_2$ | Humid $SO_2$ |
|---|---|---|---|
| ZIF-8 | 20.12 | | |
| ZIF-14 | 19.56 | | |
| ZIF-$8_{15}14_{85}$ | 19.56 | | |
| ZIF-7 | 14.86 | | |
| ZIF-11 | 14.86 | | |
| ZIF-90 | 13.58 | | |
| ZIF-$8_{50}90_{50}$ | 13.58 | | |
| ZIF-71-RHO | 11.01 | | |
| ZIF-71-SOD | 11.01 | | |
| ZIF-65 | 10.22 | | |

Georgia Tech | Energy Frontier Research Center

Burtch et al. *Chem Rev* **2014**, 114 (20).
Lu et al. *J Mater Chem A* **2014**, 2 (38).
Liptak et al. *J Am Chem Soc* **2001**, 123 (30).
Manz and Sholl. *J Chem Theory Comput* **2010**, 6 (8).



# Results: Defect Formation Energy

DFT calculated formation energies with experimental stability colour coding

| Adsorbent (SOD) | Humid air | Humid $SO_2$ | Humid $CO_2$ |
|---|---|---|---|
| ZIF-8 | 0.61 | -0.59 | -0.57 |
| ZIF-14* | 0.72 | -0.20 | 0.20 |
| ZIF-7 | 0.27 | -0.80 | -0.32 |
| ZIF-90 | 0.25 | -0.02 | 0.25 |
| ZIF-71 | -0.04 | -0.53 | -0.04 |
| ZIF-65 | 0.74 | -0.47 | -0.31 |

Negative energy (eV) indicates lower stability, favourable degradation reaction

Correlates reasonably well with experimental observation

Georgia Tech | Energy Frontier Research Center

UNCAGE ME

# Results: Defect Formation Energy, con't

Can topology explain ZIF-71 RHO stability in humid $SO_2$, $CO_2$?

Degradation reactions are local events

Same bond angles/lengths in RHO, SOD

Compare $\Delta E$ in hypothetical ZIF-71 polymorphs

$|\Delta\Delta E|$ ~ 0.02–0.26 eV could make a difference

| POLY | Humid air | Humid $SO_2$ |
|---|---|---|
| $\Delta\Delta E = \Delta E_{POLY} - \Delta E_{SOD}$ | | |
| CRB | -0.16 | -0.02 |
| DFT | -0.23 | -0.26 |
| CAG | 0.05 | 0.07 |
| GIS | -0.05 | 0.11 |

| SOD | Humid air | Humid $SO_2$ |
|---|---|---|
| ZIF-8 | 0.61 | -0.59 |
| ZIF-14* | 0.72 | -0.20 |
| ZIF-7 | 0.27 | -0.80 |
| ZIF-90 | 0.25 | -0.02 |

Georgia Tech Energy Frontier Research Center

Gee and Sholl. *J Phys Chem C* **2013**, 117.
Han et al. *Pending Publication*, **2017**.

15

# Results: Hydrophobicity

Hydrophobic linkers shield metal centers from humid adsorbates

Can hydrophobicity explain the stability trends?

ZIF-90 is the most hydrophilic, ZIF-8 is the most hydrophobic

However, ZIF-8 and ZIF-90 have similar degradation behaviour

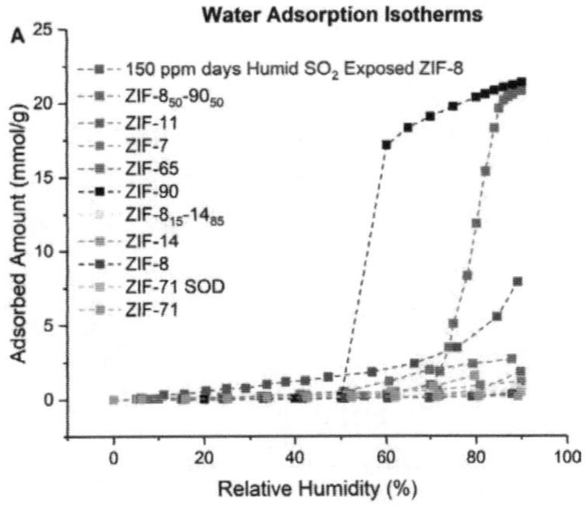


Georgia Tech Energy Frontier Research Center

Zhang et al. *J Phys Chem Lett* **2013**, 4 (21).
Calero and Gomez-Alvarez. *J Phys Chem C* **2015**, 119 (41).

16

# Conclusions

Comprehensive investigation of humid $SO_2$, $CO_2$ degradation mechanisms in ZIFs

Defect formation energy correlates reasonably well with material stability

ZIF-71 RHO alone was stable to humid $SO_2$ and $CO_2$, likely due to a combination of steric hindrance, cage size, and hydrophobicity

Previous indicators (e.g. ligand pKa) thought to predict stability in select MOFs, are not good predictors of ZIF acid gas stability



# Acknowledgement

Sholl research group

The work was supported as part of the Center for Understanding and Control of Acid Gas-Induced Evolution of Materials for Energy (UNCAGE-ME), an Energy Frontier Research Center funded by the U.S. Department of Energy, Office of Science.

**For experimental discussion of this work, see the talk Thursday at 3:30pm (MCC 103D)**

# 494942 Multiconfiguration Pair-Density Functional Theory for Computational Catalysis

**Abstract:**

Multiconfiguration pair-density functional theory[1],[2] is a generalization of Kohnâ`'Sham density functional theory in that the electron kinetic energy and classical electrostatic energy are calculated from a reference wave function, with the rest of the energy obtained from a density functional.Â By combining the advantages of wave function theory and density functional theory it provides a better treatment of strongly correlated systems. Some recent applications of the new method in computational catalysis will be presented.

[1] L. Gagliardi, D. G. Truhlar, G. Li Manni, R. K. Carlson, C. E. Hoyer, and J. Lucas Bao, Multiconfiguration Pair-Density Functional Theory: A New Way To Treat Strongly Correlated Systems, *Acc. Chem. Res*., 50, pp 66â€"73 (2017)

[2] A. M. Sand, D. G. Truhlar, and L. Gagliardi, Efficient algorithm for multiconfiguration pair-density functional theory with application to the heterolytic dissociation energy of ferrocene, *J. Chem. Phys*., 146, 034101 (2017)

# 493670 Quantum Mechanical Description of Excited-State Heterogeneous Catalysis Via Embedded Correlated Wavefunction Methods

**Abstract:**

Nanoplasmonics is recognized to be potentially transformative for the field of heterogeneous catalysis. Localized surface plasmon resonances (LSPRs) can greatly influence chemical processes occurring on the metal catalystsâ€™ surfaces, however the mechanism of enhancement is yet to be understood. First-principles quantum mechanics modeling is a widely used tool to understand heterogeneous catalysis, however, â€œgarden-varietyâ€ density functional theory (DFT) is not suitable for the accurate description of such excited-state phenomena. Additionally DFT is known to have poor predictive ability when it comes to charge-transfer and bond dissociation processes. We thus employ embedded correlated wavefunction methods[1] (*e.g.*, multi-configurational second order perturbation theory embedded in a density-functional-derived embedding potential) to explore the feasibility of light-driven catalysis on metallic nanoparticles. We evaluate if local electronic excitations (with energies at or below the LSPR frequency) centered at the surface reaction sites, can enable overcoming energy barriers for reactions that are otherwise kinetically forbidden to proceed in the ground state. $N_2$ dissociation on Fe- and Mo-doped Au(111) surfaces and $CH_4$activation on Ru-doped Cu(111) will be presented as examples.

[1] F. Libisch, C. Huang, and E. A. Carter, *Acc. Chem. Res.*, **47**, p. 2768 (2014).

# 500172 Incorporation of Linear Scaling Relations into Automatic Mechanism Generation for Heterogeneous Catalysis

**Abstract:**

To predict the selectivity and reactivity of novel catalysts at industrially relevant conditions requires a detailed microkinetic mechanism, comprising many elementary reactions. Recent advances such as such as the work of Ulissi et al [5] use a combination of scaling relations, machine learning, and DFT calculations, to gradually refine a microkinetic model until the rate limiting steps have been calculated with sufficient accuracy to be confident that they are correctly identified. However, such a system requires as input a comprehensive kinetic model containing all the possible pathways. Our recently developed Reaction Mechanism Generator for Heterogeneous Catalysis (RMG-Cat) [3], built upon the open-source RMG software primarily used for gas-phase pyrolysis and combustion [1,2], can provide such mechanisms ab inito: the user supplies just the initial conditions (eg. reactant composition, temperature, pressure) and the software predicts all the possible reactions, estimates the thermochemical and kinetic parameters, solves the governing equations, and decides which reaction pathways to include and explore further. RMG-Cat makes its decisions regarding which pathways to explore and which to ignore, using the estimated parameters, so it is important that the estimates are reasonable, even if the important parameters will be refined with more accurate calculations later in the model development process.

Linear Scaling Relations (LSRs) can provide reasonable estimates of adsorption energies in a very computationally efficient manner [4-6]. We have now implemented linear scaling relationships for the estimation of adsorption energies in the RMG-Cat software. Our database of parameters is organized in a hierarchical tree structure, enabling detailed functional group descriptions to be used when data are available and more general descriptions to be used when necessary. We include parameters to describe many adsorbates on a range of metal surfaces, and a framework to re-train the parameters whenever new data are available.

[1] Gao, C.W., Allen, J.W., Green, W.H., West, R.H. "Reaction Mechanism Generator: Automatic construction of chemical kinetic mechanisms". Computer Physics Communications, 203, 212-225, (2016) http://doi.org/10.1016/j.cpc.2016.02.013

[2] RMG - Reaction Mechanism Generator, open-source software, RMG-Py version XXX http://reactionmechanismgenerator.github.io

[3] Goldsmith, C. F., West, R. H. "Automatic Generation of Microkinetic Mechanisms for Heterogeneous Catalysis" J. Phys. Chem. C. http://doi.org/10.1021/acs.jpcc.7b02133

[4] Medford, A. J., Lausche, A. C., Abild-Pedersen, F., Temel, B., Schjødt, N. C., Nørskov, J. K., Studt, F. "Activity and Selectivity Trends in Synthesis Gas Conversion to Higher Alcohols" Topics in Catalysis (2013) 57, 135-142

[5] Ulissi, Z. W., Medford, A. J., Bligaard, T., Nørskov, J. K. "To address surface reaction network complexity using scaling relations machine learning and DFT calculations" Nature Comm. (2017) 8, 14621-14627

[6] Hummelshøj, J. S., Abild-Pedersen, F., Studt, F., Bligaard, T., Nørskov, J. K. "CatApp: A Web Application for Surface Chemistry and Heterogeneous Catalysis" Angewandte Chemie International Edition (2011) 51, 272-274

# 493221 DFT+U-Inspired Functional for Improved Modeling of Molecules and Solids

**Abstract:**

Accurate description of chemical bond energetics is a pivotal part of computational chemistry and catalysis. It ultimately determines how successful quantum mechanical models are and how reliable their predictions. Density functional theory (DFT) together with the generalized gradient approximation (GGA) is inarguably the method of choice for modeling surface phenomena in heterogeneous catalysis due to its optimal accuracy-to-cost ratio. Despite the success of GGA functionals (PBE, RPBE, PW91) in describing chemistry on metal surfaces, the GGA-inherent self-interaction error poses challenges for accurate treatment of processes on semiconductor surfaces, such as oxides.

Herein, we reveal the inadequacy of the PBE functional for describing experimental trends on metal oxide surfaces (e.g., $RuO_2$, $TiO_2$) for the vacancy-catalyzed[1] furfuryl alcohol hydrodeoxygenation, using first principles microkinetic modeling. We attribute the error to nonsystematic deviations of metal-oxygen bond energetics from experimental values. By utilizing adiabatic connection and the nth order perturbation theory, we arrive at an orbital-occupancy-dependent correction to the PBE functional that holds promise in mitigating the DFT accuracy problem. Corrections are formulated in the spirit of the DFT+U method. The new PBE-based parameter-free functional demonstrates clear improvement over pure PBE in describing dissociation energetics of molecules, NiO local magnetic moment and density of states, and cohesive energies of metal oxides, at minimal extra computational cost.

References:

1. Mironenko, A.V., Vlachos, D.G. "Conjugation-Driven “Reverse Mars–van Krevelen”-Type Radical Mechanism for Low-Temperature C–O Bond Activation." *Journal of the American Chemical Society* 138.26 (2016): 8104-8113.

# 497250 Development of "Surrogate" Hybrid Functionals Based on Electron Density Convolutions

**Abstract:**

In this work we explore the potential of a new data-driven approach to the design of exchange-correlation (XC) functionals. The approach, inspired by "surrogate" functions from optimization and convolutional filters in computer vision, utilizes convolutions of the electron density in order to form a feature space to represent local electronic environments. These features are orbital free, and provide a systematic route to including information at various length scales. Dimensional reduction and regression models are utilized to quantitatively connect these orbital-free descriptors to the exchange-correlation energies and potentials extracted from hybrid functionals. In this talk the approach will be introduced and demonstrated for the construction of a PBE-B3LYP surrogate functional that is tested for small hydrocarbon molecules. The results suggest that this is a promising new route for designing exchange-correlation functionals and embedding higher level physics into density functional theory.

# 495033 Accurate Correction of DFT Delocalization Error in Transition Metal Catalysis

**Abstract:**

Semi-local density functional theory (DFT) suffers from delocalization error that prevents the accurate treatment of systems such as transition metal oxides in the solid state and transition metal complexes. Rigorously, delocalization error is defined as deviation from piecewise linearity of the energy with respect to fractional electron removal or addition. We will describe the extent to which popular approximate treatments for alleviating self-interaction error, such as global and range-separated hybrids as well as DFT+U, alleviate delocalization error and recover the derivative discontinuity. We recently showed for the first time that DFT+U eliminates delocalization error and recovers the derivative discontinuity, but not at the value of U typically employed in most simulations. Even when the former delocalization error is eliminated through DFT+U or exchange mixing in functionals, we have recently identified residual density delocalization error, when compared to accurate correlated wavefunction theory references. We will present recent extensions to this work on representative test cases in both catalytic transition metal complexes as well as in representative bulk and slab models of transition metal oxides. Although we identified DFT+U and Hartree-Fock exchange to equivalently alter properties of the density and energetics in molecules, we will present new, periodic-table-dependent, divergent behavior in the solid state. We describe the interplay between in models and quantities relevant for heterogeneous catalysis, including vacancy formation energies, adsorption energies, surface formation energies, and properties of the electron density. We identify where these functional tuning strategies provide improvements in properties consistently between DFT+U and hybrid exchange tuning and with respect to experiment. Finally, time permitting, we present new functional development strategies we have advanced that address both delocalization error due to fractional charge and fractional spin error. Prior methodological developments, including hybrid exchange and DFT+U, have eliminated delocalization error at the cost of increasing fractional spin error, which is correlated to increased errors in transition states in catalysis.

# 492779 The Effects of Oxidation and Transition Metal-Doping on the Structure and Properties of Pt-Ni Nanoparticles

**Abstract:**

We present the use of *ab-initio* calculations and the cluster expansion method to build a direct bridge between atomic-scale structures and catalytic properties for Pt-Ni nanoparticles for the oxygen reduction reaction (ORR). Although the practical use of Pt-Ni catalysts is limited by Ni dissolution in operating conditions, it has been shown that it is possible to stabilize octahedral Ptâ€"Ni nanoparticles by doping them with a small amount of Mo (around 1.6% mole fraction). Using a cluster expansion built on nanoparticles, we show that the combination of the stabilization of sites with low coordination (e.g. edge and vertex sites), and/or the reduction of equilibrium surface Ni composition due to Mo doping greatly stabilizes the shape and composition of Pt-Ni nanoparticles. Our calculations also reveal insights into the shape evolution of Ptâ€"Ni nanoparticles: the preferential oxidation of edges can make (111) face sites more vulnerable to dissolution than edge sites, which may contribute to the observed formation of Pt-Ni nanoframes and nanoparticles with concave surfaces. To facilitate the rational design of Pt-Ni based nanoparticles with optimized structures for the ORR, we have extended our approach to model the oxygen binding energy on the surface of nanoparticles with experimentally observed sizes (~5nm), enabling the prediction of ORR current as a function of particle structure.

# 489522 The Influence of Hubbard U Parameter in Simulating Adsorption and Reactivity on CuO Surface(s): A Combined Theoretical and Experimental Study

**Abstract:**

Transition metal oxides are an important class of catalytic materials widely used in the chemicals manufacturing and processing industry owing to their low cost, high surface area, low toxicity and easily tuneable surface and structural properties. For these strongly correlated transition metal oxides, standard approximations in the density functional theory (DFT) exchange-correlation functionals fail to describe the electron localization accurately, due to the intrinsic errors arising from electron self-interactions. DFT+U method is a widely used extension of DFT, where the Hubbard U term is an onsite potential which puts a penalty on electron delocalization, successfully describing these systems at only slightly higher computational cost than standard DFT methods. The U-value is usually chosen based on its accuracy in reproducing bulk properties like lattice parameters and band structure. However, chemical reactions on transition metal oxide surfaces involve complex surface-adsorbate interactions and using the bulk properties based U-values in a locally changing surface environment may not describe reaction energetics correctly. Hence, in the current benchmarking work, using CuO as a model transition metal oxide, we perform DFT+U calculations to investigate the dissociative chemisorption of $H_2$ on it. it is observed that the U-value changes the calculated adsorption energies by over 100 kJ $mol^{-1}$. Chemisorption experiments are performed at 30-50 °C to ensure that adsorption products do not leave the system. The (dissociative) adsorption configurations along the reaction coordinate are determined by performing in-situ infrared analysis at adsorption temperatures. For the experimentally validated equilibrium configuration, adsorption energies are comupted using DFT+U for progressive U values and compared with experimental adsorption energies. We reveal that the commonly used U-value of 7 eV (fitted against CuO bulk properties) overestimates the adsorption energy by 20-40 kJ $mol^{-1}$. The U-value between 4.5-5.5 eV correctly predicts the adsorption of $H_2$ on CuO. The DFT+U benchmarking procedure elucidated in this paper, encapsulates surface-adsorbate interactions, surface reactivity and dynamic surface reaction environment and thus, provides an appropriate U-value to be used to model reactions on metal oxide surfaces.

# 499037 Simulating Solvent Effects in Catalytic Systems

**Abstract:**

Many of the catalytic transformations involved in the conversion of chemicals and biomass-derived intermediates are carried out in solvents that can significantly influence the catalytic chemistry, the kinetics and the underlying mechanisms. Solvents can act to stabilize or destabilize reaction intermediates and transition states and, in some cases, they are intimately involved in co-catalyzing reactions. We will discuss methods to simulate solvent structure and its influence on the free energy for catalytic transformations that occur in the solution phase for mixed solvent systems, on metal surfaces as well as within nanoporous architectures.

# 497801 Electronic Excitations in Thermal Heterogeneous Catalysis

**Abstract:**

Most studies of thermal catalysis assume that the system remains in the electronic ground state. However, electronic excitations are easily induced in metal surfaces, due to the lack of a band gap. Here, we show that processes on catalytic surfaces can induce electronic excitations, and we examine the possible effects of these processes on surface chemistry.

We performed nonadiabatic dynamics using real-time, time-dependent density functional theory, propagating the nuclei using Ehrenfest dynamics. We study several trajectories of an $N_2$ molecule interacting with Ru nanoparticles ranging in size from 13 to 147 atoms. These simulations show that, during the adsorption and dissociation processes, a significant amount of energy can be dissipated into electronic degrees of freedom, comparable to the amount of energy dissociated into ionic vibrations. This suggests that the electronic structure of catalytic surfaces may, transiently, be quite different from the ground state. We characterize the electronic structure and spatial character of the excitations, and we examine how they can affect the surface chemistry of other species reacting on the nanoparticle.

# 500631 Large-Scale Nonadiabatic Molecular Dynamics Enabled By Machine Learning

**Abstract:**

Transferring energy from solid surfaces to chemical bonds of adsorbed species is a fundamental process in heterogeneous catalysis. To a great extent, the dynamics of molecule-surface interactions driven by thermal energy, i.e., heat, can be well described within the framework of the Born-Oppenheimer approximation. New reaction channels[1] often open up in response to electronic excitations via surface-mediated energy transfer. In this process, the energetic electrons or holes can scatter into surface species and heat up vibrational modes of adsorbates. This process involves not only the electronic ground state but also the excited state, and it is nonadiabatic with respect to the nuclear motions[2].

In the weak coupling limit, the nonadiabatic energy transfer on metal surfaces can be modeled using only the ground-state potential energy surfaces and the nonadiabacity is taken into account with electronic friction contributions in Langevin dynamics by solving Newtonian equations with a thermal fluctuation term[2], [3]. The electronic friction via partial populating and depopulating of the adsorbate ground state right above the Fermi level is governing the energy transfer. The crucial parameter here is the electronic friction of an excited charge with nuclear motions of surface adsorbates. However, it is extremely time consuming to compute electron-phonon coupling strength for large systems.

In this talk, we will discuss our most recent development of a machine-learning nonadiabatic molecular dynamics approach that uses predicted forces and electronic friction coefficients by 'learning from data'. Machine learning algorithms, such as the artificial neural networks[4] can use past trajectories as training datasets for fast and accurate prediction of forces and electronic friction coefficients, thus allows us to perform statistical analysis of many trajectories. We benchmark the approach for activation of oxygen from hollow to bridge because of strong interests in understanding laser-induced chemistry of CO oxidation on metal surfaces. By comparing the Head-Gordon/Tully electronic friction with the Local Density Friction Approximation (LDFA), we found that the LDFA is mode unselective, thus giving identical coefficients along the lateral and vertical directions to surface and no oxygen activation is observed, while the Head-Gordon/Tully friction can differentiate the electron dragging force in vertical and lateral modes and capture the laser-induced oxygen activation on Ru [2].

[1] M. Bonn *et al.*, "Phonon- Versus Electron-Mediated Desorption and Oxidation of CO on Ru(0001)," *Science*, vol. 285, no. 5430, pp. 1042–1045, Aug. 1999.

[2] M. Head-Gordon and J. C. Tully, "Molecular dynamics with electronic frictions," *J. Chem. Phys.*, vol. 103, no. 23, pp. 10137–10145, Dec. 1995.

[3] M. Brandbyge, P. Hedegård, T. F. Heinz, J. A. Misewich, and D. M. Newns, "Electronically driven adsorbate excitation mechanism in femtosecond-pulse laser desorption," *Phys. Rev. B*, vol. 52, no. 8, p. 6042, 1995.

[4] A. Khorshidi and A. A. Peterson, "Amp: A modular approach to machine learning in atomistic simulations," *Comput. Phys. Commun.*, vol. 207, pp. 310–324, Oct. 2016.

# 500076 Modeling of Segregation on Au-Pd (111) Surfaces with Monte Carlo Simulations and Neural Network Atomic Potentials

**Abstract:**

The simulation of segregation in multi-component alloy surfaces is challenging with atomistic approaches because of the need to model a very large number of possible configurations with a high degree of accuracy. Density functional theory (DFT) is too expensive to use directly, and atomistic potentials are often a compromise between accuracy and computational speed. Neural networks (NN) are a promising alternative to traditional atomistic potential. Using the Behler-Parrinello framework, we were able to create a potential capable of predicting highly-accurate energies for any configuration of a AuPd(111) slab. The fully trained NN spanning all configurations and lattice constants of a AuPd binary alloy is trained from only 3,914 DFT calculations, each with only 7-21 atoms. From these relatively small slabs, Monte-Carlo simulations can be performed on 10 x 10 x 15 atom slabs with average errors under 5 meV/atom.

Mean compositions from the Monte-Carlo simulations were then used to construct segregation profiles spanning bulk compositions between 10 and 90% Au, and at temperatures ranging from 700-1000 K. These simulations result in excellent agreement with available experimental low-energy ion scattering spectroscopy data, which is sensitive only to the composition of the top layer of a metal surface. Site distributions were computed and compared to random distributions, indicating the presence of some short-range ordering favoring the formation of Au-Pd surface bonds. These trends in site distribution are also in excellent agreement with available scanning tunneling microscopy data, further validating our model. These profiles can also be fit to the Langmuir-McLean formulation of the Gibbs-isotherm with a model for the enthalpy of segregation. Based on the trends observed in the calculated enthalpy of segregation across all bulk compositions, it is not clear how one would derive the same trend using more traditional techniques, such as course-grained models based on dilute limit segregation energies alone. The techniques derived in this work are easily implemented, even with limited computational recourses. Due to the flexible nature of the NN, results from this work are also useful in application to more complex systems with adsorbates.

# 488959 Accelerating Electronic Structure Calculations with Machine Learning

**Abstract:**

Electronic structure calculations have produced mechanistic insights and have led to systematic design principles for catalysis. However, a typical atomistic calculation involves much computational "waste": often, thousands of single-point electronic structure calculations are performed to produce each one or two configurations which are ultimately published. Recently, methods have been developed which can "learn" from the output of electronic structure calculations, allowing for the prediction of the energies and forces of atomic configurations that they have not been exposed to. In this talk, we will discuss new algorithms that can use such machine learning techniques to strategically accelerate and extend electronic structure calculations, through such routines as transition-state searches and hybrid quantum-mechanics/molecular-mechanics approaches. These methods will be introduced in terms of our open-source machine-learning package, "Amp".

# 499604 A Data-Driven *in silico* Research Paradigm for the Rational Design of Catalyst Systems and the Exploration of Chemical Space

**Abstract:**

Trial-and-error research approaches are increasingly ill equipped to meeting the complex challenges involved in the discovery and design of next-generation chemistry. Our work recognizes the great opportunities that are arising with the shift towards data-driven *in silico* research and a rational design paradigm. These approaches are poised to mitigate the inefficiencies, shortcomings, and limitations of traditional trial-and-error research. However, the notion to utilize modern data science in the chemistry context is so recent that much of the basic infrastructure has not yet been developed, or is still in its infancy. The existing tools and expertise tend to be in-house, specialized, or otherwise unavailable to the community at large. Data science is thus in practice beyond the scope and reach of most researchers in the field. Our work aims to chart new paths in this area by creating an open, general-purpose software ecosystem designed to overcoming this situation, filling the prevalent infrastructure gap, and thus making data-driven research a viable and widely accessible proposition. Our software ecosystem fuses *in silico* modeling (in particular computational quantum chemistry), high-throughput screening techniques, and Big Data analytics into an integrated research infrastructure. We will demonstrate the utility of this approach in the context of the development of molecular catalysts for solar water splitting.

UCL

# Overcoming the Compromise between Accuracy and Efficiency in Modelling Catalytic Kinetics

Miguel Pineda and Michail Stamatakis
Department of Chemical Engineering
University College London

New Developments in Computational Catalysis II session
*2017 AIChE Annual Meeting*
*Tuesday October 31, 2017*

## Motivation

A popular and successful formalism used to analyse catalytic systems is the so-called chemical kinetic approach (deterministic and stochastic)

### Micro-kinetic modelling (MKM)

Efficiency but inaccuracy

$$\frac{d\theta}{dt} = r_+ - r_-$$

Well-mixed

Mean-field

- Not correlations (thermodynamically or kinetically driven)
- Not stochasticity

### Kinetic Monte Carlo (KMC) method

Accuracy but inefficiency

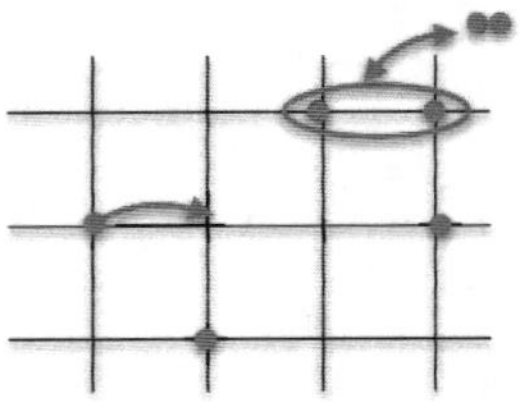

$$\frac{dP_\sigma}{dt} = \sum_\beta \left[k_{\sigma\beta}P_\beta - k_{\beta\sigma}P_\sigma\right]$$

- Correlations (thermodynamically or kinetically driven)
- Stochasticity and detailed models of complex catalysis

Developing accurate and efficient methods that take into account the best of MKM and KMC

## Spatial Correlations

- Cluster mean-field approximation (CMA) adapted for use in Chemical Kinetics
- Reduced model for $NO$ oxidation and $NO_2$ reduction on $Pt(111)$

+ a cloud of atoms

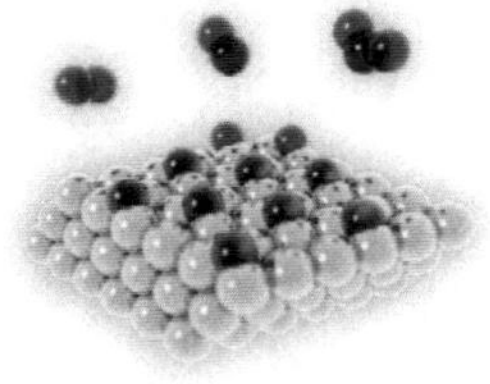

## Stochasticity

- Reflected chemical Langevin equations (RCLE)
- Validation in a representative reaction mechanism with a single species

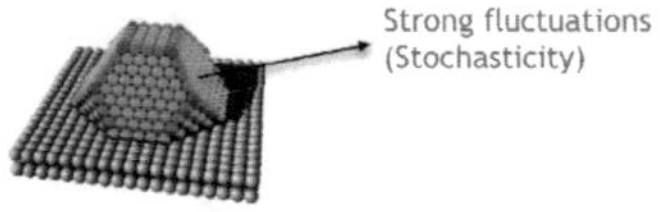


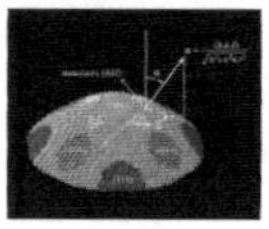

Field emitter tips

McEwen et. al. PNAS., **2009**, *106*, pp 3006

## • Spatial correlations

•**Cluster mean-field approximation:** Lattice gas Hamiltonian formulation (LGHF) or cluster expansion

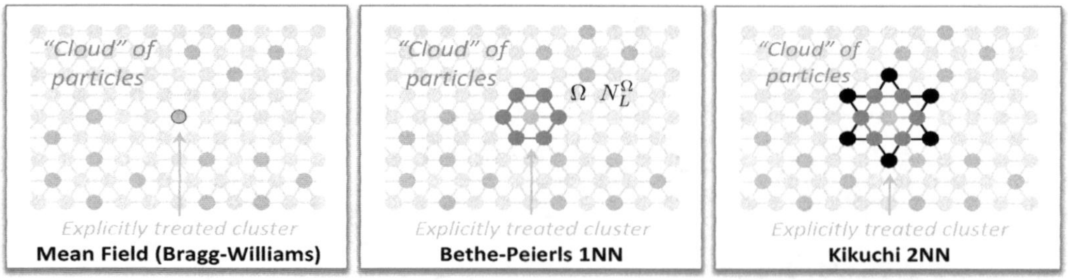


$\sigma_i = 0, 1 \qquad \sigma = \left\{\sigma_1, \sigma_2, \cdots, \sigma_{N_L^\Omega}\right\}$

$$H_C(\sigma) = H_\Omega(\sigma) + \sum_{i=2}^{N_L^\Omega} a_i\sigma_i + \frac{1}{2!}\sum_{i=2}^{N_L^\Omega}\sum_{\substack{j=2\\ j\neq i}}^{N_L^\Omega} b_{ij}\sigma_i\sigma_j + \cdots$$

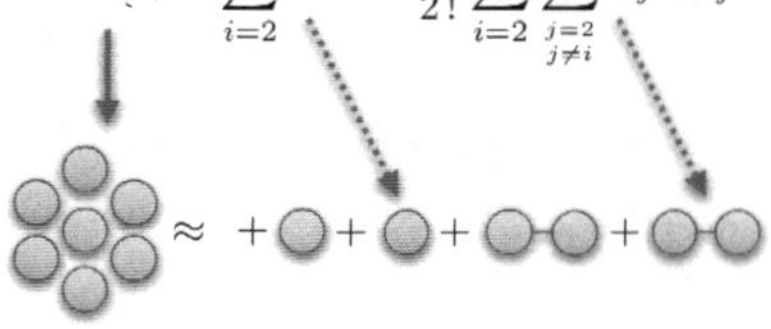

$a_i, b_{i,j} \longrightarrow$ Effective fields determined self-consistency by imposing:

- Equality of coverages of all sites
- Equality of correlation functions for symmetrically equivalent cluster of sites

- **Benchmark system**

Reduced model for $NO$ oxidation and $NO_2$ reduction on $Pt(111)$

1) $NO_{(gas)} + O^* \underset{k^{red}}{\overset{k^{oxi}}{\rightleftarrows}} NO_{2(gas)} + *$

2) $O_{2(gas)} + 2* \underset{k^{des}}{\overset{k^{ads}}{\rightleftarrows}} 2O^*$

3) $O^* + * \underset{k^{diff}}{\overset{k^{diff}}{\rightleftarrows}} * + O^*$

- Equilibrated interconversion between $NO$ and $NO_2$ allows us to assume

$$\mu_{O^*} = \mu_{NO_2} - \mu_{NO}$$

→ Kinetic observations

- Strong repulsive interactions and very fast diffusion of oxygen atoms led to oxygen adlayer quasi-equilibration (superlattice ordering)

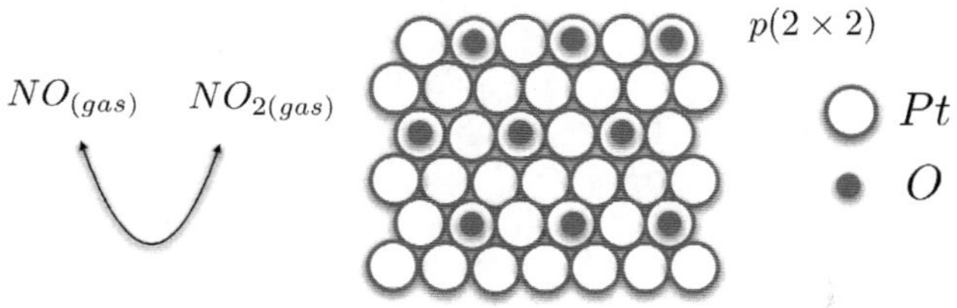


Getman et. al. Physical Review Letters., **2009**, *102*, pp 076101
Nielsen et. al. Journal of Chemical Physics., **2013**, *139*, pp 224706
Wu et. al. J. Cata., **2012**, *286*, pp 88

Rate-limited process ⟶ $O_{2(gas)} + 2* \underset{k^{des}}{\overset{k^{ads}}{\rightleftarrows}} 2O^*$

$NO$ oxidation rate (TOF)

$$TOF = R_{O_2,ads} - R_{O_2,des} \qquad \beta \longrightarrow \sigma$$

$$R_{O_2,ads} = \frac{2}{N_L^{\Omega}} \sum_{\sigma\beta} P_\beta k_{\sigma\beta} \qquad R_{O_2,des} = \frac{-2}{N_L^{\Omega}} \sum_{\sigma\beta} P_\beta k_{\sigma\beta}$$

Probability distribution:

$$P_\sigma = \frac{1}{Q} e^{-\frac{(H_C(\sigma) - \mu_{O^*} N_\sigma)}{k_B T}}$$

The reaction rates:

$$k_{\sigma\beta} = k_{\sigma\beta}^{ads} = A e^{-\frac{E_a^\ddagger(\beta)}{k_B T}}$$

$$k_{\sigma\beta} = k_{\sigma\beta}^{des} = A' e^{-\frac{E_d^\ddagger(\beta)}{k_B T}}$$

BEP relationship adsorption barrier:

$$E_a^\ddagger(\beta) = E_a^\ddagger(0_a) + \omega(\Delta E_{rxn}(\beta) - \Delta E_{rxn}(0_a))$$

$$\Delta E_{rxn}(\beta) = H_C(\sigma) - H_C(\beta) + \Delta E_{O_2}$$

$$\Delta E_{O_2} = 002eV \qquad \omega = 1$$

Nielsen et. al. Journal of Chemical Physics., **2013**, *139*, pp 224706

1st nearest neighbour O*···O* interactions

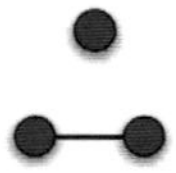

$h_1 = -1.200eV$

$J_{ij} = J = 0.300eV$

$h_0 = -0.027eV$

K2NNC1: first nearest neighbour corrections

K2NNC2: second nearest neighbour corrections → "Ghost interactions"

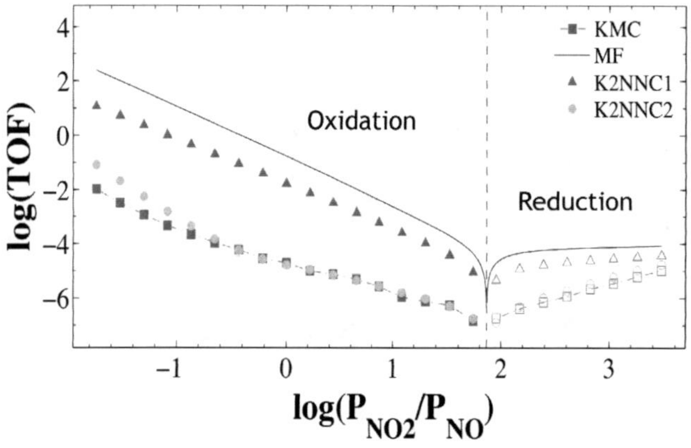


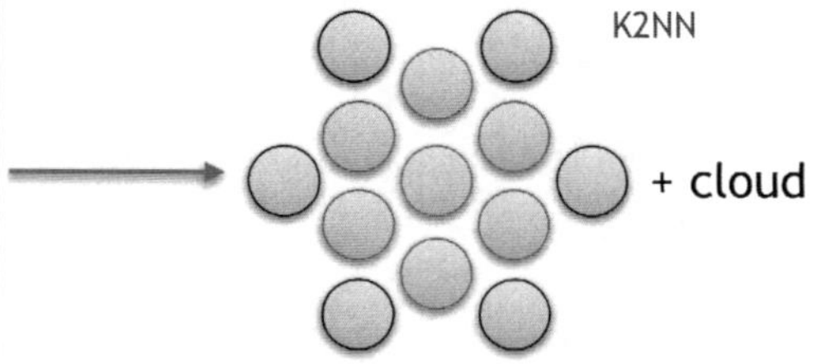


Computational efficiency

| Method | Cost in CPU seconds |
|---|---|
| KMC | $2.1 \times 10^{7}$ |
| MF | $4.3 \times 10^{-1}$ |
| K2NNC1 | 1.5 |
| K2NNC2 | 2.6 |

**Tremendous computational gains!**
**Negligible loss of accuracy.**

M. Pineda and M. Stamatakis.
Journal of Chemical Physics., **2017**, *147*, pp 024105

## • Stochasticity

Let us neglect repulsive interactions and consider very fast diffusion of adsorbates

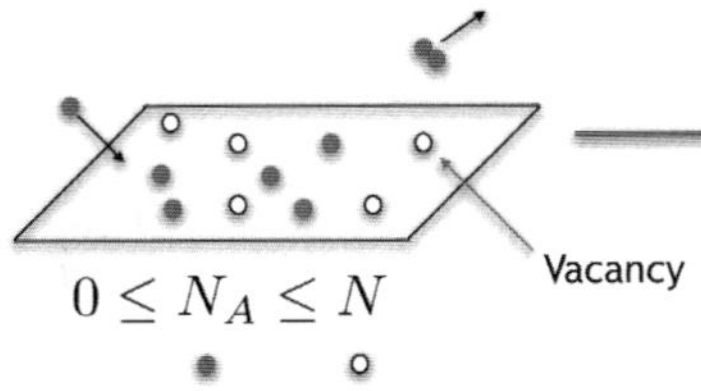


$0 \leq N_A \leq N$

Chemical master equations (CME)

$$P(N_A; t)$$

Discrete-state continuous-time description

Computationally expensive

Trajectories and probability distributions

In the limit of $N$ very larger

$$y = \epsilon N_A \qquad \epsilon = \frac{1}{N}$$

$$0 \leq y \leq 1$$

Chemical Fokker-Plank equation (CFPE)

$$P(y; t)$$

Continuous-state continuous-time description

Computationally efficient

Only probability densities and average values

The chemical Langevin equations (CLE)

$$dy = a(y)dt + \sqrt{\epsilon b(y)}dW$$

Numerical solution: Euler-Maruyama (EM) scheme

$$y_{i+1} = y_i + a(y_i)\Delta t + \sqrt{\epsilon b(y_i)\Delta t}\mathcal{N}(0,1)$$

Gaussian

$$y_i \approx 0 \longrightarrow y_{i+1} < 0$$

$$y_i \approx 1 \longrightarrow y_{i+1} > 1$$

Solutions of the CLE not comparable to solutions of the CFPE

Reflected chemical Langevin equation (RCLE)

$$y = x + k$$

$x \in (0, 1)$

Minimal process that force the solution to remain in the unit interval

$$dy = a(y)dt + \sqrt{\epsilon b(y)}dW + dk$$

Solution of the CFPE

$$P_{st}(y) = \frac{C}{b(y)} exp\left\{\frac{2}{\epsilon}\int_0^y dx \frac{a(x)}{b(x)}\right\}$$

## • Benchmark system

1) $A_{(gas)} + *_s \underset{k^{des}}{\overset{k^{ads}}{\rightleftarrows}} A_s$

2) $A_{(s)} \xrightarrow{k^{rxn1}} *_{(s)} + B_{(gas)}$

3) $A_{(s)} + A_{(n)} \xrightarrow{k^{rxn2}} *_{(s)} + *_{(n)} + A_{2(gas)}$

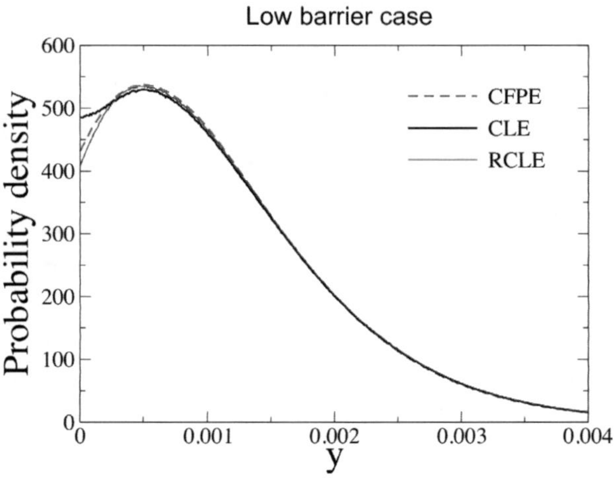


Euler-Lepingle (EL) scheme for two-side barrier

$$y \in [0, 1]$$

- Gaussian variable and a independent exponential variable

D. Lepingle, Mathematics and Computers in Simulations., **1995**, *38*, pp 119

## Comparison of the RCLE with the CME and CFPE

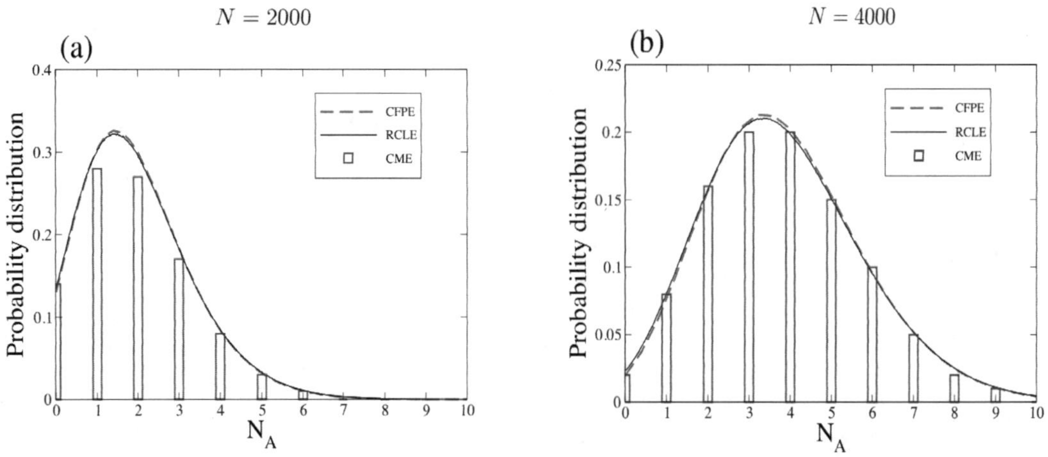


CME solutions are obtained solving the evolution equation for $P(N_A; t)$ at steady state

## Computational efficiency

CME solutions are obtained using the stochastic simulation algorithm (SSA) and stopped at t=1

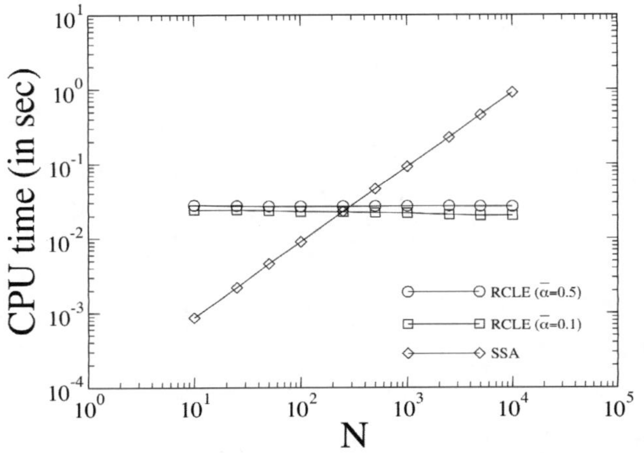


Low barrier case

M. Pineda and M. Stamatakis. Submitted for publication., **2017**

**Computational gains!**

**Reproduce very good the well-mixed CME and CFPE.**

## Ongoing work

- Coupling the two methods in order to have an efficient and accurate approach involving spatial correlations and stochasticity

## Funding

- Leverhulme Trust: RPG-2014-161

## Computational resources

- Legion@UCL

14

# 496191 CO Adsorption on Platinum and Cobalt: Site Preference and Coverage Effects

**Abstract:**

**CO Adsorption on Platinum and Cobalt: Site Preference and Coverage Effects**

G. T. Kasun Kalhara Gunasooriya,† Mark Saeys†

†Laboratory for Chemical Technology, Ghent University, Technologiepark 914, 9052 Gent, Belgium


Calculation of correct site preference, adsorption energies and distribution of the adsorbates forming stable phases remain important challenges for DFT.[1] In this presentation we demonstrate that van der Waals functionals[2] provide an accurate description for these challenges on platinum and cobalt catalyst surfaces. Platinum catalysts are widely used in CO oxidation in automotive exhausts and as anode catalysts for direct methanol fuel cells. Cobalt catalysts are the preferred choice for Fischer-Tropsch synthesis (FTS), the conversion of synthesis gas, a mixture of CO and $H_2$, to long-chain hydrocarbons.

At low CO coverage, CO adsorbs at the top sites on Pt(111) and Co(0001) facets. Low coverage CO vdW-DF adsorption energy agrees well with the experimental data and the site preference is correctly predicted as top site. PBE functional overestimates CO adsorption energy and favors hollow over top site as most stable configuration. Interestingly, low coverage CO BEEF-vdW adsorption energy at the top site is close to the respective vdW-DF adsorption energy. However, the BEEF-vdW indicates bridge site to be the most favorable over hollow and top sites on Pt(111) facet. With increase in CO coverage, CO adsorbs both on top and bridge sites on the Pt(111) and Co(0001) facets. The observation of change in site preferences with increase in CO coverages is remarkable. To elucidate the change in stabilities of top and bridge sites with increase in CO coverage, a Bader charge and Natural Bond Orbitals (NBO) analysis was conducted. Charge analysis indicate that, CO adsorption on top site is more sensitive to surface charge compared to bridge site. Bonding analysis indicate that increase in surface charge, increases Pauli repulsion between metal $dz^2$ states and the filled CO 5σ orbital. Thus indicating that charge is the essence in determining the preferred adsorption site. This effect was further experimentally observed on $TiO_2$supported Pt clusters.[3]

CO adsorption is often described by a Langmuir isotherm, sometimes accounting for a gradual decrease in the adsorption energy with coverage. Here, we show that CO adsorption on cobalt does not follow this typical behavior. Instead, adsorption on Co(0001) is dominated by two surface phases. At low pressures, the (√3×√3)R30°–CO structure is the stable phase, and CO forms (√3×√3)R30°–CO islands for coverages below 1/3 ML because of attractive CO–CO interactions. However, a similar attractive interaction was not observed on Pt or with PBE and RPBE functionals. Charge and bonding analysis indicate that, these attractive interactions are due to decrease in charge at the diagonal Co atoms, hence reducing the Pauli repulsion between the partially filled Co $dz^2$ state and the CO 5σ orbital. Increasing the pressure does not gradually increase the coverage beyond 1/3 ML. Instead, a transition to a high coverage (2√3×2√3)R30°–

7CO surface structure occurs at 0.1 mbar at room temperature and at 21 bar at 500 K. First principle phase diagram calculations with the van der Waals functional confirm the stability of these two phases and show a true first-order phase transition between the two phases.[4,5] It is important to note that CO adsorption introduces five new low frequency modes, i.e., vibrations associated with frustrated rotation and translation, with frequencies in the range of 40 to 400 $cm^{-1}$. Those modes contribute significantly to the adsorption entropy and need to be computed accurately. For comparison, CO adsorption on Pt(111) was also studied, where a similar phase transition is not observed.

**References**

[1] Schimka, L., Harl, J., Stroppa, A., Grüneis, A., Marsman, M., Mittendorfer, F., Kresse, G. *Nat. Mater.* **2010**, 9, 741

[2] Dion, M.; Rydberg, H.; Schroder, E.; Langreth, D. C.; Lundqvist, B. I. *Phys. Rev. Lett.* **2004**, *92*, 246401

[3] Gunasooriya, G. T. K. K., Seebauer, E. G., Saeys, M. *ACS Catal.*2017, 7, 1966

[4] Beitel, G. A.; Laskov, A.; Oosterbeek, H.; Kuipers, E. W. *J. Phys. Chem.* **1996**, *100*, 12494

[5] Gunasooriya, G.T.K.K.; van Bavel, A.P.; Kuipers, H.P.C.E.; Saeys, M. *Surf. Sci.* **2015**, 642, L6

## The 10th Industrial Fluid Properties Simulation Challenge

# The Pressure-Viscosity Relation

**Scott Bair**
**Regents' Researcher**
**Center for High Pressure Rheology**
**George W. Woodruff School of Mechanical Engineering**
**Atlanta, Georgia 30332-0405**
**scott.bair@me.gatech.edu**

**October 31, 2017**



The 10th Industrial Fluid Properties Simulation Challenge
The Pressure-Viscosity Relation

One of many wind turbines whose overheated gearboxes caused the lubricating oil to catch fire. From "Wind Turbine Gearbox Technologies," Adam M. Ragheb and Magdi Ragheb, Department of Aerospace Engineering and Department of Nuclear, Plasma and Radiological Engineering, University of Illinois at Urbana-Champaign, 216 Talbot Laboratory

The importance of tribology is often not appreciated until failures occur.

# Introduction to Full Film Lubrication

Hydrodynamic bearings are conformal and support a load on a liquid film at low pressure, p<300 MPa.

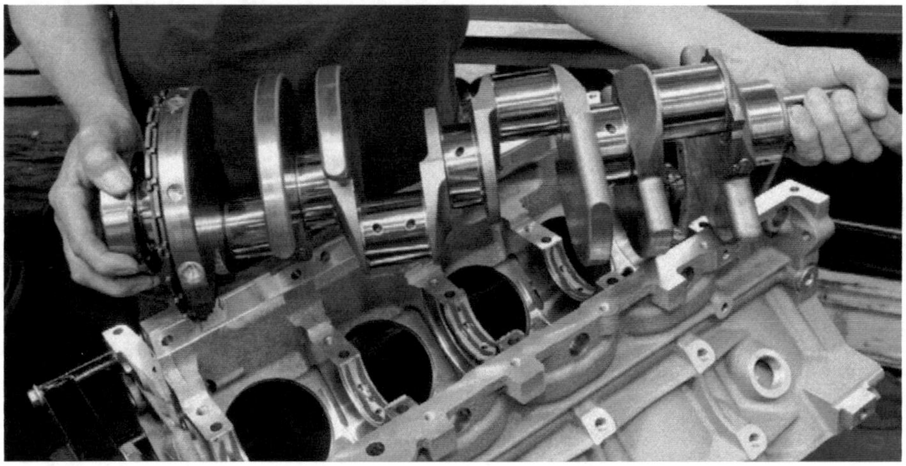

3

# Introduction to Full Film Lubrication

Roller bearings, gears, and cams are non-conformal and support a load on a liquid film at very high pressure, 500<p<3000 MPa. This is Elastohydrodynamic lubrication.

4

The 10th Industrial Fluid Properties Simulation Challenge
The Pressure-Viscosity Relation

## Hydrodynamic Lubrication of Rolling / Sliding Contacts

Conformal Contact

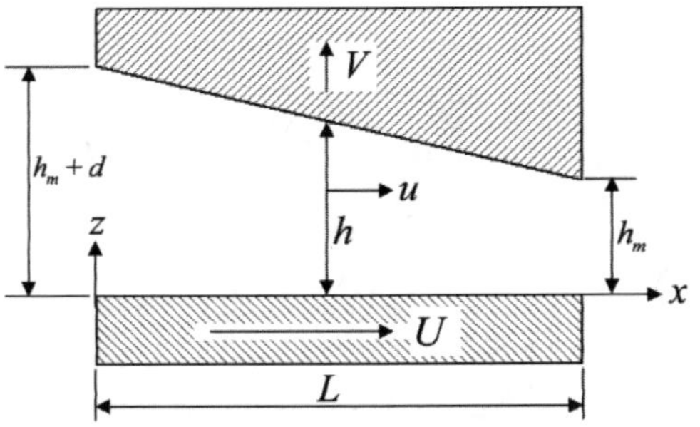


Hydrodynamic Lubrication
p < 300 MPa

Non-Conformal Contact

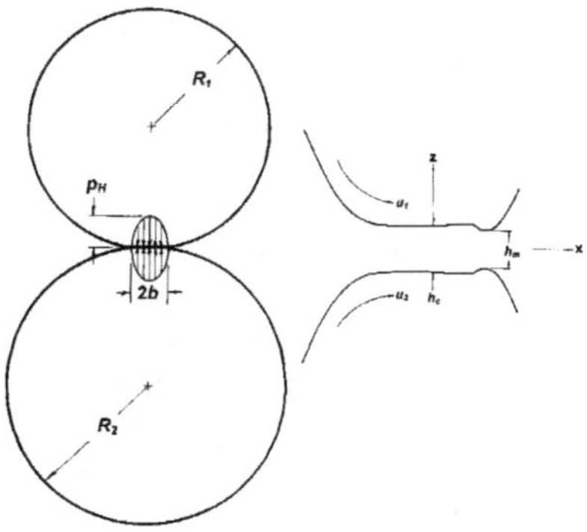


Elastohydrodynamic Lubrication
p > 300 MPa

Georgia Tech Center for High Pressure Rheology 5

The 10th Industrial Fluid Properties Simulation Challenge
The Pressure-Viscosity Relation

# Fluid Film Lubrication

For hydrodynamic lubrication, the contact load is supported by a pressure distribution which is calculated from the Reynolds equation.

$$\frac{\partial}{\partial x}\left(\phi_x \frac{\rho h^3}{12}\frac{\partial p}{\partial x}\right) + \frac{\partial}{\partial y}\left(\phi_y \frac{\rho h^3}{12}\frac{\partial p}{\partial y}\right) = \bar{u}\frac{\partial(\rho h)}{\partial x} + \bar{v}\frac{\partial(\rho h)}{\partial y} + \frac{\partial(\rho h)}{\partial t}$$

$$\phi_x = \frac{12}{h^3\left(\partial p/\partial x\right)}\int_{-\frac{h}{2}}^{+\frac{h}{2}} z\left(\tau_{mx} + z\frac{\partial p}{\partial x}\right)\frac{1}{\eta\left(T, p, \tau^*\right)}\mathrm{d}z$$

$$\phi_y = \frac{12}{h^3\left(\partial p/\partial y\right)}\int_{-\frac{h}{2}}^{+\frac{h}{2}} z\left(\tau_{my} + z\frac{\partial p}{\partial y}\right)\frac{1}{\eta\left(T, p, \tau^*\right)}\mathrm{d}z$$

$$\tau^* = \sqrt{\left(\tau_{mx} + z\frac{\partial p}{\partial x}\right)^2 + \left(\tau_{my} + z\frac{\partial p}{\partial y}\right)^2}$$

Two fluid properties appear directly in the Reynolds equation, $\eta$, the generalized Newtonian viscosity, and $\rho$, the mass density. If the temperature variations within the liquid film affect viscosity, then the energy equation must be solved simultaneously and thermal conductivity becomes quite important as well.

Georgia Tech Center for High Pressure Rheology 6

# Only consider elastohydrodynamic lubrication (EHL) from here on.

The central film thickness is established in the inlet region where viscosity rises to such values that the liquid cannot escape entrainment.

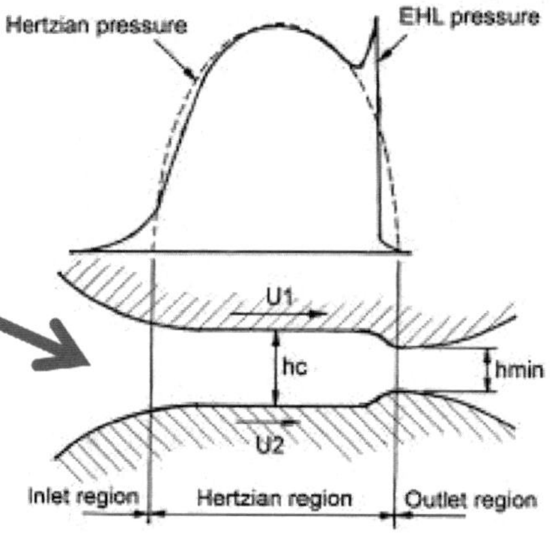


Film pressures cause significant elastic deformation of the solid surfaces and order-of-magnitude changes in viscosity. Friction is primarily generated in the Hertz region.





## Aircraft, automobiles, industrial machines and many appliances rely on lubricated concentrated contacts to function. A fundamental requirement of elastohydrodynamic lubrication (EHL) is the pressure dependence of viscosity.

## There are now two very different approaches to studying EHL

1. 1950 – Present **Classical EHL** with two essential assumptions regarding rheology
   a. The liquid in the inlet zone responses in Newtonian fashion.
   b. A friction vs sliding velocity curve has the same functional form as a shear stress vs shear rate curve at some pressure.

   The pressure-viscosity relation takes any form which reconciles these assumptions with experimental measurement of film thickness and friction. Viscometers are not employed.

2. 2006 – Present **Quantitative EHL** The temperature, pressure and shear dependence of viscosity is obtained from viscometer measurements and applied as the liquid properties using correlations.





# The first full EHL numerical simulation of film thickness to use a viscometer measurement was only published in 2006.

**Full EHL numerical simulations solve for both the film pressure and the elastic deformation.**

Georgia Institute of Technology

**The 10th Industrial Fluid Properties Simulation Challenge**
**The Pressure-Viscosity Relation**

## The first quantitative EHL film thickness analysis was published in 2006 using viscometer data for the pressure dependence of squalane.

Liu, Y., Wang, Q. J., Wang, W., Hu, Y., Zhu, D., Krupka, I., & Hartl, M. (2006). EHL simulation using the free-volume viscosity model. *Tribology Letters*, *23*(1), 27-37.

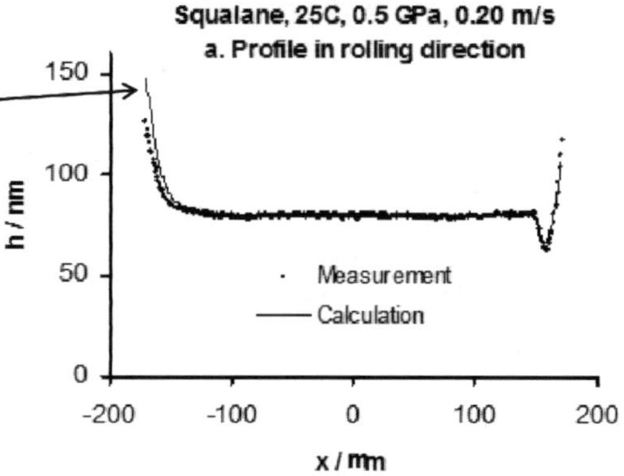


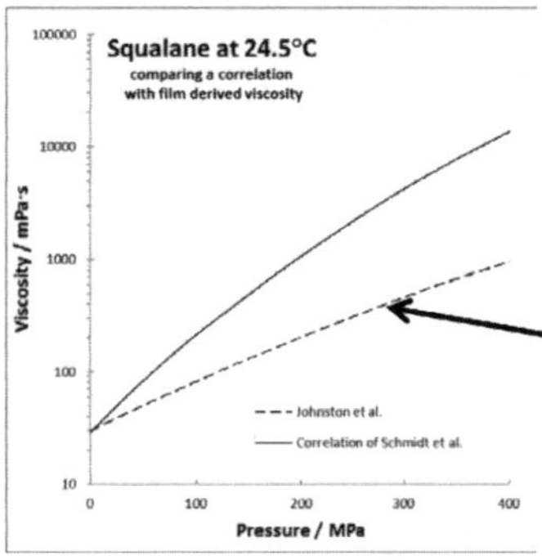


**A previous classical EHL analyses adjusted the pressure dependence of squalane to explain an observation of an anomalous feature in the film thickness.**



Georgia Institute of Technology

**The 10th Industrial Fluid Properties Simulation Challenge**
**The Pressure-Viscosity Relation**

## The lack of agreement with viscometers has not gone unnoticed.

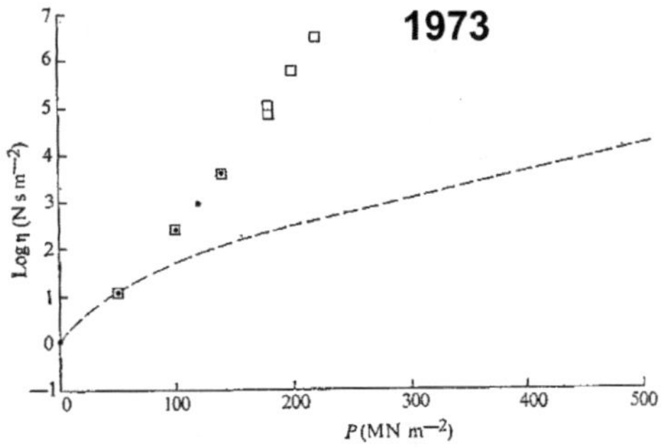


Viscosity-pressure results for polyphenyl ether.

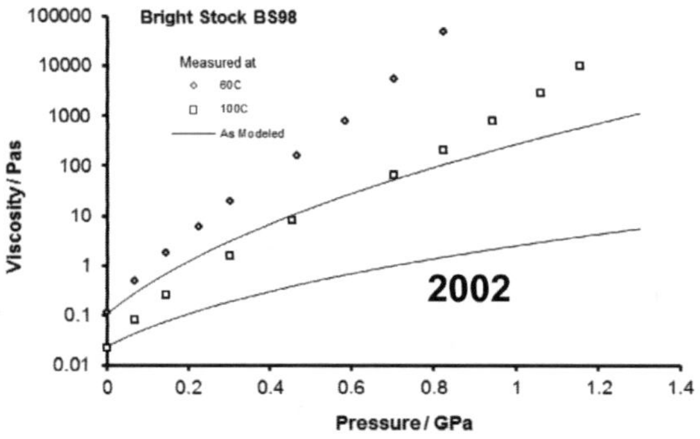


The curve is from: Paul, GR and Cameron, A, Proc Roy Soc A 331, 171, 1971.
Viscometer data points: Hutton, JF and Phillips, MC, High Pressure Viscosity of a Polyphenyl Ether Measured with a New Couette Viscometer, Nature Phys Sci, Vol.245, 1973, 15-16.

Kaneta, M., Kawashima, R., Masuda, S., Nishikawa, H., Yang, P. and Wang, J., "Thermal Effects on the Film Thickness in Elliptic EHL Contacts with Entrainment along the Major Contact Axis", *ASME, J. Tribology*, Vol.124, No.2, 2002, pp. 420-427.

The 10th Industrial Fluid Properties Simulation Challenge
The Pressure-Viscosity Relation

# The first full EHL numerical simulation of both film thickness and friction using viscometer measurements was only published in 2007.

**Full EHL numerical simulations solve for both the film pressure and the elastic deformation.**



The 10th Industrial Fluid Properties Simulation Challenge
The Pressure-Viscosity Relation

## Applications in Quantitative Elastohydrodynamics

**The low shear viscosity was measured to high pressure for a simulation.**

**PAO's have a large value of viscosity at the inflection pressure.**

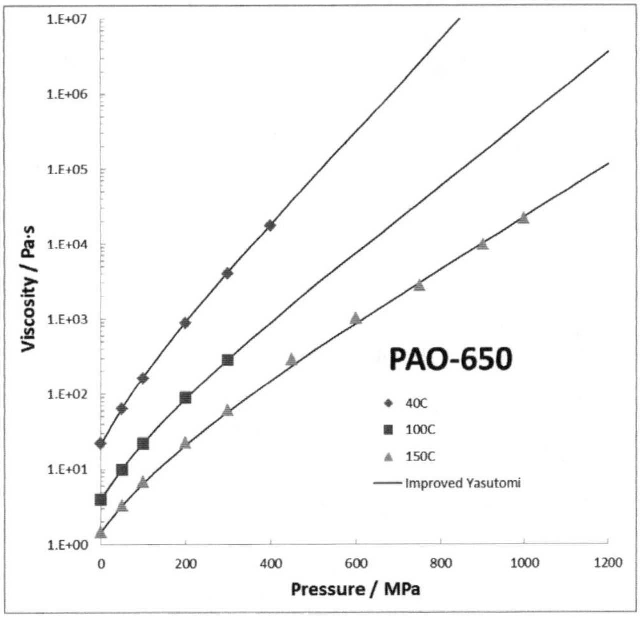

## Applications in Quantitative Elastohydrodynamics

**The shear dependent viscosity was measured to high pressure and stress for a simulation.**

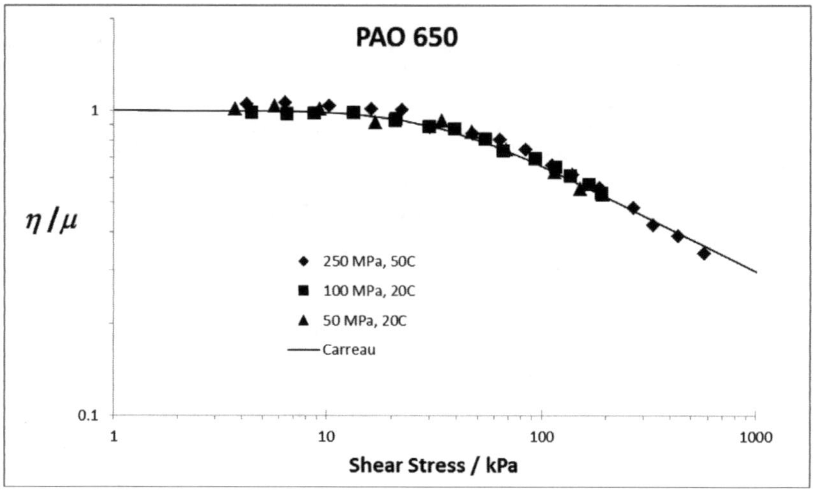






## Numerical simulation by Yuchuan Liu at Northwestern University of the film thickness measurements by Philippe Vergne at INSA Lyon.

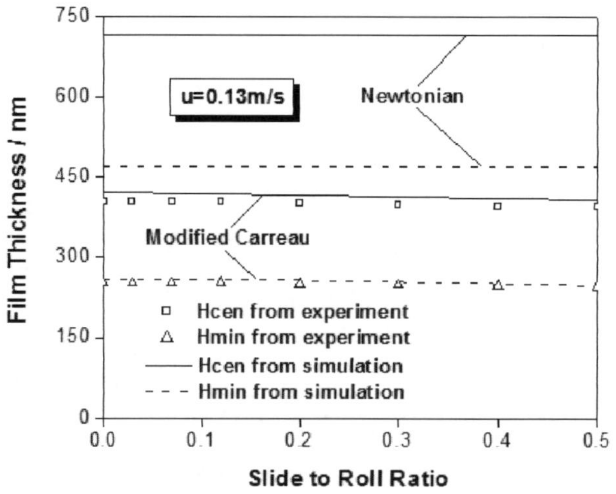

The 10th Industrial Fluid Properties Simulation Challenge
The Pressure-Viscosity Relation

**Numerical simulation by Yuchuan Liu at Northwestern University of the friction measurements by Philippe Vergne at INSA Lyon.**

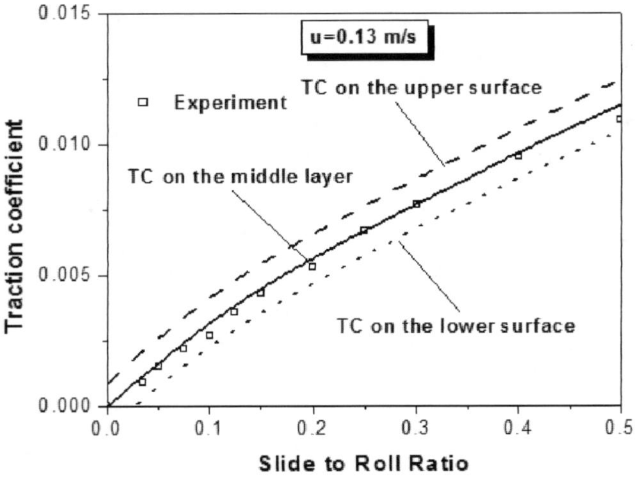




The 10th Industrial Fluid Properties Simulation Challenge
The Pressure-Viscosity Relation

# The temperature and pressure dependence of the low-shear viscosity, especially the inflection, is essential to understanding friction.

The 10th Industrial Fluid Properties Simulation Challenge
The Pressure-Viscosity Relation

## The faster-than-exponential part cannot be predicted from low pressure data.

The POE generates significantly greater friction than the PAO and the reason can be clearly seen in the viscosity at high pressure. However the pressure sensitivity is opposite that at low pressure.

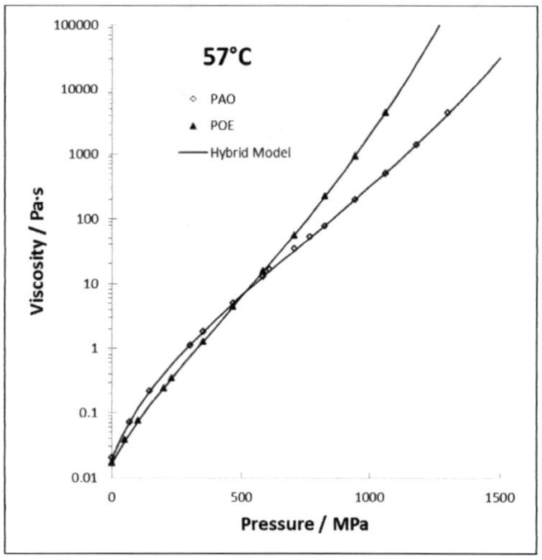


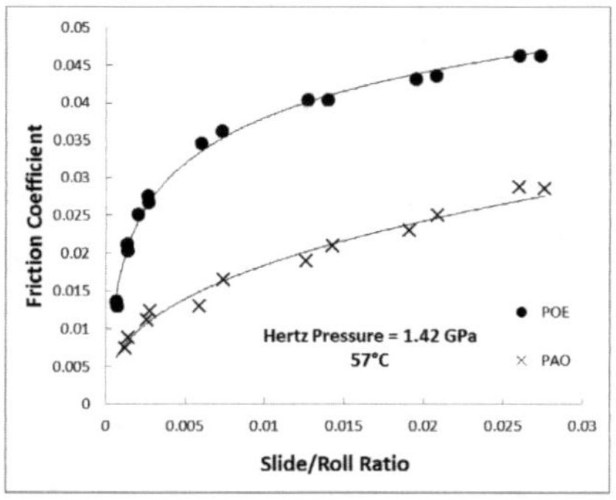




The 10th Industrial Fluid Properties Simulation Challenge
The Pressure-Viscosity Relation

# High Pressure Instruments

91 years ago Bridgman published a complete and accurate description of the pressure-dependence of organic liquids similar to lubricating oils.

Incredibly, the description of pressure response in tribology text books, using fictional accounts of measurements by Barus and Roelands, ignore Bridgman's accurate description.

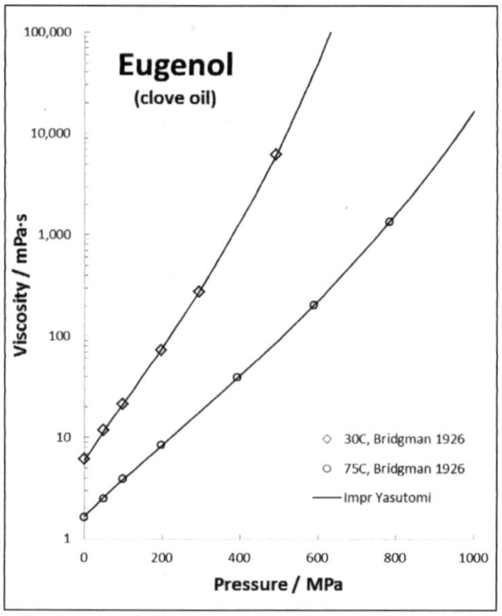

The 10th Industrial Fluid Properties Simulation Challenge
The Pressure-Viscosity Relation

# Measuring the Low-Shear Viscosity

## Techniques for high pressure

Capillary

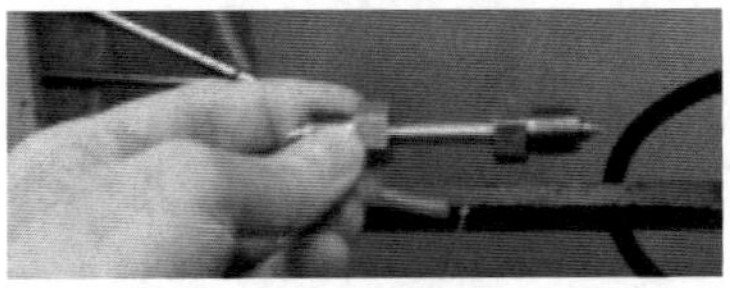

Circular Couette

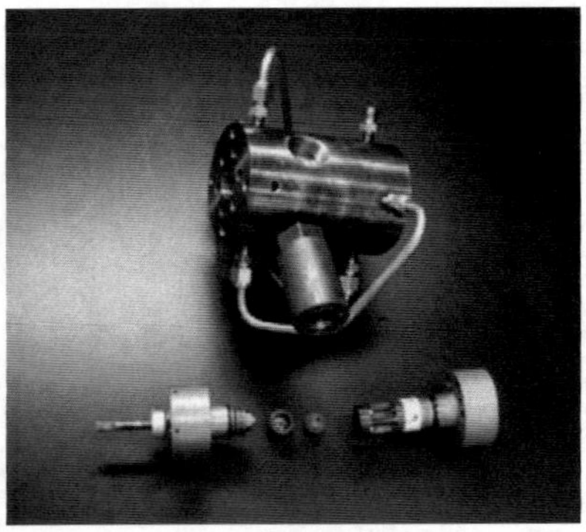

Falling cylinder

The 10th Industrial Fluid Properties Simulation Challenge
The Pressure-Viscosity Relation

# Measuring the Low-Shear Viscosity

**The four different pressurized viscometers provide consistent measurements over a range of pressure. This is important since the capillary and rotating viscometers are also used for the shear dependent viscosity measurements.**

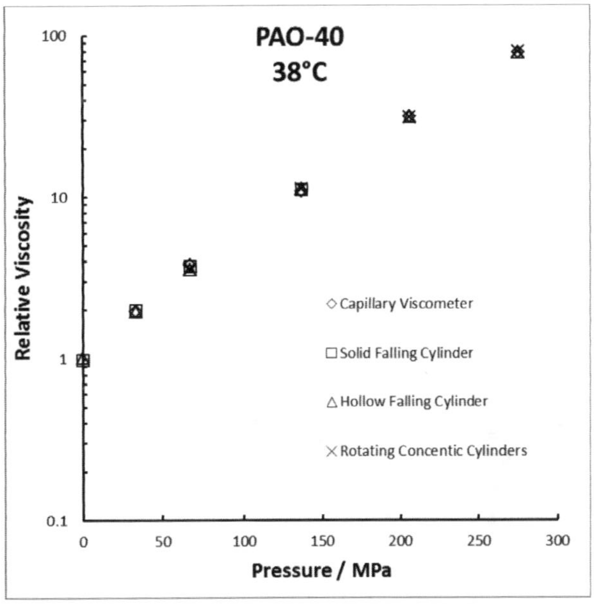

The 10th Industrial Fluid Properties Simulation Challenge
The Pressure-Viscosity Relation

# Measuring the Low-Shear Viscosity

The falling cylinder is preferred. Using a special transformer, an LVDT, measurements can be made to 1.4 GPa with good accuracy and high throughput. The pressure vessel shown below accepts various viscometer cartridges.

The 10th Industrial Fluid Properties Simulation Challenge
The Pressure-Viscosity Relation

## Measuring the Low-Shear Viscosity

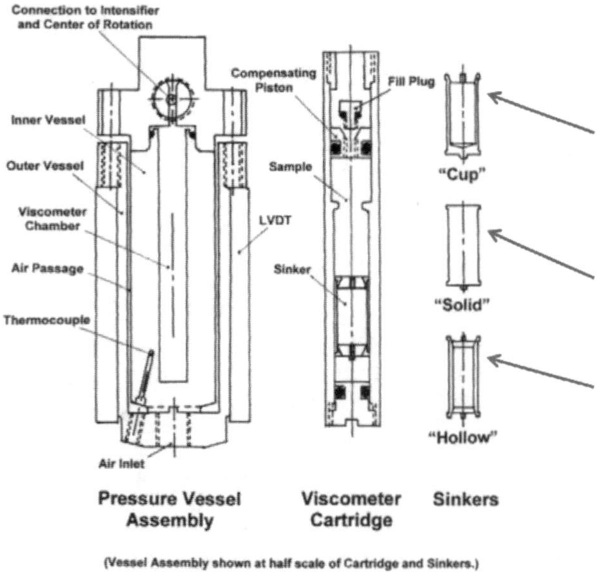


Guiding the cylinder with side lugs minimizes the falling distance required to reach steady velocity.

Lightweight version of "Solid"
Principally Poiseuille flow
Shear Stress ≈ 1 Pa

Principally Poiseuille flow
Shear Stress ≈ 5 Pa

Principally Couette flow
Shear Stress ≈ 50 Pa

## Measuring the Low-Shear Viscosity

The cartridge is assembled in the order shown and inserted into the pressure vessel. The pressure medium is di(2-ethylhexyl) adipate or sebacate.

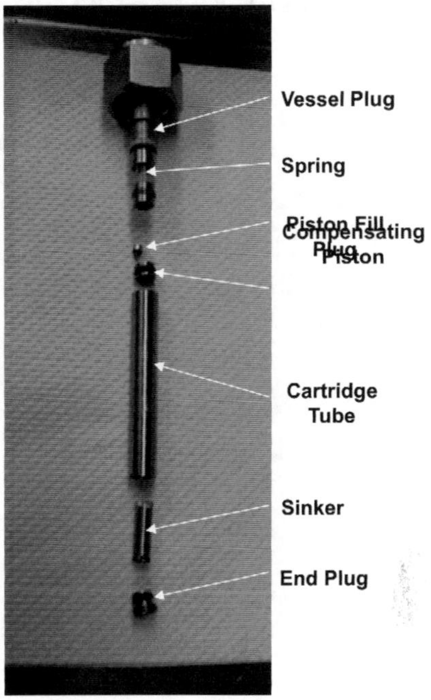




The 10th Industrial Fluid Properties Simulation Challenge
The Pressure-Viscosity Relation

# Measuring the Low-Shear Viscosity

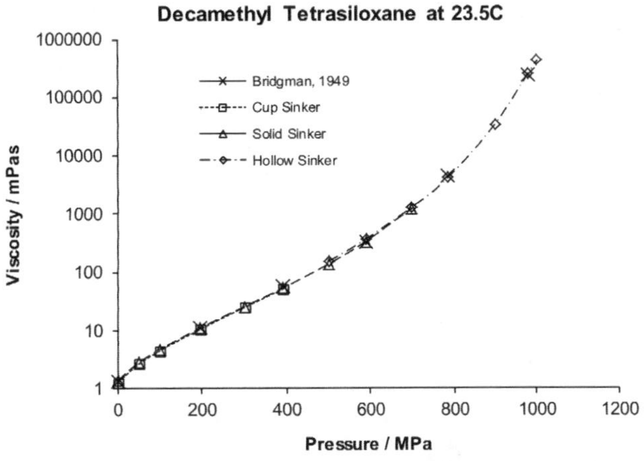


**The viscosity of $Si_4O_3(CH_3)_{10}$ by Bridgman in 1949 using a falling cylinder and a swinging vane viscometer. Viscosities from the LVDT-based viscometer obtained with three sinkers.**

Georgia Institute of Technology

The 10th Industrial Fluid Properties Simulation Challenge
The Pressure-Viscosity Relation

## Temperature and Pressure Dependence of the Low-Shear Viscosity of a Reference Liquid, **Dibutyl Phthalate**, by 3 labs

## Pressure to 1.6 GPa

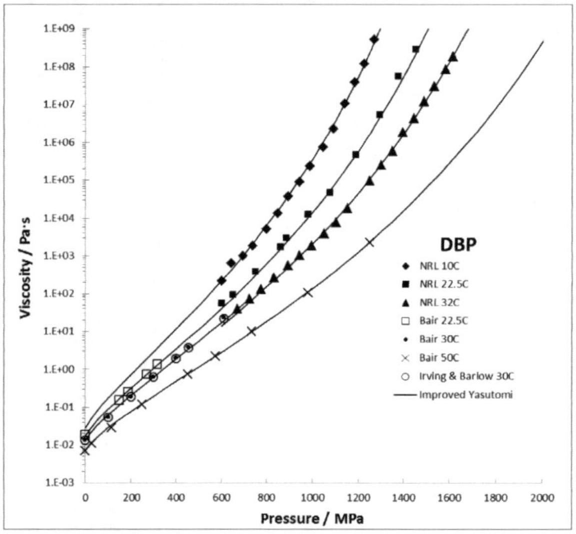




Georgia Institute of Technology

The 10th Industrial Fluid Properties Simulation Challenge
The Pressure-Viscosity Relation

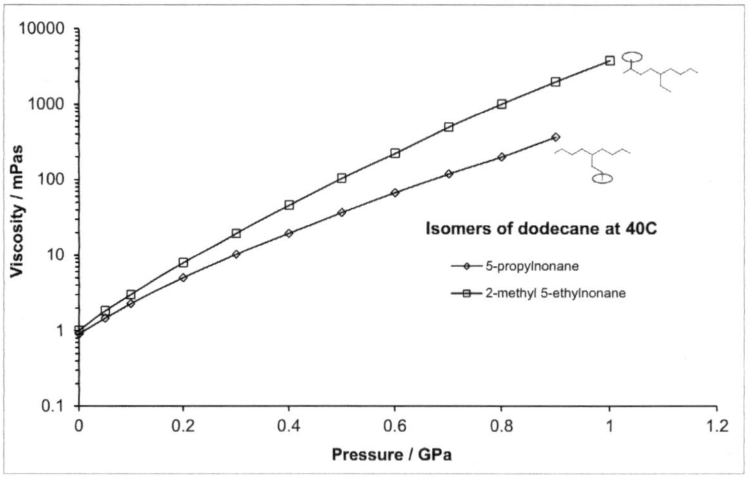


The effect of pressure on the viscosity of two isomers of dodecane measured with a falling cylinder viscometer. Subtle changes in chemical structure such as moving a carbon from the central branch to the backbone can result in order-of-magnitude change in viscosity at high pressure with little effect at low pressure.

The 10th Industrial Fluid Properties Simulation Challenge
The Pressure-Viscosity Relation

# Determining the Equation of State

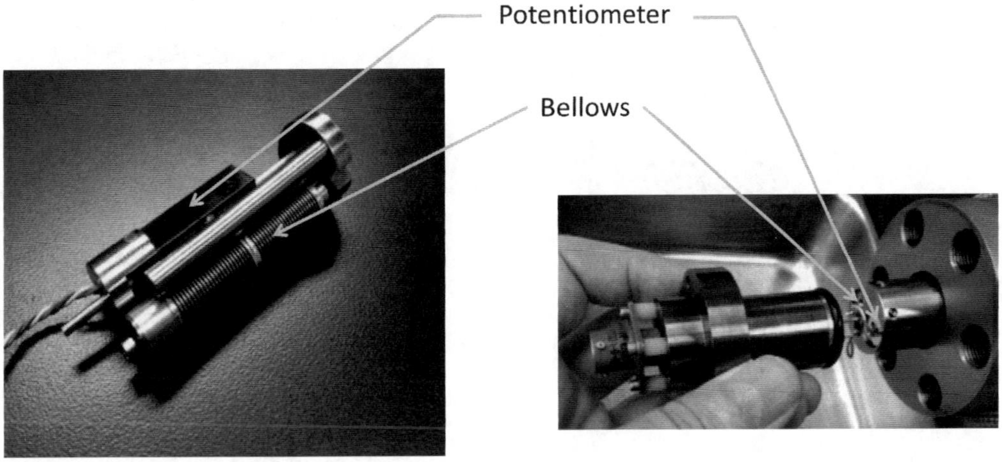




The 10th Industrial Fluid Properties Simulation Challenge
The Pressure-Viscosity Relation

# The rest of the Story
## Rate Independent Shear Response (Limiting Shear Stress)

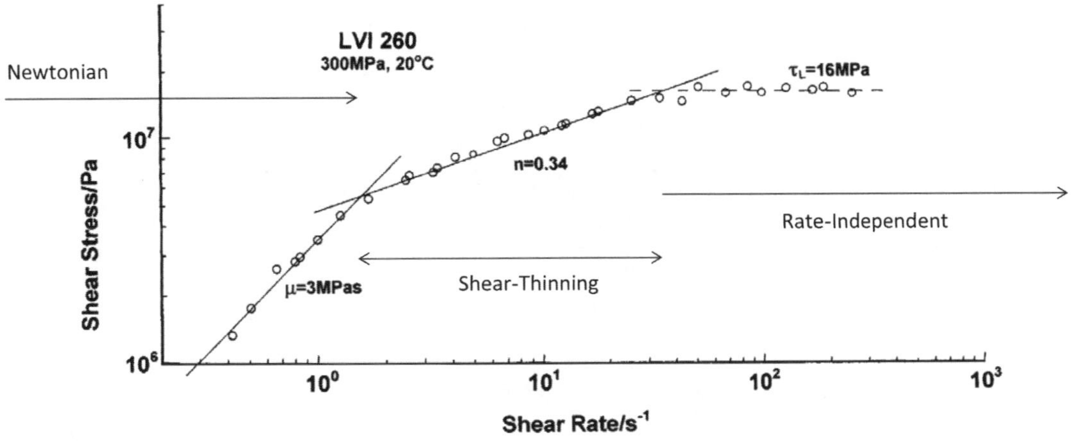

The 10th Industrial Fluid Properties Simulation Challenge
The Pressure-Viscosity Relation

## PRIMARY NORMAL STRESS DIFFERENCE

When a shear stress, $\tau_{yx}$, is applied to a Newtonian liquid, the normal stresses are equal. That is, $\sigma_y=\sigma_x$. Here, positive normal stress is tensile. For a non-Newtonian liquid, $\sigma_y<\sigma_x$, defined such that $\sigma_x$ is more tensile than $\sigma_y$.

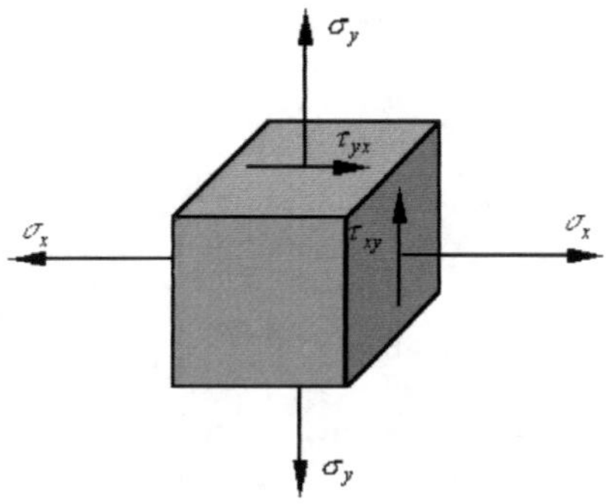


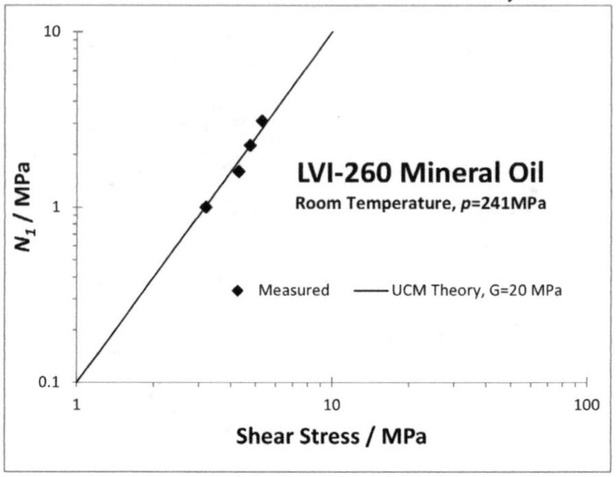


**The maximum shear stress occurs on a plane which is not oriented along the flow direction.**

The 10th Industrial Fluid Properties Simulation Challenge
The Pressure-Viscosity Relation

## Flow Visualization

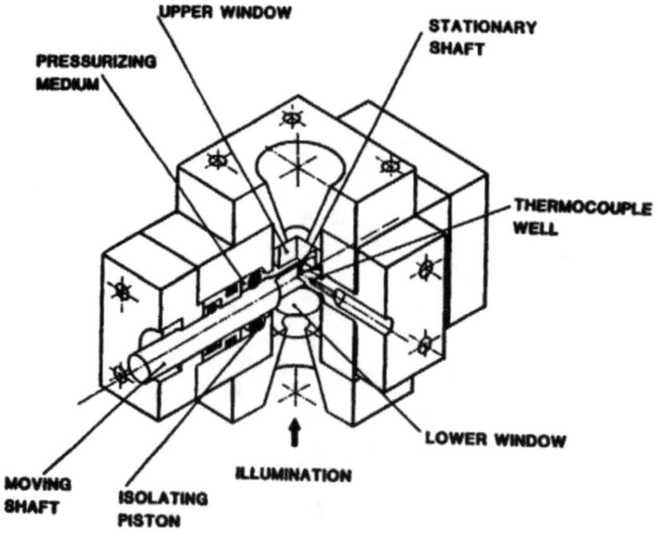

The 10th Industrial Fluid Properties Simulation Challenge
The Pressure-Viscosity Relation

**The maximum shear stress occurs on a plane which is not oriented along the flow direction. If slip occurs, it should be along such planes and it is!**

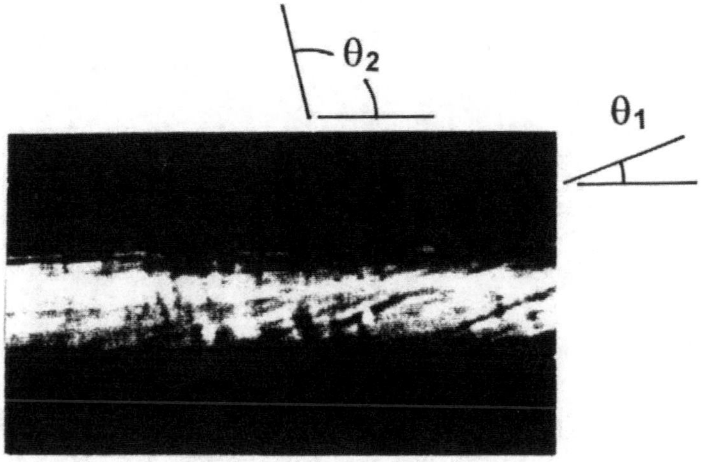




The 10th Industrial Fluid Properties Simulation Challenge
The Pressure-Viscosity Relation

# Conclusions

1. **Data for the pressure dependence of viscosity are not only welcome now, for quantitative understanding and prediction, they are essential.**

2. **Viscometer measurements require a physical sample which may not be easily produced.**

3. **MD may provide the essential piezoviscous response if the results can be experimentally validated.**

The 10th Industrial Fluid Properties Simulation Challenge
The Pressure-Viscosity Relation

# The challenge molecule, 2,2,4 Trimethylhexane

Pressures of 0.1, 25, 50, 100, 150, 250, 350, 400, 500, 600, 700, 800, 900, and 1000 MPa were characterized at a single temperature of 20°C (293K).

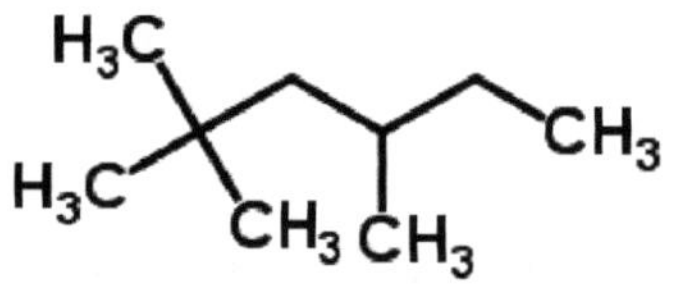

# 511509 Viscosity Predictions using Nonequilibrium Molecular Dynamics Simulations

**Abstract:**

Having accurate viscosity-pressure data for alkanes is key for a large number of applications that rely on the use of these molecules as lubricants. Molecular simulation methods have emerged over the past 20 years as a means to predict the dependence of viscosity upon pressure and temperature for these systems, when experimental data is difficult to obtain e.g. under high pressure. These methods also allow us to gain insight into the microscopic mechanisms, at the molecular level, occurring in the fluid during viscous flow. Here, we discuss how equilibrium and nonequilibrium molecular dynamics (NEMD) simulations have been developed to determine the viscosity of a wide range of fluids. In particular, we focus on two nonequilibrium approaches, based either on the steady-state response or on the transient response of the fluid to an applied shear rate. We show how these two methods can be applied to determine the viscosity of fluids, with the Transient-Time Correlation Function (TTCF) approach providing a bridge between the equilibrium molecular dynamics simulation results, based on the Green-Kubo formalism, and the steady-state NEMD results. We also discuss several applications of these methods, and compare their performance.

# 511510 Elastohydrodynamic Lubrication and Glassy Flow: Linking Experiments and Simulations at High Rates and Pressures

**Abstract:**

Elastohydrodynamic lubrication (EHL) is important in many practical devices and produces extreme pressures (> 1 GPa) and shear rates ($10^5$ $s^{-1}$ to $10^7$ $s^{-1}$). As a result, viscosity of EHL fluids rises by orders of magnitude leading to a glass-like rheological response. Further, the high strain rates lead to large shear stresses and shear-thinning in these fluids. This talk will present nonequilibrium molecular dynamics (MD) simulations of the model EHL fluid squalane from $10^5$ $s^{-1}$ to $10^{10}$ $s^{-1}$ and a wide range of temperatures and pressures. The results are consistent with experimental data for a broad range of equilibrium and nonequilibrium stress and viscosity measurements and allow shear response to be correlated to molecular orientation and dynamics. At high temperatures and low pressures, where the Newtonian viscosity is low, the fluid shows simple shear-thinning associated with molecular alignment which can be described by power-law fluid models. As the Newtonian viscosity rises above ~1 Pa-s, shear-thinning is increasingly dominated by activated hopping. Over more than 10 decades in strain rate, the stress rises logarithmically with rate as implied by Eyring theory. Extrapolating Eyring fits to simulations at $10^6$ $s^{-1}$ and above yields Newtonian viscosities that are consistent with available low-rate experiments, and allows predictions to higher pressures and lower temperatures. Implications of our results to glass transition and EHL phenomena are discussed. Perspectives in the use of molecular simulation to extract accurate viscosities of glassy systems are offered.

# 494281 The Surprising Accuracy of Dispersion-Corrected Ggas in the Prediction of Dissociation Barriers on Transition Metal Surfaces

**Abstract:**

The development of density functional theory (DFT) for accurate description of molecule-surface interactions, while essential for heterogeneous catalysis and energy conversion, is severely limited by the absence of well-defined, chemically accurate benchmark databases for reactions on surfaces. Therefore, unlike gas phase barriers for which density functionals have been extensively benchmarked, limited information is available on computational accuracies for surface reactions. We have developed a database of ten experimental apparent barrier heights (SBH10) for dissociation of small molecules such as $H_2$, $N_2$, and $CH_4$ on transition metal surfaces. The barriers are carefully chosen from combined molecular beam experiments and quantum dynamics simulations, laser assisted associative desorption (LAAD) experiments, and thermal rate measurements.

We have also benchmarked the performance of a dispersion-corrected GGA, BEEF-vdW, a metaGGA, MS2, and a screened hybrid functional, HSE06. In the gas phase, GGAs systematically underestimate barriers due to self-interaction errors that lead to incorrect charge separation at the transition state. By including exact exchange, hybrid functionals correct for these errors and lead to better barrier estimates. The trends are reversed for reactions occurring on transition metal surfaces. Our benchmarking study shows that the BEEF-vdW functional (mean error $< 0.2$ eV) outperforms MS2 and HSE06, which underestimate barriers by 0.3 eV, and 0.5 eV, respectively. This is because transition states for dissociative adsorption closely resemble the final chemisorbed states, and errors in barrier heights mirror errors in chemisorption energies. In other words, dissociation transition states are bound to the surface, while gas phase transition states are isolated species. The key driver for functional accuracy for reactions on surfaces, therefore, is the description of surface-adsorbate interactions, and not charge separation or self-interaction correction. A functional that can accurately predict chemisorption energies, like BEEF-vdW, is also reliable for estimating barrier heights on surfaces.

Catalysis Center for Energy Innovation

UNIVERSITY OF DELAWARE

# Adsorbate Vibrations on Transition Metal Surfaces: Applications and Theory


**Joshua Lansford** and D. G. Vlachos


2017 AIChE Annual Meeting, Minneapolis, MN
Computational Catalysis I: Fundamentals


University of Delaware


October 31, 2017


Catalysis Center for Energy Innovation

UNIVERSITY OF DELAWARE


## Advances in First-Principles Modeling

- Rapid advances over the past 20 years
  - Computer power and accurate methods
- Insights into mechanisms and kinetics using Density Functional Theory (DFT)
  - Adsorption
  - Spectroscopy (vibrational frequencies)
  - Reaction barriers
  - Rate constants
- **Microkinetic models enable direct comparison to experimental data**
- **Catalyst discovery using microkinetics is impractical**
- **The materials gap (actual catalyst structure, and active site) prohibits predictive modeling**

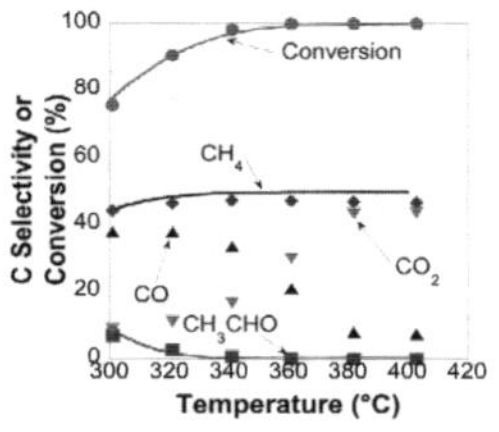


2

## Adsorption Energy Linear Scaling Relations

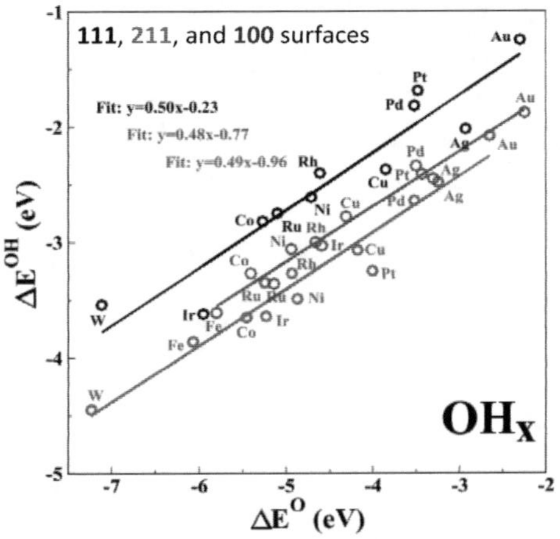


$$\Delta E^{AHx} = m_E \Delta E^A + b_E$$

Linear scaling relations improve our understanding of surface reactions and **enable catalyst screening**[1]

**Extension of these relations to entropies is lacking**

[1] F. Abild-Pedersen et al., Phys. Rev. Lett. 99, 4 (2007).





## Spectroscopy is Very Surface Sensitive Technique

CO on **Pt (A)** and **Pd (B)** electrodes[1]

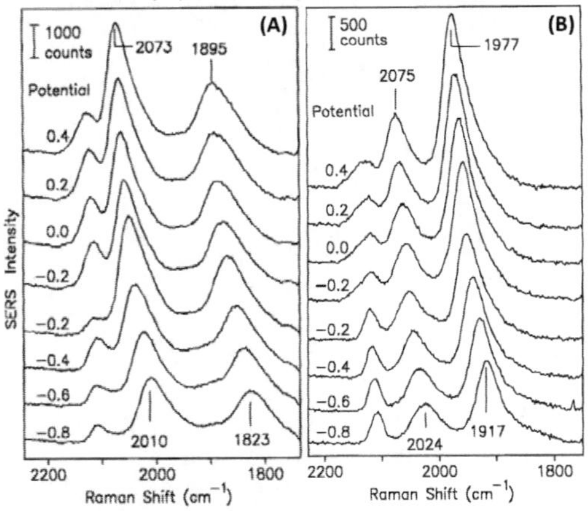


| **Exp. CO Frequency on Pt(111)**[2] | |
|---|---|
| **Site** | **Freq. [cm$^{-1}$]** |
| **atop** | 2070 |
| **bridge** | 1830 |
| **fcc** | 1760 |

**Frequencies depend on metal and site**

[1] Zou et al., J. Phys. Chem. 100 (1996)
[2] Dabo et al., J. Am. Chem. Soc. 129 (2007)

# Objectives

- **Exploit scaling relations** from experiments and first-principles calculations
- **Develop** theory for vibrations
- **Assess effect of entropy** on coverages and microkinetics
- **Deduce adsorption sites and ambiguous species**





# Experimental Evidence for Frequency Scaling

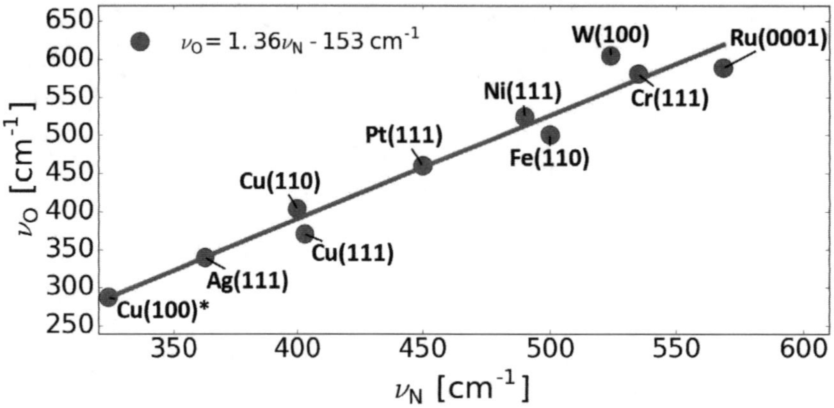


- Experimental HREEL data for chemisorbed atomic O and N

**Data from 40+ different HREEL studies suggest frequencies scale across metals**

J.L. Lansford, A.V. Mironenko, D.G. Vlachos, Nat. Comm. Scaling Relationships and Theory for Vibrational Frequencies of Adsorbates on Transition Metal Surfaces (Accepted)

Catalysis Center for Energy Innovation

UNIVERSITY OF DELAWARE

## Computational Methods

- Periodic Density Functional Theory (DFT) using the Vienna Ab initio Simulation Package (VASP)[1]
- RPBE functional with D3 dispersion corrections[2,3]
- 4-layer slabs with the bottom 2 layers fixed and 16 atoms per layer
- All vacuum space 20 Å
- 400 eV cutoff with 3x3x1 k-point sampling
- Force cutoff 0.02 eV
- Frequencies determined by diagonalizing a mass-weighted Hessian under the harmonic approximation
- **Algorithmic criteria developed for assigning frequencies to normal modes**

[1] G. Kresse and J. Furthmüller, Phys. Rev. B. 54, 11169 (1996).
[2] B. Hammer, Phys. Rev. B. 59, 7413 (1999).
[3] S. Grimme, J. Antony, S. Ehrlich, and H. Krieg, J. Chem. Phys. 132, 154104 (2010).

7

Catalysis Center for Energy Innovation

UNIVERSITY OF DELAWARE

## Scaling of Normal Mode Vibrations from DFT

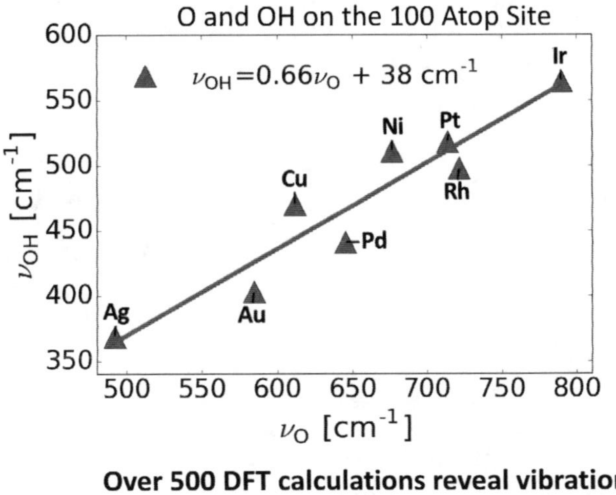

**Over 500 DFT calculations reveal vibrational scaling relations (VSRs)**

J.L. Lansford, A.V. Mironenko, D.G. Vlachos, Nat. Comm. Scaling Relationships and Theory for Vibrational Frequencies of Adsorbates on Transition Metal Surfaces (Accepted)

8

## Linear Scaling Relationships (LSRs)

$$\Delta E^{AHx} = m_E \Delta E^{A} + b_E$$

### Energy Is Additive for Transition Metals

$$\Delta E = \Delta E_{sp} + \Delta E_{d}$$

Effective Medium Theory (EMT)[1,2]

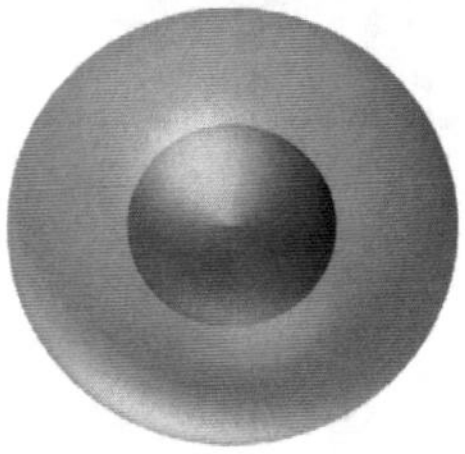

Linear Muffin Tin Orbital Theory (LMTO)[3,4]

[1] J. K. Nørskov and N. D. Lang, Phys. Rev. B. 21, 2131 (1980).
[2] J. K. Norskov, J. Chem. Phys. 90, 7461 (1989).
[3] O. K. Andersen and O. Jepsen, Phys. Rev. Lett. 53, 2571 (1984).
[4] B. Hammer and J. K. Nørskov, Surf. Sci. 343, 211 (1995).





## Competitive Nature of sp and d Bands

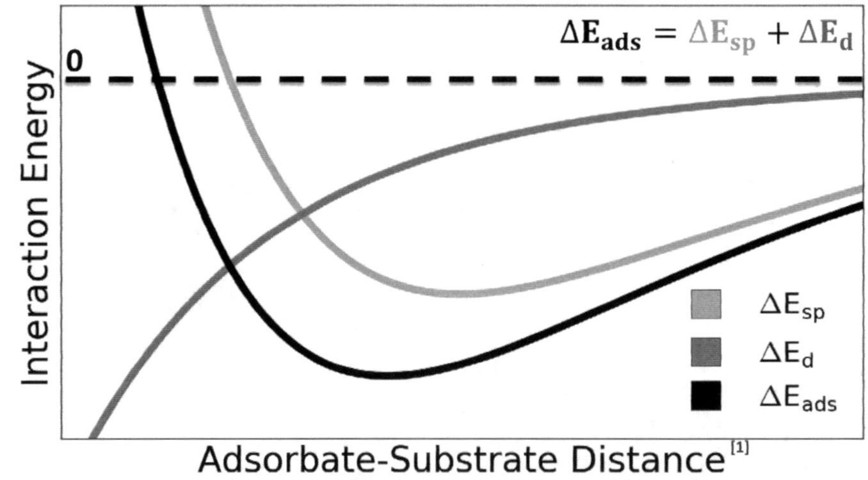

- $\Delta E_d$ at equilibrium taken from the Newns-Anderson model[2]; curves produced from EMT and LMTO

[1] J.L. Lansford, A.V. Mironenko, D.G. Vlachos, Nat. Comm. Scaling Relationships and Theory for Vibrational Frequencies of Adsorbates on Transition Metal Surfaces (Accepted)
[2] B. Hammer, Faraday Discuss. 110, 323 (1998).

Catalysis Center for Energy Innovation

UNIVERSITY OF DELAWARE

## Extending Adsorption Scaling to Vibrations

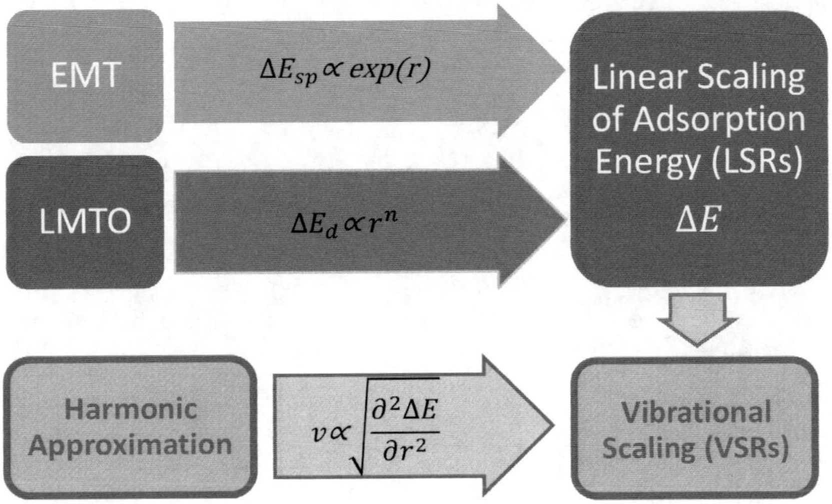


11

Catalysis Center for Energy Innovation

UNIVERSITY OF DELAWARE

## Vibrational Scaling Relation (VSR)

$$(v^{AHx})^2 = m_v (v^A)^2 + b_v$$

## Deriving the VSR in terms of the LSR

$$m_v = m_E \frac{\mu^A}{\mu^{AHx}} \left( \frac{R_n^{AHx}}{R_n^A} \right)$$

where

$$R_n = f(bond\ distance, orbital\ character)$$

**We can predict $m_v$ from a single DFT calculation!**

J.L. Lansford, A.V. Mironenko, D.G. Vlachos, Nat. Comm. Scaling Relationships and Theory for Vibrational Frequencies of Adsorbates on Transition Metal Surfaces (Accepted)

12

Catalysis Center for Energy Innovation

UNIVERSITY OF DELAWARE

## Predicted vs. Actual Slopes of Correlations

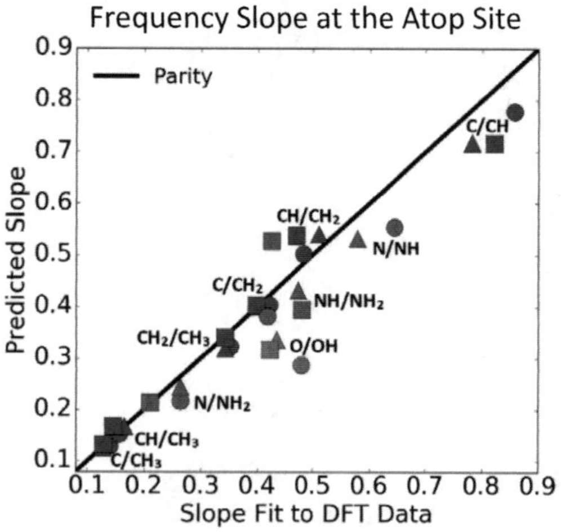


**Predicted Slope**

$$m_v = m_E \frac{\mu^A}{\mu^{AHx}} \left( \frac{R_n^{AHx}}{R_n^A} \right)$$

**Slope Fit to DFT Data**

$$(v^{AHx})^2 = m_v (v^A)^2 + b_v$$

| Standard Error in Regressed Slope | | |
|---|---|---|
| Species | LSR | VSR |
| O/OH | 0.13 | 0.06 |
| $NH/NH_2$ | 0.03 | 0.05 |
| $CH_2/CH_3$ | 0.05 | 0.05 |

J.L. Lansford, A.V. Mironenko, D.G. Vlachos, Nat. Comm. Scaling Relationships and Theory for Vibrational Frequencies of Adsorbates on Transition Metal Surfaces (Accepted)

13

Catalysis Center for Energy Innovation

UNIVERSITY OF DELAWARE

## Impact of Vibrational Correlations on Modeling

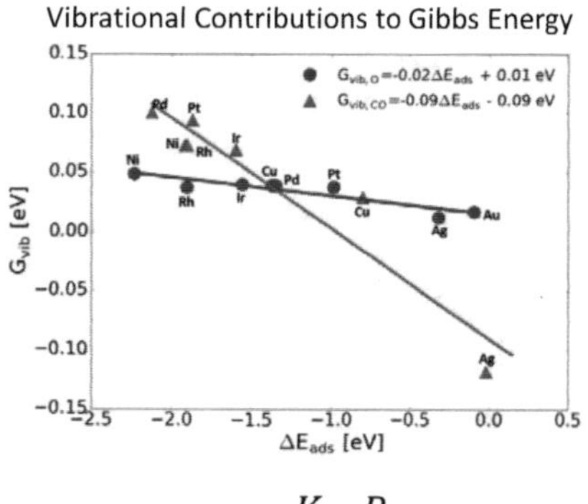


$$\theta_{CO} = \frac{K_{CO}P_{CO}}{1 + K_{CO}P_{CO} + \sqrt{K_O P_{O_2}}}$$

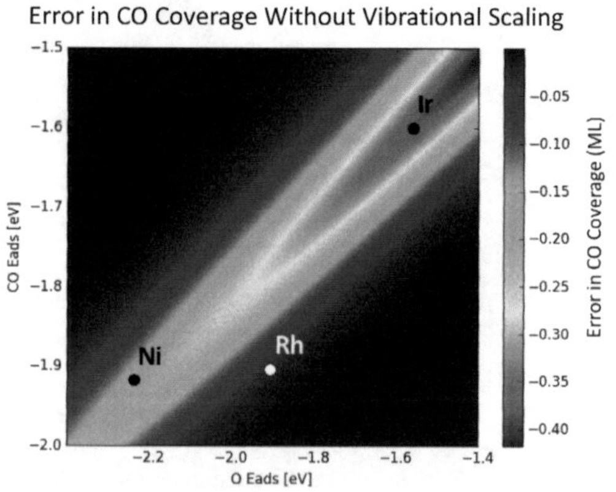


*400 K, 1 bar and 100 ppm CO in $O_2$*

J.L. Lansford, A.V. Mironenko, D.G. Vlachos, Nat. Comm. Scaling Relationships and Theory for Vibrational Frequencies of Adsorbates on Transition Metal Surfaces (Accepted)

14

# Interpreting Complex Spectra on Bimetallics

NH$_X$ on the 100 Atop Site

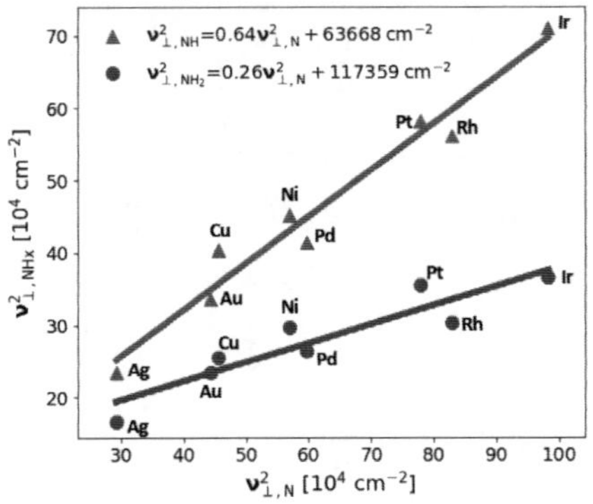


Scaling from DFT[1]

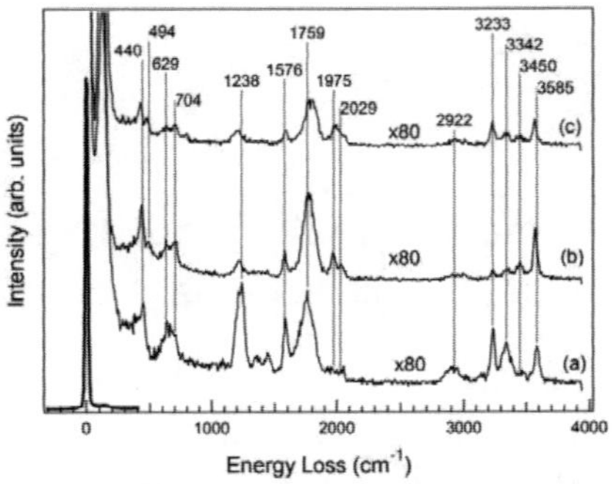


Ammonia decomposition on Ni-Pt-Pt at various temperatures[2]

[1] J.L. Lansford, A.V. Mironenko, D.G. Vlachos, Nat. Comm. Scaling Relationships and Theory for Vibrational Frequencies of Adsorbates on Transition Metal Surfaces (Accepted)
[2] D. A. Hansgen, D. G. Vlachos, and J. G. Chen, Nat Chem 2, 484 (2010).





# Identifying the $O_2$ Adsorption Site and Normal Modes

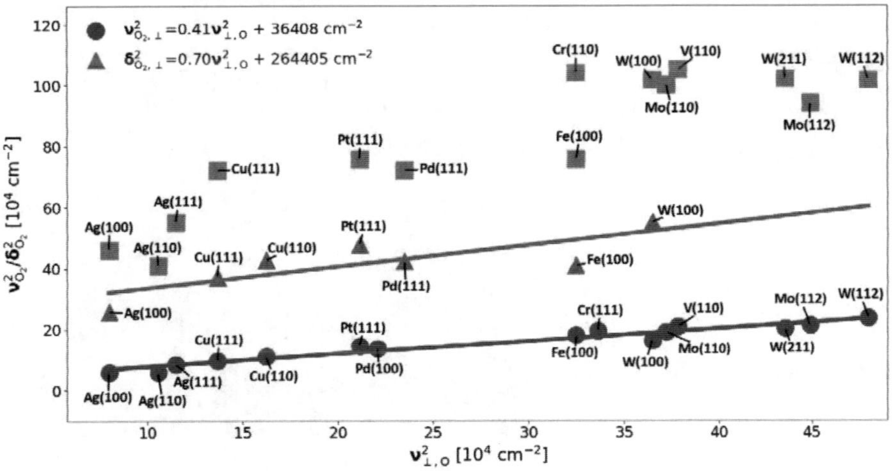


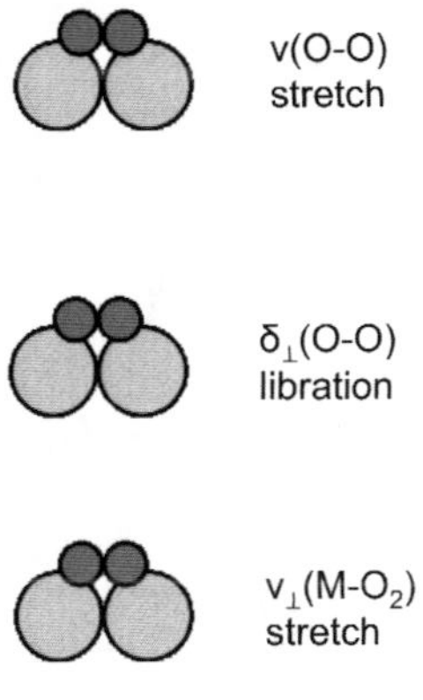


- Experimental HREEL data for $O_2$ and O

J.L. Lansford, A.V. Mironenko, D.G. Vlachos, Nat. Comm. Scaling Relationships and Theory for Vibrational Frequencies of Adsorbates on Transition Metal Surfaces (Accepted)

## Summary

- We develop vibrational scaling relations **(VSRs)** and **corresponding theory**
- We **demonstrate** the **predictive power** of our theory and **impact** of vibrational scaling on thermodynamics
- **Scaling frequencies** can be used to **identify** adsorption site and species



Catalysis Center for Energy Innovation

## Acknowledgements


Funding from DARPA

Professor Dionisios G. Vlachos for advisement

The Vlachos group

# 494955 Genetic Algorithm Enhanced By Atomistic Neural Network: Pt Clusters at the $H_2$ Atmosphere as an Example

**Abstract:**

Revealing the catalyst structures at the atomic level is essential to rationally improve or design new efficient catalyst. Accordingly, it is a key goal for theoretical catalysis studies. Although theoretical calculations can reveal the catalyst structures in atomic details, which is beyond the capability of most current experimental techniques, there are still some challenges. First, theoretical calculations based on first-principles calculations are very expensive, therefore the simulations are normally limited within several hundred atoms; Secondly, realistic catalyst structures are quite diverse and they are easily affected by reactants, products et al[1]. The catalyst might be associated to several competing low energy structures, and not to a single unique structure, which in turn needs more computations. Finally the exploration of potential structures on a system with a large number of degrees of freedom is computationally complex[2]. All of those factors hamper the understanding of the relationship between reaction activity and catalyst structures.

We proposed two strategies to address above obstacles. Firstly, in order to investigate all the possible catalyst structures, we devise a scheme to enhance the capability of normal genetic algorithm (GA)[3]. Considering that the most stable structure is not necessary the most active catalyst structure, a comprehensive exploration of many low energy catalyst isomers is vital. With this in mind, our enhanced GA is aimed at finding all the possible structures that could have a reasonable population in catalytic temperatures rather than only locating the global minimum. Secondly, since the genetic algorithm will exploit a large number of structure optimizations, it is time-consuming to carry out all optimizations by DFT directly. Therefore, we accelerated the computationally expensive structure optimizations with an atomistic neural network (ANN)[4], which is trained before performing GA optimizations. An iterative scheme is also utilized to efficiently find the optimal training set structures during the ANN training procedures.

With those tools, we studied the structures of platinum clusters covered with hydrogen adsorbates ($Pt_nH_x$). For every composition, we can find all the possible structures whose energy lies between the $E_{gm}$ (gm=global minimum) and $E_{gm}$+0.5 eV. Those structures cover all the catalytically relevant structures that may appear during catalytic reactions, therefore we can locate the most important catalytic species from the structure population. The influence of the supports is in consideration. In the end, we can improve our understanding of Pt clusters at more realistic conditions.

All in all, we presented a combined routine, which takes advantages of two cutting-edge techniques: global optimizations and neural-network methods. This method can efficiently find out the low energy structures for a large system without loss of energy accuracy vs DFT. Then the method is utilized for studying a realistic system-Pt cluster at the atmosphere of $H_2$. Besides the example in this study, this scheme can be easily applied to similar catalytic problems benefiting from the generality of genetic algorithm and atomistic neural network.

1. Mager-Maury, C.; Bonnard, G.; Chizallet, C.; Sautet, P.; Raybaud, P., H  -induced reconstruction of supported Pt

2

clusters: metal-support interaction versus surface hydride. *ChemCatChem* **2011,** *3* (1), 200-207.

2. Zhai, H.; Alexandrova, A. N., Ensemble-Average Representation of Pt Clusters in Conditions of Catalysis Accessed through GPU Accelerated Deep Neural Network Fitting Global Optimization. *J Chem Theory Comput* **2016,** *12* (12), 6213-6226.

3. Vilhelmsen, L. B.; Hammer, B., A genetic algorithm for first principles global structure optimization of supported nano structures. *J Chem Phys* **2014,** *141* (4), 044711.

4. Artrith, N.; Hiller, B.; Behler, J., Neural network potentials for metals and oxides - First applications to copper clusters at zinc oxide. *Physica Status Solidi (b)* **2013,** *250* (6), 1191-1203.

# 500606 Prediction of Chemisorption Energies By Gaussian Processes

**Abstract:**

Design of heterogenous catalysts is going through a revolution due to the development of quantum chemical computations and descriptor based approaches, which reduce the dimensionality of the design problem based on a detailed physical understanding.[1]

Chemisorption energies of reaction intermediates are most often the descriptors of choice. Meanwhile, a bottleneck is currently caused by the CPU time needed to calculate energetics of these intermediates and transition states from atomic structures. It is therefore feasible to use machine learning methods to predict chemisorption energies, particularly as the amount of data grows.[2]

For catalyst screening purposes, a Gaussian Process (GP) has advantageous properties, the biggest of which is the capability of error estimates using Bayesian statistics.[3] Simple linear models are already sufficient for describing correlation between adsorbate energies, when they bind through the same atom to various surfaces. Simple GP models may therefore be made with the advantage of being highly interpretable, which provides physical insight. Typical numbers of datapoints in calculated chemisorption energy databases has yet to reach the regime of "big data" and GP's may therefore also prove to be sufficiently efficient.

We present our framework for modeling chemisorption energies using a Gaussian Process and we present applications in elucidating trends and designing metal catalyst surfaces for relevant problems in modern thermal catalysis.

1. Jens K. Nørskov et al. “Towards the computational design of solid catalysts”. In: Nature chemistry 1.1 (2009), pp. 37–46.

2. Zachary W. Ulissi et al. “To address surface reaction network complexity using scaling relations machine learning and DFT calculations”. In: Nature Communications 8 (2017).

3. Carl E. Rasmussen and Christopher K. I. Williams. “Gaussian processes for machine learning. 2006”. In: The MIT Press, Cambridge, MA, USA 38 (2006), pp. 715–719.

# 494916 New Computational Tools for High-Throughput Discovery in Transition Metal Catalysis

**Abstract:**

Virtual high throughput screening has emerged as a powerful tool for the discovery of new materials. Nevertheless, screening of inorganic complexes is presently limited by the high cost of accurate property evaluation and the large, complex search space. To address these challenges, we have recently developed molSimplify[1], an open-source toolkit for the automated generation of highly accurate inorganic molecular structures and property analysis to accelerate high throughput screening of inorganic complexes. In conjunction with our structure generation toolkit, we have also developed and implemented a robust and automated workflow involving: (1) selection of a test set of ligands maximizing relative molecular diversity[2], (2) prediction of electronic structure outputs with a neural network prior to simulation for pre-screening and (3) screening of multi-million molecule organic libraries according to the discovered design rules to discover experimentally accessible ligand candidates[2]. We demonstrate the broad utility of our integrated approach to inorganic complex discovery across the following applications: design of redox active functionalized ferrocenium complexes for selective ion separation[3], discovery of octahedral Fe(II/III) redox couples with nitrogen ligands[4], and design of single-site Fe(II) catalysts for oxygen activation. We also demonstrate the potential for molSimplify to accelerate catalytic property evaluation by automatically generating high-quality transition state guesses for common reactions in inorganic chemistry.

[1]E. I. Ioannidis, T. Z. H. Gani, and H. J. Kulik "molSimplify: A toolkit for automating discovery in inorganic chemistry" J. Comput. Chem. 37, 2106-2117 (2016).

[2]J. Y. Kim, A. H. Steeves, and H. J. Kulik "Harnessing Organic Ligand Libraries for First-Principles Inorganic Discovery: Indium Phosphide Quantum Dot Precursor Design Strategies" Chem. Mater., just accepted (2017).

[3]T. Z. H. Gani, E. I. Ioannidis, and H. J. Kulik "Computational Discovery of Hydrogen Bond Design Rules for Electrochemical Ion Separation" Chem. Mater., 28 6207-6218 (2016).

[4]J.P. Janet, T. Z. H. Gani, A. H. Steeves, E. I. Ioannidis, and H. J. Kulik "Leveraging Cheminformatics Strategies for Inorganic Discovery: Application to Redox Potential Design" Ind. Eng. Chem. Res., just accepted (2017).

# 493380 More Accurate Depiction of Adsorption Energy on Transition Metals Using Work Function As One Additional Descriptor


**Abstract:**

Transition metals (TMs), benefiting from their active interactions with chemical species owing to partial occupancy of the d bands, represent one most important category of catalysts and have been widely used in many reactions. The catalytic properties of TMs in a chemical reaction are primarily determined by adsorption energy ($E_{ads}$) of the involved species. In this regard, it is of high research significance to get accurate $E_{ads}$ values, however, which is often challenging be measured experimentally. Active research interests have been devoted to determine Eads by establishing an exact relationship with measurable catalyst parameters like the d-band (εd) theory, which was based on the Newns-Anderson model and suggested for use by Hammer and Nørskov. It describes a simple $E_{ads}$–$\varepsilon_d$ linear correlation and has been widely adopted for application, which turns out to work well for certain adsorption systems, for instance $H_2$ and CO adsorption, but becomes less effective for many other systems, such as O, OH and OOH adsorption. These controversial results suggest the current $E_{ads}$–$\varepsilon_d$linear correlation could be oversimplified to describe these adsorption systems and εd might not be the only influencing parameter, thus improvements are needed.

Considering the fact that band hybridization and electron transfer could occur simultaneously when a molecule adsorbs to a TM surface, we propound that $E_{ads}$ contains a mixture of covalent and ionic contributions (i.e., $E_{ads}=E_{covalent} + E_{ionic}$). The $E_{covalent}$ term would be describable using the $\varepsilon_d$ parameter as it accounts for the band hybridization, whereas the $E_{ionic}$ term is determined by the charge transfer between TM and adsorbing molecules and thus would be related to the work function ($W$) parameter. Herein by taking both the ionic and covalent contributions into account, we suggest a new description model for $E_{ads}$ by dividing it into $E_{covalent}$ and $E_{ionic}$ components and adding $W$ as one additional descriptor. We study the adsorption of O, OH, and OOH groups on TM surfaces using the new $E_{ads}$–($\varepsilon_d$, $W$) model with the explicit formula as $E_{ads} = f(\varepsilon_d, W) = z_0 + a \times \varepsilon_d + k \times (W - W_0)^2$ and achieve significantly better goodness of fit compared to the current $E_{ads}$–$\varepsilon_d$ model. We further demonstrate the new $E_{ads}$–($\varepsilon_d$, $W$) model can better predict activity property of TMs in oxygen reduction reaction (ORR) benefiting from a more accurate description of the oxygen binding energy ($E_O$).

# 496405 Molecular Simulation Study of How the Structure of Liquid Water Affects the Free Energies of Adsorption and Reaction in Aqueous Phase Heterogeneous Catalysis

**Abstract:**

There are many applications of aqueous phase heterogeneous catalysis, including aqueous phase reforming (APR), Fischer-Tropsch synthesis, electrocatalytic oxygen reduction, etc., and a grand challenge in catalysis research is understanding how water influences the chemistry, mechanism, thermodynamics, and kinetics of reactions occurring on metal catalyst surfaces under liquid $H_2O$. Recent work by our group and others has shown that $H_2O$ molecules form hydrogen bonds (HB) with reaction intermediates, alter their adsorption energies, co-catalyze certain reactions, affect kinetic barriers and reaction mechanisms, amongst other phenomena. We are interested not only in the roles that liquid $H_2O$ molecules play, but also how these phenomena influence the rates and equilibria of catalytic surface reactions. For example, we previously found that hydrophilic adsorbates form stronger and more stable HB with $H_2O$ molecules than hydrophobic adsorbates, which we postulated would lead to greater degrees of order in the interfacial water structure around hydrophilic adsorbates than for hydrophobic ones. We further postulated that this would give rise to a free energy effect in reactions where hydrophilic adsorbates are converted into less hydrophilic adsorbates or even hydrophobic ones, e.g., in APR reactions where sugar alcohol molecules are sequentially deprotonated down to $CO_2$ and $H_2$ (and other analogous reactions). In this work, we use density functional theory (DFT) and methods in molecular dynamics (MD) and statistical mechanics to study these effects in the context of the decompositions of $CH_3OH$ and $NH_3$ over Pt catalysts. Specifically, we compute the free energies of different surface intermediates along the pathways for these reactions under liquid $H_2O$ using the following methods: potential of mean force (PMF), calculated from the radial distribution function (RDF), which gives the free energy due to configurational fluctuations in the liquid $H_2O$ structure, and free energy perturbation (FEP), thermodynamic integration (TI), and implicit solvation method for surfaces (iSMS), which give the free energies of solvation for the different adsorbates. A key step in the PMF/RDF, FEP, and TI methods is producing an explicit structure of liquid $H_2O$ at the catalyst interface, in order to identify how it influences catalytic thermodynamics. We provide values for these interfacial free energy contributions relative to the free energies of reaction and discuss how they influence the thermodynamics of adsorption to the surface from aqueous phase and reaction on the surface under liquid $H_2O$. We summarize by discussing how these contributions influence the equilibrium and rate constants in aqueous phase heterogeneously catalyzed reactions.

# 496237 Top Ten Mistakes Applying Computational Chemistry in Industry

**Abstract:**

In this talk, I will present mistakes that are often made in managing, directing, and applying computational chemistry in an industrial R&D or problem-solving context. The focus is not on the practical details of doing calculations, but on the decisions that are made regarding when and how computational chemistry (CC) and molecular simulations are used and what is expected of them. Each mistake will be paired with positive messages for how to avoid the mistake and how to best integrate computational chemistry into an R&D program. Topics include what happens when the limits or, alternatively, the power of computational chemistry are overemphasized, when to bring CC into a project, who should be doing CC, how CC relates to the experimental program, what ancillary activities might be useful, and how to choose what to work on.

# 495072 Atomic-Level Insights into Chemical Additives Used for Silicon-Containing Film Removal during Integrated Circuit Manufacturing

**Abstract:**

The evolution of the semiconductor industry over the past several decades has been greatly impacted by effective surface preparation and cleaning formulations. During integrated circuit (IC) manufacturing, cleaning is the most common operation, but the challenges in preparing the wafer surface for the next process step continue to increase as contamination tolerance is lowered, material removal limits decrease, and threeâ€‘dimensional fine features become more common. The advancement of novel formulation additives for removing contaminants and residues from earlier process steps have led to higher yields and enabled new integration schemes.Â First-principles analysis of etching mechanisms and additive function can significantly decrease experimental costs through initial virtual screening of potential substrates and additives. This presentation will focus upon the removal of silicon-containing materials used in ICs, such as $SiO_2$ and $Si_3N_4$, and relevant formulation additives used to remove these materials. Particularly, in this collection of studies, Density Functional Theory (DFT) was employed to examine the underlying thermochemistry and kinetics of etching mechanisms and additive function on substrates found in common ICs. Our studies have employed both cluster and periodic surface slab models, as well as mixed implicit and explicit solvation schemes, in efforts to balance computational time and accuracy. Modeling results were found to be in agreement with experiments and have provided insights on the parameters which should be considered in the formulation design for selective thin film removal. These results demonstrate how theoretical investigations are currently impacting formulation additive discovery and surface cleaning technology in industry.

# 492731 Thiohypoiodous Acid (HSI) Formation and Its Role As an Intermediate in $H_2$ Production


**Abstract:**

In 1979, S. W. Benson observed a room temperature liquid formed by the reaction of Hydrogen Sulfide ($H_2S$) and Iodine ($I_2$) and postulated that this liquid was Thiohypoiodous acid (HSI), a never before observed molecule.[1] However, Benson had only a little bit of experimental evidence and no theoretical evidence to support this interpretation of this experiment. Aside from some proposed potential atmospheric relevance,[2] this molecule has attracted little attention because of the absence of a definitive characterization, and also because it was not obvious that it would be useful even if it did exist. No one has studied its reaction kinetics.

Very recently, a method for converting $H_2S$ into $H_2$ + S has been proposed which would have significant environmental and industrial impact if successful. The first step in this process is the reaction of $H_2S$ with $I_2$. We hypothesize that HSI is a key intermediate in the formation of HI from $H_2S$ and $I_2$ in aqueous solutions. It is therefore important to quantitatively understand the thermochemistry and reaction kinetics of HSI.

Optimized geometries, energies and frequencies of HSI and the reactants and products in proposed HIS reactions have been calculated at the CCSD(T) level using Dunning's correlation consistent triple zeta basis with added diffuse functions. The computed equilibrium constants are consistent with experimental observations of Benson et al. Also, the energetics of HSI's secondary reactions with $H_2S$, $I_2$, HSI and HI have been studied and will be reported.


---


[1] R. J. Hwang and S. W. Benson, "Thermochemistry of the Reaction of Hydrogen Sulfide with Iodine and the Heat of Formation of HSI – a New Sulfur Compound," *J. Am. Chem. Soc.*, vol. 101, no. 10, pp. 1615–1617, 1979.

[2] W. Hermoso and F. R. Ornellas, "Predicting new molecular species of potential interest to atmospheric chemistry: The isomers HSI and HIS," *Chem. Phys. Lett.*, vol. 459, no. 1, pp. 77–81, 2008.

# 498059 Computational Design of Thermodynamically Stable Metal Nanoparticles

**Abstract:**

The physicochemical properties of metal nanoparticles (e.g. catalytic behavior), as well as their thermodynamic stability, depend on the nanoparticle metal composition and morphological characteristics, such as their size and shape. For this reason, the application performance of metal nanoparticles is commonly a fine balance between their stability and optimal property functionality (e.g. catalytic activity). As a result, the computational design of nanoparticles that maximize their application properties should be performed utilizing criteria for the nanoparticle thermodynamic stability. In this work, we blend first-principles calculations with scientific computing to introduce a novel computational framework able to predict the stability of bimetallic nanoparticles of different morphology and metal composition. Our nanoparticle stability predictions are tested against experimentally synthesized nanostructures with excellent agreement.

# 498222 Understanding the Geometrical and Electronic Properties of Imidazolium-Based Ionic Liquids in the Presence of Amino Acid Substituted Metal Porphyrins

**Abstract:**

Ionic liquids have proven themselves to be potential candidates for driving industrial processes in a sustainable fashion due to their environmentally benign properties. These liquids have extremely low vapor pressures that translates to the fact that they have negligible role in air emissions as compared to conventional volatile organic solvents. They are also highly design flexible and their physical, chemical and biological properties may be tuned easily by adapting a suitable combination from a wide array of cations and anions. Out of the myriad of a wide range of possibilities, imidazolium-based ionic liquids have been successfully utilized in diverse applications including catalysis, extractive distillation, liquid-liquid extraction among many others. Although being unique and highly effective in driving complex processes, experimental work on these ionic liquids have raised questions on their toxicity and environmental degradability. Some of them have been shown to be water soluble and may leave a large ecological footprint that may lead to detrimental changes in the ecosystem. Thus, it becomes imperative to include biodegradability in the rational design of ionic liquids. Limited experimental work focused on biodegradation of ionic liquids has suggested that long alkyl chain imidazolium-based ionic liquid cations can undergo hydroxylation in the presence of microbial community while the smaller cation resist any biotransformation. The molecular level details for such transformations have not yet been elucidated.

Cytochrome P-450's have been identified and widely studied for their role in the oxidation of a wide variety of molecules in aerobic and anaerobic environments and can act as a possible enzyme implicated in the hydroxylation of ionic liquids. In order to determine the effect of cytochrome P-450 on the 1-n-alkyl-3-methylimidazolium $[C_nmim]^+$ class of ionic liquids, protein environment was modeled with cysteine ligated iron porphyrin (FePCys). Imidazolium cations $[C_nmim]^+$ (n=2,4,6,8, and 10) were considered as potential substrates that interact with different variants of FePCys. These variants may be oxygenated (FePOCys) or non-oxygenated referring to the enzymatic cycle of the P-450 enzyme in focus. All the systems considered ($[C_nmim]^+$FePCys or $[C_nmim]^+$FePOCys) were treated quantum mechanically for understanding the effect of the porphyrin receptor on the electronic environment of imidazolium cations considering their properties in the gas phase as the datum for the same. To include the regioselective and stereoselective aspects of the system, geometrical analyzes were also conducted to identify key cationic orientations. Electronic properties such as frontier orbital energies i.e. Highest Occupied and Lowest Unoccupied energy levels (HOMO and LUMO) were obtained to determine the reactivity of these molecules. Additionally, the results of HOMO and LUMO locations, hybridization of molecular orbitals of FePCys and the cations, and FePOCys and the cations will be presented along with the thermodynamic properties such as binding energy and free energy change upon binding to aid in inferring molecular mechanism of ionic liquid biodegradation.

# 496397 Towards the Prediction of the Liquid Phase Oxidation of Aromatics – an Experimental and Modeling Study for Toluene Autoxidation

**Abstract:**

## Introduction

Due to adequate physicochemical properties as well as high octane number, aromatics are key components for fuel formulation. They limit abnormal combustion, that is knock for gasoline fuel, and also ensure seal swell and lubrication in jet fuels . However, the same aromatics are also subject to different discussions regarding their impact on other types of abnormal combustion such as low speed pre-ignition, where the reactivity in the liquid phase for fuel/lubricant mixes could be an important parameter. Literature also focuses on their effect on particulate formation in combustion as well as deposits in liquid phase in an autoxidation regime. Due to an important need for better simulation tools regarding fuel autoxidation as well as experimental data that can be simulated, this work focuses on these two aspects. Toluene being the simplest alkyl aromatic, its autoxidation at high temperature in liquid phase is investigated in this study.

## Experimental work

The experimental work relies first on a PetroOxy which is employed in a standardized oxidation test method. It is designed for the assessment of automotive fuels according to test methods ASTM D7545, D7525 and EN 16901. This device has been previously used for a similar analysis [1]. This is a closed batch reactor that includes a gold plated dish on which a thin film of liquid is placed. The head-space is pressurized with oxygen to 7 bar and heated to the targeted temperature. The pressure is then recorded while the system evolves. Fuel can then be discriminated according to their induction period (IP) which correspond to a 10% pressure drop. In this study, two types of measurements are employed : (1) single induction period measurements at three individual temperatures of 120°C, 140°C and 160°C to characterize the temperature dependence of the system; (2) The experiment is stopped at half IP so that a liquid sample is analyzed and the toluene consumption can be monitored.

Secondly, an autoclave is set-up to enable a systematic investigation of intermediate autoxidation products at the highest temperature studied in the PetroOxy. A 250 $cm^3$ reactor is filled with 50 mL of sample and agitated with pure oxygen bubbling through the liquid. The system is pressurized at 10 bar to the targeted temperature and the liquid is sampled at different times to follow products build-up.

The oxidation products analysis is carried out using gas chromatography coupled with mass spectrometry for identification. A gas chromatograph CP3800 Varian with a FFAP column of 30 m length, 250 µm and 0.25 µm film thickness is used. Quantification is performed with a flame ionization detector and a different gas chromatograph. It is an Agilent 6890N with the same type of FFAP column. Finally, qualitative measurements are also performed with a HP5 column to assess the feasibility of additional product identification.

## Modeling approach

The Reaction Mechanism Generator RMG is an automated code, designed to propose chemical kinetics mechanisms for both the gas and liquid phase [2]. The generation requires a set of initial conditions including reactant species, their structure and concentration, the temperature as well as, for liquid phases, the solvent used. The mechanism generation is based on a library of reaction types such as hydrogen abstraction or radical recombination, with relevant kinetic parameters, for defined molecular graphs. For every molecule in the mechanism, possible reactions from the library, based on structural information of the molecule are evaluated. The highest reaction flux enables the code to select the next species that moves into the core mechanism. This process is repeated until the defined end criteria is reached. This can be a certain time or reactant conversion for example. Where no kinetic data is available in the database, an average, based on related parameters, is calculated in order to obtain an estimate for the reaction. Besides the kinetic estimation, thermochemical parameters are also estimated based on Benson group additivity for the gas phase. A liquid phase system implies solvation effects, RMG use corrective methods in order update the thermochemical parameters. The first one is based on a linear solvation energy relationship (LSER) which presents a parametrized description of solvent effects to estimate the Gibbs free energy of solvation. This method takes aspects such as viscosity and polarity into account as well as solute descriptions. This approach has been shown to provide a good accuracy for most solvents at low computational cost [3]. The second correction applied in RMG uses Mintz model which aims to obtain the enthalpy of formation in the liquid phase [4]. This model is quite similar in principle to the LSER parametrization method and has been evaluated for a large set of closed shell species. The remaining liquid phase entropy is then estimated by RMG using the two corrections cited above. Finally, the code also ensures that reactions do not exceed their diffusion limit. This is achieved by estimating an effective rate constant that takes into account the radii of the involved molecules as well as the diffusion rates in the solvent.

In the context of this work, quantum chemical calculations have been performed to asses and improve the quality of the RMG-predicted thermochemical parameters for individual species in the liquid phase. The functional used in this study is M06-2X which has proven its capabilities in terms of accuracy since, including in the application of organic chemistry. A modern Aldrich basis set and the RI-J auxiliary basis set is coupled for lower cost calculations to evaluate a large number of species with reasonable accuracy. Indeed, in the GMTKN30 benchmark, M06-2X performed best for basic properties with a weighted total mean absolute deviation of 3.2 kcal/mol and is one of the best performing hybrid functional [5]. Finally, solvation is considered using the SMD model [6] which has shown to perform well when compared to other implicit solvation methods. Explicit solvation may provide better predictions however such a calculation is beyond the scope of this study.

## Results and conclusion

From an experimental side, the IP of toluene has been measured at three different temperatures using the Petrooxy apparatus. The toluene species profile has been measured by sampling the liquid at different times. A more systematic and robust sampling and analysis method has been applied with the autoclave experiments. This second set of data supports the previous findings for major oxidation products of toluene autoxidation at a moderate temperature, that is to say benzaldehyde, benzyl alcohol, benzoic acid as well as benzyl hydroperoxide. In addition, a detailed analysis of termination species for toluene autoxidation is also given. That way a set of potential deposit precursors is established.

These experimental results are used to generate and improve a detailed kinetic mechanism with RMG. This generator provides a first skeletal mechanism for which the thermochemistry is updated with quantum chemical calculations for thermochemistry parameters. Major changes applied to the thermochemistry impacted the equilibrium of key reactions such as the well-known ROO + RH = ROOH + R propagation step, that controls the system at these temperatures. To evaluate the modifications applied to the mechanism, the resulting rate constant for that key step is compared to Denisov's data. The update of the thermodynamic parameters decreases the rate constant by one order of magnitude relative to RMG, leading to a better agreement with experimental data from literature [7]. This contributes to the fact that the updated mechanism reproduces within less than an order of magnitude the IP obtained with the PetroOxy reactor. On the speciation aspect, species observed experimentally are included in the mechanism. The toluene species profile is well simulated by the model and the qualitative trends for major products are also reproduced [8,9]. Quantitatively, additional work is ongoing to improve the simulation of the species profile.

This methodology demonstrates the feasibility of autoxidation global reactivity modeling with a detailed kinetic mechanism for a simple system. Potential improvements of RMG are targeted through the quantum calculations performed for the liquid phase mechanism generation. In addition, due to its interest for many aspects, this study with toluene provides a first building block to evaluate the capability of detailed kinetic mechanisms in specific co-oxidation phenomena as well as deposit formation that is of interest for many applications.

**References**


[1] Chatelain, K. et al., Energy & Fuels, 2016, 30 (2), 1294 – 1303

[2] Gao, C. W. et al., Computer Physics Communications, 2016, 203, 212 – 225

[3] Jalan, A. et al., The Journal of Physical Chemistry B, 2013, 117 (10), 2955 – 2970

[4] Mintz, C. et al., QSAR & Combinatorial Science, 2008, 27, 179 – 186

[5] Goerigk, L and Grimmie, S., Physical Chemistry Chemical Physics, 2011, 6670 – 6688

[6] Marenich, A. et al., The Journal of Physical Chemistry B, 2009, 113, 6378 – 6396

[7] Denisov, E. T. and Afanas'ev; I. B., Taylor & Francis, CRC Press, 2005

[8] Hoorn, J. et al., International Journal of Chemical Reactor Engineering, 2005, 3, 1 – 19

[9] Hermans, I. et al., ChemPhysChem, 2007, 8, 2678 – 2688

# 490860 Multiscale Molecular Dynamics Simulations of Asphaltenes in Crude Oils Based on the SAFT-γ Mie Force Field

**Abstract:**

The modelling of crude oil behaviour is particularly challenging due to inherent uncertainty concerning the chemical characterization of the system, and also due to the diverse length and time scales involved. This work focuses on the particular problem of the characterization of the asphaltene aggregation process. Here we describe a framework that spans multiple time, size and complexity scales to provide a richer description of the problem, ending with a large scale Molecular Dynamics simulation of model asphaltene molecules in a synthetic discrete crude oil at conditions which mimic oil reservoirs.

At the bottom of the size/complexity scale is an experimentally-based description of plausible molecular structures of asphaltenes and resins; the Quantitative Molecular Representation (QMR) approach [1] was used to generate structures based on experimental data from crude oils. These structures were optimised to select a small subset of molecules that give the best match with the experimental data input, consisting of elemental analysis parameters, and both 1H and 13C NMR-spectroscopy parameters.

The behaviour of the model QMR-generated asphaltene structures in heptane, toluene and heptol solvents was then determined using a fully atomistic description employing the OPLS-AA force field in classical molecular dynamics (MD) simulations. The simulations shed light on the incipient clustering mechanisms and the size and distribution of asphaltenes in simple solvents. While simulations at this level of detail are enlightening, it has been shown [2] that no level of current computational resources is sufficient to cover the real times required to represent the clustering process, hence the need to recourse to a lower level of fidelity.

At the next level of scale, coarse-grained (CG) models were built, in which beads representing multiple atoms were used, significantly reducing the number of particles and calculations involved in the simulation. We used here the SAFT-γ Mie approach [3,4], where the parameters describing the intermolecular forces are obtained from fitting the properties of molecules and molecular segments to macroscopic thermophysical properties, and produced CG models of the asphaltenes, resins and different components (aromatics, waxes, non-condensables, etc.) of a typical live oil. The output of selected atomistic MD simulations was used to validate the robustness of the CG models.

Complex systems composing of realistic mixtures of solvents, resins and multiple asphaltenes in wide ranges of pressures and temperatures were then explored using CG models in large-scale parallel MD simulations, in which more than 120,000 molecules or roughly 2M atoms were involved. Similarly, time scales of fractions of μs were explored. It was observed that only at these larger scales can one fully appreciate the effect of clustering and aggregation in the system. These results suggest that most of the reports in the literature, based on atomistic modelling of systems an order of magnitude smaller, are riddled by system size effects and by observation of un-equilibrated states.

Results suggested that the polydispersity and complexity of the crude helped maintain the larger asphaltene molecules from segregating and precipitating. The effects of the concentrations of aromatics and saturated compounds on asphaltene aggregation were investigated, and the bubble points were determined.

### References

[1] Neurock, M.; Nigam, A.; Trauth, D.; Klein, M. T. (1994). Chem. Eng. Sci., 49, 4153–77 . for a recent application see Boek, E.S., Yakovlev, D.S., Headen, T. (2009). Energy & Fuels 23, 1209-1219.

[2] Headen, T. F. et al. (2017). Energy & Fuels, 31, 1108–1125.

[3] Müller, E. A., Jackson, G. (2014). Annu. Rev. Chem. Biomolec. Eng., 5, 405-427

[4] Herdes, C., Totton, T. S., Müller, E. A. (2015). Fluid Phase Equilibria, *406*, 91–100.

# 491228 Evaluating the Consistency and Accuracy of COSMO-RS Based Free Energy Predictions

**Abstract:**

Since its first publication in 1995, COSMO-RS[1] has found wide-spread use and has developed a reputation[2] as an efficient approach for the qualitative and quantitative prediction of many thermodynamic properties in liquid systems. The commonly predicted properties include activity coefficients, solubilities, phase separation, vapor pressures and two phase partitioning.

In this contribution, we evaluate the accuracy and consistency of COSMO-RS predictions on the basis of two different cases: a recent benchmark [3,4] comparing different methods for predicting the free energy of solvation of simple organic liquids and the 2016 SAMPL[5] blind challenge for water / cyclohexane distribution of drugs.

The presented data show that COSMO-RS provides the most accurate predictions in both cases. A good consistency over different properties and different classes of molecules is also observed as all predictions were conducted with the same parameters and without prior modifications or adjustments.

References:

1) A. Klamt, “Conductor-like screening model for real solvents: a new approach to the quantitative calculation of solvation phenomena,” *The Journal of Physical Chemistry*, vol. 99, no. 7, pp. 2224–2235, 1995.

2) P. Deglmann, A. Schäfer, and C. Lennartz, “Application of quantum calculations in the chemical industry - an overview,” *International Journal of Quantum Chemistry*, vol. 115, no. 3, pp. 107–136, 2015.

3) J. Zhang, D. Tuguldur, D. van der Spoel, “Force Field Benchmark of Organic Liquids. 2. Gibbs Energy of Solvation.” *Journal of Chemical Information and Modeling*, vol. 55, no. 6, pp. 1192-1201, 2015

4) J. Zhang, D. Tuguldur, D. van der Spoel, “Correction to Force Field Benchmark of Organic Liquids. 2. Gibbs Energy of Solvation” *Journal of Chemical Information and Modeling*, vol. 56, no. 5, pp. 819-820, 2016

5) https://drugdesigndata.org/about/sampl

# 495434 Support Effects in Heterogeneous Catalysis Using Au Nanoparticles on Oxides: A DFT Analysis

**Abstract:**

Supported Au nanoparticles have been found to be active for numerous heterogeneously catalyzed reactions [1]. In addition, these catalysts have been found to exhibit a bifunctional behavior, where the combined Au/oxide system works in a synergistic manner, giving lower reaction barriers than those of either the Au or the oxide in isolation [2]. The identity of the support plays a key role in determining the activity of Au/oxide systems, with Au nanoparticles showing different turnover rates when supported on different oxides [2,3]. Although support effects have been observed for individual systems, a fundamental understanding, especially in terms of a common unifying descriptor for the activity of these systems, is poorly understood. In this work, we have studied model Au nanowire supported on a range of oxides â€" MgO, ZnO, $Al_2O_3$, $Fe_2O_3$, $ZrO_2$ and $TiO_2$ - to understand how the catalytic properties at the interface of these systems are influenced by the choice of the oxide support. The geometric effect of strain in these systems was minimized by orienting the Au nanowire in non-epitaxial registries on the oxide supports. Water Gas Shift (WGS) is known to be promoted at the interfaces of supported Au/oxide systems [2,4] and have, therefore, been used as a probe reaction in this study. Rigorous DFT calculations were performed for reactive intermediates usually associated with WGS, at the interfaces of the various model Au/oxide systems. Our results correlate well with the experimental trend in WGS reactivity on these systems. In particular, we found the WGS rates to correlate well with experimental water orders and calculated adsorption energies of dissociated water at the interface of Au/oxide, suggesting dissociated water to be a common intermediate involved in kinetically significant reaction steps on all these systems. This combined theoretical and experimental analysis hints at the possibility of a common mechanistic basis for trends in support effects.

References

[1] Rodriguez, J.A., *Catal. Today*. **2011**, 1, 160

[2] Zhao, Z.-J., Li, Z., Cui, Y., Zhu, H., Schneider, W. F., Delgass, W. N., Ribeiro, F. and Greeley, J., *J. Catal.* **2017**, *345*, 157

[3] Shekhar, M., Wang, J., Lee, W.S., Williams, W.D., Kim, S.M., Stach, E.A., Miller, J.T., Delgass, W.N. and Ribeiro, F.H. JACS 2012, 134, 4700

[4] Williams, W.D., Shekhar, M., Lee, W-S., Kispersky, V., Delgass, W.N., Ribeiro, F.H., Kim, S.M., Stach, E.A., Miller, J.T. and Allard, L.F. JACS, 2010, 132, 14018

# 501423 Coordination-Based Descriptors for Rational Design of Metal Nanocatalysts

**Abstract:**

One major challenge for designing metal nanocatalysts is to know a priori appropriate surface structure for a given catalytic reaction. Understanding the effects of particle morphology and composition on the interaction of surface atoms with adsorbates is of pivotal importance for identifying optimal active sites and engineering nanoparticles with maximized fraction of such sites. Many efforts have been made aiming to predict the binding energy of an adsorbate at metal surfaces using the electronic and/or geometric factors of the adsorption site (termed descriptor).[1] While these models have been successfully used as reactivity descriptors for either extended-metal surfaces or simple metal nanoparticles, their extension to complex particle systems with varying strains and/or metal ligands remains elusive due to a formidable computational cost or the lack of an explicit consideration of interatomic interactions.

In this work, we used the previously developed reactivity descriptor, i.e., the orbital-wise coordination number ($CN^a$), for understanding catalytic effects of size/shape/composition of metal nanocatalysts[2]. The $CN^a$, quantified by interatomic coupling matrix elements between the site of interest and its all neighbors within a certain cutoff radius, provides a robust description of CO, $O_2$, and O adsorption energies on Au metal nanoparticles of varying size and shape attributed to its explicit consideration of broken-bond strains. Furthermore, the $CN^a$ shows promise as a general descriptor for predicting adsorption properties of core-shell alloyed structures of Au nanoparticles with d10 metal ligands (e.g., Cu and Ag). We further tailored the metal ligands interacting with surface Pt atoms in a core-shell type alloy structure for a rapid screening of metal nanoparticle sites that have optimized activity toward $O_2$ reduction. The approach can be readily extended to understand and predict reactivity trends of large and more complex metal catalysts with defects, impurities, transition-metal additions, supports, etc, and thus provides a general basis for rational design of metal nanocatalysts[3]

1. Hammer, B. & NÃ¸rskov, J. K. Electronic factors determining the reactivity of metal surfaces. *Surf. Sci.* **343,** 211â€“220 (1995).

2. Ma, X. & Xin, H. Orbitalwise Coordination Number for Predicting Adsorption Properties of Metal Nanocatalysts. *Phys. Rev. Lett.* **118,** 36101 (2017).

3. Wang, S. & Xin, H. Size Effects of Au Nanocatalysts: The Role of Local Coordination of Surface Atoms. Submitted (2017).

# 501459 Theoretical Investigation of the Ring Opening Mechanism of Cyclohexanes on Ir Surfaces


**Abstract:**

Diesel engines have higher fuel economy and produce less greenhouse gas than equivalent gasoline engines. An important requirement in diesel production is to reduce the amount of aromatic compounds, which have poor ignition quality (low cetane numbers, or CN) and form soot during burning. Traditional hydrocracking produces a large fraction of low-molecular weight products that cannot be used in diesel fuel. A better approach would be to convert aromatics to saturated cyclic compounds, followed by selective ring opening (SRO). Although catalyst support, metal dispersion, and reaction conditions all make a difference, iridium compared to other metals possesses much higher activity for ring-opening vs. cracking, but it tends to favor branched isoparaffins, which have lower CN than linear paraffins [1-4]. The dicarbene mechanism, which cleaves secondary-secondary C-C bonds, is postulated to dominate over those that open rings at substituted carbon positions (e.g., the metallocyclobutane mechanism), but neither the details of these mechanisms nor the reason why the dicarbene mechanism is favored by Ir is fully understood.

The focus of this talk will be on the effects that surface morphologyand substitution have on the ring opening reaction on Ir. We are theoretically investigating the ring opening mechanism of cyclohexane and substituted cyclohexanes on the close-packed Ir(111) and stepped Ir(211) surfaces as a basis for understanding the fundamental ability of Ir to promote SRO. We perform detailed density functional theory calculations to identify the more facile reaction channels for each reaction intermediate in order to construct the minimum-energy reaction pathway for the ring opening of a given cycloalkane species. We find that in the low coverage limit, repeated dehydrogenation of cyclohexane at positions closer to the surface is kinetically highly facile, so that neither cyclohexene nor benzene is a major product. Instead, C-C cleavage ensues after 4~5 sequential dehydrogenation steps depending on the surface, producing a metallocycle. This stands in contrast to the decomposition of cyclohexane on Pt, where benzene is a major product [5,6]. The reaction intermediates generally adsorb more strongly on Ir(211) than on Ir(111), but the reaction activation barriers are not always lower on the step edge. Our findings provide clues for enhancing the ability of Ir catalysts to more selectively cleave substituted C-C bonds, thereby increasing the CN of the resultant hydrocarbons.



[1] G. B. McVicker et al., *J. Catal.* **2002**, *210*, 137.

[2] P. T. Do, W. E. Alvarez, D. E. Resasco, *J. Catal.* **2006**, *238*, 477.

[3] H. Shi, X. Li, G. L. Haller, O. Y. Gutiérrez, J. A. Lercher, *J. Catal.* **2013**, *295*, 133; H. Shi, O. Y. Gutiérrez, G. L. Haller, D. Mei, R. Rousseau, J. A. Lercher, *J. Catal.* **2013**, *297*, 70.

[4] D. W. Flaherty, A. Uzun, E. Iglesia, *J. Phys. Chem. C* **2015**, *119*, 2597.

[5] J. A. Rodriguez, C. T. Campbell, *J. Phys. Chem.* **1989**, *93*, 826.

[6] D. H. Parker, C. L. Pettiette-Hall, Y. Li, R. T. McIver, Jr., J. C. Hemminger, *J. Phys. Chem.* **1992**, *96*, 1888.

# 498309 $CO_2$ Reduction on the Nickel Surface

**Abstract:**

Nickel is a typical catalyst for carbon dioxide reforming. In this study, we investigate the mechanism of reducing $CO_2$ on the Ni(110) surface in the presence of the subsurface or impinging hydrogen using spin-polarized periodic density functional theory and ab initio molecular dynamics simulations. In the case of the subsurface H atoms, they act as both spectator species and reactants of the reactions and hence provide the extra energy for overcoming the reaction barriers, in which the major contribution comes from the energetics of subsurface H emerging to the surface as a reactant. While in the case of the impinging hydrogen, the simulation results show direct theoretical evidence for both associative and redox mechanisms in the reaction of atomic hydrogen with $CO_2$. Because $H_2$ is dissociatively chemisorbed on Ni(110) with nearly unit probability, the mechanisms we find are also relevant to the reverse water-gas shift reaction ($H_2$ with adsorbed $CO_2$). Furthermore, we provide the first real-time demonstration of both Eley-Rideal and hot atom mechanisms when H impinges on adsorbed $CO_2$.

# 497771 Designing Cu-Based Bimetallic Nanoparticles for $CO_2$ Activation

**Abstract:**

Copper-based catalysts are good candidates for the conversion of carbon dioxide into useful chemicals. Using Density Functional Theory calculations, we investigate the adsorption and activation of carbon dioxide on CuM (M = Au, Cu, Mn, Mo, Ni, Pd, Rh, Ru, Sc, V, Zn, and Zr) nanoparticles. Candidate heteroatoms (M) for $CO_2$ adsorption were selected based on the heteroatom preference to occupy a surface site of the nanoparticle. We identified two significant descriptors for carbon dioxide chemisorption on the nanoparticles: (1) the heteroatom local d-band center must be positioned higher in energy than the LUMO of carbon dioxide, and (2) the electropositivity of the heteroatom must be higher compared to Cu. Both descriptors lead to an effective charge transfer from the nanoparticle to $CO_2$. Using these criteria, bimetallic nanoparticles can be screened rapidly for the ability to chemisorb (and activate) $CO_2$, which is an important step for the subsequent catalytic conversion of $CO_2$. Our work highlights the importance of generating binding sites on a nanoparticle surface based on stability and electronic structure properties, which can lead to designing bifunctional catalysts.

# 497290 Understanding Heterogeneous Catalyst Deactivation By Biogenic Impurities on Ni (111) Surface and Bimetallic Alloy

**Abstract:**

Interaction of biogenic impurities with the Ni(111) catalyst surface was studied to understand the important role played by them in inhibiting catalysis. Non-bonding interactions of amino acids (Alanine (Ala), Cysteine (Cys), Methionine
(Met), Tryptophan (Trp), Histidine (His), Lysine(Lys), Glutamic acid (Glu) and Threonine (Thr)) in aqueous enivronment were examined using classical molecular dynamics (MD) simulations. The potential of mean force (PMF) profiles of amino acids revealed qualitative differences, resulting into altered orientation and preferential interacting site with the metal surface. Most of the preferential interacting sites were observed to be in direct contact with the surface except Lys which interacted with the backbone nitrogen oriented away from the surface. The strength of nonbonded
interactions varied from -13 to -71 kJ/mol with Ala as the weakest and Trp as the strongest adsorber. The interaction energies scaled linearly with the solvent accessible surface area of the amino acids. DFT calculations showed strong binding for both Cys (-123 kJ/mol) and Met (-102 kJ/mol), undergoing dissociation to form atomic S with an intrinsic activation barrier of 133 kJ/mol and 60 kJ/mol respectively for the C-S bond cleavage on the Ni (111) surface. However, the mechanistic routes for dissociation followed by the two amino acids were a bit different. The mechanistic insights thus obtained lead to an explorative theoretical framework for the design of a bimetallic catalyst which may be used in such an integrated bio and chemo-catalytic process with reduced chances of deactivation. The NiAu alloy surface was calculated to show both reduced Met (-88 kJ/mol) and S binding (132 kJ/mol), as compared to pure Ni surface, indicating promising prospects for the activity of such an alloy catalyst.

# 492352 Determination of Iridium Alloy Properties for Catalytic Applications

**Abstract:**

Alloys can have a number of structures, such as ordered, fully mixed, segregated, shell-core, or near surface. The alloy structure in turn determines the electronic and catalytic properties of the alloy. Iridium-based alloys have shown promise in a number of catalytic applications, including ethanol oxidation[1, 2], oxygen reduction[3], and hydrogen generation[4]. Iridium alloys possess high melting point, high corrosion and oxidation tolerance, as well as excellent mechanical properties. Our goal was to provide models and predictions of realistic iridium alloys, including their stability and electronic properties at various temperatures and synthesis conditions. Identifying how structure affects stability and electronic properties is key to predicting catalyst performance. We used density functional theory (DFT) calculations to model these alloys. Statistical methods, such as the cluster expansion approach, were also combined with DFT to model a large number of possible bimetallic configurations. We studied iridium-platinum alloys as a prototypical bimetallic system. We have identified viable structures of these alloys at various composition levels ($Ir_xPt_{1-x}$; $0 < x < 1$), and shown how these metals prefer to segregate at low temperature. While identification of the most thermodynamically stable alloy 0 K structures can be straight forward, real working catalysts often have not reached equilibrium. Monte Carlo simulations at finite temperature (600 K+) indicate that disordered iridium-platinum structures occur with larger degree of mixing. We have examined how the electronic structure (e.g. density of states) changes at various compositions and structures (e.g. segregated versus mixed), and potential implications of these changes on catalysis. We have also modeled a large number of other iridium alloys with various metals, including Pd, Rh, Sn, Cr, Fe, etc. This systematic study of iridium alloys provides details on which bimetallic systems are stable, their expected electronic properties, and how to realistically model these catalysts from various synthesis conditions. Our results show how theoretical methods can advance alloy development and lead to potential new catalysts.

1. Cao, L.; Sun, G.; Li, H.; Xin, Q., Carbon-supported IrSn catalysts for direct ethanol fuel cell. Fuel Cells Bulletin 2007, 2007 (11), 12-16.

2. Du, W.; Wang, Q.; Saxner, D.; Deskins, N. A.; Su, D.; Krzanowski, J. E.; Frenkel, A. I.; Teng, X., Highly active iridium/iridiumâ€“tin/tin oxide heterogeneous nanoparticles as alternative electrocatalysts for the ethanol oxidation reaction. J Am Chem Soc 2011, 133 (38), 15172-15183.

3. Ioroi, T.; Yasuda, K., Platinum-iridium alloys as oxygen reduction electrocatalysts for polymer electrolyte fuel cells. Journal of the Electrochemical Society 2005, 152 (10), A1917-A1924.

4. Singh, S. K.; Xu, Q., Bimetallic nickel-iridium nanocatalysts for hydrogen generation by decomposition of hydrous hydrazine. Chemical Communications 2010, 46 (35), 6545-6547.

# 489005 Elucidating Molecular Details of Protein Liquid-Liquid Phase Separation By a Coarse-Grained Model

**Abstract:**

Recently, much interest has been generated in the biological phenomenon of of phase-separated assemblies composed of protein and RNA. These assemblies are ubiquitous within the eukaryotic cell, making up structures such as membraneless organelles and stress granules, and have been shown to display liquid-like properties including dripping, rapid internal rearrangement, and relaxation to a spherical shape. Due to the scale of these systems, atomistic simulations would be severely limited in scope, thus motivating the development of appropriate coarse-grained models. We propose a CÉ'-based model in which relative inter-residue interactions are tuned based on their hydrophobicity and parameterized to capture dimensions of intrinsically disordered proteins. We show that our model is able to capture concentration and temperature dependent phase separation between dilute and dense liquid states. The model is further used to obtain thermodynamic phase diagrams for many systems of interest and changes in the phase behavior due to changes in the protein sequence or/and length.

# 489243 Real-Time Electron Dynamics of Large Complex Systems from a Density Functional Tight Binding Approach

**Abstract:**

Light-harvesting systems and plasmonic materials continue to garner significant attention due to their importance in both energy conversion and sensitive detection technologies. However, several challenges exist in improving their performance: these electronic multifunctional systems are complex, coupled, many-body systems with both structural and electronic interactions with their surrounding environments. All of these processes occur at different time and length scales, and span an immense multi-scale space (i.e., chemical interactions within their environment, all in tandem with external light and electric fields). To this end, we have developed a new computational approach based on density functional tight binding (DFTB) theory to directly probe and calculate these complex materials. Our approach significantly differs from conventional DFT methods in that we can directly calculate real-time dynamical effects in the presence of strong, non-perturbative couplings with the surrounding environment. Furthermore, and most importantly, our implementation allows us to calculate the electronic and structural properties of large systems (~10,000 atoms), whereas conventional DFT approaches are typically limited to only hundreds of atoms. Using this new DFTB-based capability, I will highlight our recent work on photo-initiated dynamics in complex heterogeneous solvated systems[1] and large plasmonic nanoantennas. By treating these large systems at a quantum-mechanical level of detail, we show that the electron dynamics in these chemical systems are surprisingly rich and complex. Most importantly, these time-domain studies provide an intuitive approach to probe the microscopic details of time-dependent electron dynamics for understanding and controlling these complex chemical systems.

[1]M. B. Oviedo and B. M. Wong, *â€œReal-Time Quantum Dynamics Reveals Complex, Many-Body Interactions in Solvated Nanodroplets.â€* Journal of Chemical Theory and Computation, **12**, 1862 (2016).

# 499569 Computational Investigation of Ionic Liquids Nanostructure Formation at a Mesoscale

**Abstract:**

Ionic Liquids (ILs) are a new class of materials with great potential to impact many areas in industry and science. In particular, ILs are good candidates for battery electrolytes due to low volatility, moderate reactivity, low flammability, and a wider liquid range than most organic solvents, thus, enabling safe and high-energy batteries. ILs are characterized with a number of unique properties, and one of the main feature for all ILs is a self-assembling of ions into nanostructures. Existence of such solvent structure is a cause for a strong interest to ILs. Computer modeling and simulation of the ionic liquid systems with the molecular dynamics (MD) approach yields an accurate prediction of the systemsâ€™ structure, dynamics, and thermodynamic properties. However, self-assembling of IL occurs at larger than atomic scale (at meso- to macroscale) and MD application to these systems is prohibited due to extremely high computational cost. In this work a new Probability Distribution Function Coarse Grain (PDF-CG) method has been used for accurate prediction of ionic liquids nanostructure formation together with accurate prediction of thermodynamic, dynamic, and transport properties. For the first time, formation of nanostructures in bulk ionic liquid is observed at mesoscale level using computational approach. Application of PDF-CG method to ILs allows to establish structure-property relationships critical for a complete ILs characterization.

# 494272 Ultra-Coarse-Grained Modeling of ATP Hydrolysis in an Actin Filament

**Abstract:**

Actin is a critical component of the cellular cytoskeleton in which it forms polar filaments that are dynamic in nature. The actin sub-unit in a filament is bound to the nucleotide ATP that hydrolyzes as the filament ages. Aging of the filament, and hence ATP hydrolysis, modulates the long length scale physical network properties and dynamics critical to the function of the cytoskeleton. ATP hydrolysis occurs at a timescale of a few seconds, when ATP is converted to an intermediate ADP with a bound inorganic phosphate. The release of inorganic phosphate is even slower, and occurs at a timescale of a few minutes. These timescales are too slow to be studied using atomistic simulations.

We present a systematic multi-scale ultra-coarse-graining (UCG) approach that provides a computationally efficient way to simulate a long actin filament undergoing ATP hydrolysis, while rigorously taking into account relevant atomistic details. A systematic procedure is used to construct a heterogeneous elastic network model for the filament including the implicit representation of sub-unit specific states of the bound nucleotide, using information from underlying atomistic simulations of short actin filaments. The elastic network is enforced to be short-ranged by restricting the harmonic bonds between coarse-grained sites to two neighboring actin sub-units. The slower conformational and chemical changes associated with the hydrolysis of ATP and the phosphate release are simulated with the UCG modeling approach by assigning internal states to the coarse-grained sites. Transitions between the three states, each represented by a unique potential surface of a local heterogeneous elastic network, are simulated via stochastic transitions coupled to the positions of the coarse-grained sites, and hence to the conformations of the underlying reference atomistic system. The UCG model guarantees detailed balance and is tuned to reproduce experimental reaction rates. The model accurately reproduces long length scale properties such as persistence length of the filament in spite of the local nature of interactions, and is used to systematically study the mechanism of ATP hydrolysis cooperativty in the filament.

# 494756 Validation of Biomolecular Force Fields Regarding Structural and Thermodynamic Properties of Cyclodextrins and Their Complexes

**Abstract:**

Cyclodextrins, CyDs, are macrocyclic oligosugars most commonly composed of 6, 7, or 8 glucosidic units bearing the names Î±-, Î$^2$-, and Î$^3$-CyD. Other, usually smaller, molecules (called guests) can enter their cavity forming inclusion complexes with these hosts. As a result of this complexation CyDs are widely used in many industrial products, technologies and analytical methods. [1]

CyDs have also been extensively studied by means of molecular dynamics simulations, concentrating on structural properties and / or free energies of complex formation. In the present work CyDs as well as their complexes with small organic molecules, i.e. alcohols and ketones, are studied with different biomolecular force fields in aqueous solution [2]. The evaluation is based on structural properties from crystallographic X-ray and solution NMR data as well as on experimental binding constants. Two approaches were employed to compute standard binding free enthalpies, the alchemical double decoupling method and the potential of mean force method, respectively [3]. Based on this analysis recommendations for biomolecular force fields and simulation protocols are provided that allow for a robust and efficient calculation of binding affinities.

[1] H. Dodziuk, Rigidity versus flexibility. A review of experimental and theoretical studies pertaining to the cyclodextrin nonrigidity, J. Mol. Struct., 614 (2002), pp 33-45.

[2] J. Gebhardt, N. Hansen, Calculation of binding affinities for linear alcohols to Î±-cyclodextrin by twin-system enveloping distribution sampling simulations, Fluid Phase Equilib., 422 (2016), pp 1-17.

[3] Y. Deng, B. Roux, Computations of standard binding free energies with molecular dynamics simulations, J. Phys. Chem. B 113 (2009) 2234-2246.

# 490924 Optimal Probes: A Machine Learning Platform for Design of Experimental Probes for Protein Dynamics

**Abstract:**

In this talk, we present our recent work focused on filling the void that exists in the study of membrane protein dynamics, by a fruitful combination of computation and experiment. Despite the ubiquitousness and the diversity in their function in the physical biology of the cell, study of membrane proteins has been limited owing to huge sizes and the difficulty to probe them in their native environment within the lipid bilayer. Membrane proteins play crucial roles in signal transduction, variety of substrate transport, enzymatic activity, immune response and are popular drug targets. Few crystal structures are available limiting the structural information, is mostly restricted to only few functionally diverse intermediates and metastable state that the protein adopts. We show how machine learning can be used to combine computational and biophysical experimental methodologies for studying membrane proteins, with the vision that in such a scenario the result is often greater than the sum of the parts. First, we present algorithms that use molecular dynamics (MD) simulation data, to predict the optimal reaction coordinates of an experimental technique, in order to perform experiments that are optimized to study membrane protein conformation dynamics. When there are multiple experiments on a target protein, it is useful if the they provide overlapping yet, distinct insights into it's dynamics. Second, we determine which experimental dataset adds orthogonal information to what the theoretical/experimental studies already provide and update the model accordingly. Finally, we demonstrate a cloud-based GUI platform, which would act as a bridge between the available experimental and theoretical information; and provide the most appropriate future set of studies for a given protein of interest. The cloud platform is based on National Data Service's BROWN DOG methodology, which provides the web-based data access proxy (DAP) for file conversions and Data tilling service (Clowder) for metadata extraction. These services and tools aid in the curation, accessing and indexing of data as well as the software that might be leveraged for this purpose. Researchers would be able to use this resource to aid in the design of experiments in their studies.

# 501219 Deep Learning and Atomistic Simulations in High-Throughput Material Discovery


**Abstract:**

We applied molecular deep learning cheminformatics technology in virtual screening of thousands of candidates for Organic Photovoltaic (OPV) material and processing conditions. We built a predictive model using virtual screening. We have first collected and combined experimental datasets from literature sources. From these datasets, we extracted chemical species and processing conditions and used them to seed exploration of the parameter space. We then identified the most important chemical and experimental condition descriptors to lead fundamental research toward the engineering of molecular-scale features that contribute to higher Power Conversion Efficiencies (PCE) in OPVs. By examining the feature space of the model, we then determined which inputs lead to the largest changes in performance metrics. This feature space might include the chemical identities of the active molecules (i.e. donor and acceptor), processing method (e.g. spin-coating, inkjet printing, roll-to-roll printing), and processing conditions (e.g. solvent identity, annealing temperature and time, additive content). We also generated thousands of hypothetical acceptors and donors and screened them using our model. We performed electronic structure calculations on these hypothetical materials.

In another effort, we combined molecular dynamic simulation (MD), density functional theory (DFT) and deep learning to map the interaction map and bandgap of the solar cell devices. Acceptors and donors interactions at heterojunction interface modulate different bandgaps. The bandgap highly depends on the relative conformation and relative orientation of acceptors and donors. We performed MD simulations to generate thermodynamically stable donor-acceptor couples and computed the bandgaps of each couple using DFT. We used the molecular snapshots as an input and the calculated interaction energy and bandgap as descriptor in our deep learning model. Our model predicts the bandgap and the interaction energy with high accuracy.

# 499852 Advancing Molecular Simulation Methods with Machine Learning

**Abstract:**

In this presentation, we will demonstrate how modern data science techniques can be used to advance molecular modeling and simulation methods, how the results of traditional methods can be calibrated, and how we can devise data-derived models that are predictive and that can replace simulations at a fraction of the computational cost.

# 495084 Saddle Point Searches in Electrochemical Reactions

**Abstract:**

Electrochemical reactions have gained considerable interest as potential solutions for renewable fuels and energy storage. An accurate computational model of electrochemical systems remains challenging, due in part to the presence of a solvent and the need for charged interfaces. Moreover, electrode potentials are governed by an external circuit, which keeps the electrochemical potential of the electrode constant throughout the reaction. Conventional density functional theory (DFT) is carried out within a canonical framework, where the number of electrons is fixed. This constant-charge approach results in a potential bias in the course of an electrochemical reaction since the surface charge of the electrode is depleted [1], and practical models to overcome this constant-electron situation have no obvious energy-referencing scheme.

The nudged elastic band (NEB) method has proven useful in evaluating minimum energy pathways (MEP) in chemical reactions. One can devise methods that continuously vary the number of electrons or the local field in the unit cell to keep the chemical potential of the electrode constant. This presents another difficulty: on-site energies of systems with different numbers of electrons cannot be directly compared since the energy reference varies between images. To circumvent this issue, we introduce a force integration method within the NEB method. This method relies solely on canonical force components parallel to the MEP in each image in the band. As the resolution of images increases, the potential energy surface approaches that of a grand-canonical system, and the path-independence of the potential energy means that the interpolated images need not lie on the minimum energy pathway. We employ this method to study the kinetics of the proton discharge step of the hydrogen evolution reaction.

[1] Skúlason, E.; Tripkovic, V.; Björketun, M. E.; Gudmundsdóttir, S.; Karlberg, G.; Rossmeisl, J.; Bligaard, T.; Jónsson, H.; Nørskov, J. K. J. Phys. Chem. C 2010, 114 (42), 18182–18197.

# 495535 Enhanced Activity for Oxygen Reduction Reaction By Gold at Step/Edge Sites of Ni@Aupt Core-Shell Nanoparticles: A DFT Investigation

**Abstract:**

**Enhanced Activity for Oxygen Reduction Reaction by Gold at Step/Edge Sites of Ni@AuPt Core-Shell Nanoparticles: A DFT Investigation**

Wei An*, Hao Wang

*College of Chemistry and Chemical Engineering, Shanghai University of Engineering Science, Songjiang District, Shanghai 201620, China.*

*E-mail: weian@sues.edu.cn

**ABSTRACT**

Modifying Ni@Pt nanoparticle (NP) with Au has been proved to be an efficient strategy to enhance not only the activity of Ni@Pt nanoparticles for oxygen reduction reaction (ORR) demonstrated by previous theoretical study,[1] but also the stability of cathode from experimental observations.[2-3] Still, there is an urgent need for understanding the ORR behaviors at step/edge sites of a NP considering the large portion of surface Pt atoms consumed by step/edge sites for the closure of a NP buildup besides the key roles of step/edge ¡®defect¡¯ sites that play in promoting a catalytic reaction.

In view of the remarkable effect of Au, we investigate ORR at step/edge sites of Ni@AuPt NPs employing periodic spin-polarized density functional theory (DFT) method. We built high-index (221) facets as model catalysts and found that the ORR activity can be greatly enhanced by replacing the surface Pt atoms at edge site with Au atoms. Our results show that ORR activity at *edge* site is much higher than that at *step* site. Inert Au atoms localized at the surface edge sites of Pt-based NPs can effectively promote the elementary steps of ORR. At working potential (U=0.8V), the rate-determining step of ORR at step/edges sites of Ni@AuPt core-shell NPs is likely to be $O_2$ protonation to *OOH, compared with *O protonation to *OH on Pt NPs. Furthermore, the BEP relations are held for $O_2$ dissociation in the form of the linear correlation between the transition-state and final-state energies on potential energy surface, and are held for *O protonation in the form of the linear correlation between the activation barrier and reaction energy in which the rate of *O protonation is increased as the reaction becomes more exothermic. We identified that the surface activity of Pt atoms of core-shell NPs can be well explained by either d-band model or coordination-number model, which have the same origin of chemical nature. Our study provides fundamental understanding of functionality of step/edge sites of Pt-based core-shell NPs for ORR that can be used for engineering of practical cathode catalyst with

atomic efficiency and precision by maximizing occupancy of gold atoms at edge sites.

**REFERENCES**

1. An, W.; Liu, P. Rationalization of Au Concentration and Distribution in AuNi@Pt Core-Shell Nanoparticles for Oxygen Reduction Reaction. *ACS Catal.* **2015,** *5*, 6328-6336.

2. Zhang, J.; Sasaki, K.; Sutter, E.; Adzic, R. R. Stabilization of platinum oxygen-reduction electrocatalysts using gold clusters. *Science* **2007,** *315*, 220-222.

3. Gatalo, M.; Jovanovic, P.; Polymeros, G.; Grote, J. P.; Pavlišic, A.; Ruiz-Zepeda, F.; Šelih, V. S.; Šala, M.; Hocevar, S.; Bele, M.; Mayrhofer, K. J. J.; Hodnik, N.; Gaberšcek, M. Positive Effect of Surface Doping with Au on the Stability of Pt-Based Electrocatalysts. *ACS Catal.* **2016,** *6*, 1630-1634.

# 492128 Ammonia Synthesis Using Plasma Assisted Catalysis: Understanding Rate Enhancements By Excited Species

**Abstract:**

The industrial Haber-Bosch process for catalytic ammonia synthesis from nitrogen and hydrogen is carried out at high pressures (100-200 bar) and temperatures (400-500 °C). The robust N-N triple bond is the fundamental barrier for this reaction, and there has been a longstanding desire to develop ways to activate this bond under less extreme conditions. A few reports in the literature [1-4], as well as recent experiments performed by us, have demonstrated that is possible to produce ammonia at atmospheric pressure and temperatures between 100-300°C by coupling the catalyst with a non-thermal plasma. It is speculated that the plasma assists in activating the source gas by generating reactive species, such as vibrationally or electronically excited N2*, ions, and radicals, which interact on the catalyst surface to produce NH3. In this talk, we explore the limits of ammonia synthesis rate enhancements on transition metal surfaces by plasma activated intermediates using a simple microkinetic model based on density functional theory calculations. Our model predicts that plasma-assisted catalysis could lead to low temperature and pressure reaction rates that are competitive with the Haber-Bosch process, but the optimal metal catalysts and active-site motifs are distinct from those predicted to be most active for thermal catalysis. The models are useful for interpreting literature results and guiding new experiments.

References

1. Peng, P., Li, Y., Cheng, Y., Deng, S., Chen, P., & Ruan, R. Plasma Chemistry and Plasma Processing, 36(5), 1201â €“1210 (2016).

2. Xie, D., Sun, Y., Zhu, T., Fan, X., Hong, X., & Yang, W. RSC Advances, 6(107), 105338â€“105346 (2016).

3. Kim, H., Teramoto, Y., Ogata, A., Takagi, H., & Nanba, T. Plasma Processes and Polymers, Published online: DOI:10.1002/ppap.201600157 (2016).

4. Akay, G., and Zhang, K. Industrial & Engineering Chemistry Research, 56(2), 457â€“468 (2017).

# 504649 Systematic Multiscale Models and Physics Using the Relative Entropy

**Abstract:**

Many molecular process involve ranges of length and time scales that cannot be tackled with atomic-scale simulations. Instead, coarse-grained molecular models are necessary to interrogate such systems from a theoretical point of view. We discuss a fundamental approach to the identification of such models and more generally of emergent physical behavior from many-body systems. We proposed that a quantity called the relative entropy measures the information lost upon coarse graining and is the natural thermodynamic metric for such tasks: its minimization provides a universal variational principle for coarse-graining. This broad statistical-mechanical framework can improve and even detect both analytical and simulation models of complex systems. We discuss conceptual and numerical aspects of this approach and illustrate its use in several case studies, including our recent efforts that create greatly improved coarse models with unusual interaction potentials.

# 504650 Development of a Top-Down Coarse-Grained Model for Protein Assemblies

**Abstract:**

In this talk, I will describe ongoing efforts in my group aimed at developing an accurate simulation model to study the thermodynamics and kinetics of multiprotein assembly. As opposed to using a "bottom-up" coarse-graining approach such as the relative entropy framework of Shell and co-workers, we use a "top-down" approach for constructing a Calpha-based (one interaction site per amino acid) protein model. Development of the proposed model involves comparisons with experimental data available from the recent literature as well as comparisons with atomistic simulations of a single protein chain. The usefulness of our approach will be demonstrated by discussing results on a couple of systems. Of particular importance is the formation of liquid-like assemblies of disordered proteins that have been found to be important for the physiological function of membraneless compartments in living cells including the nucleolus and ribonucleoprotein (RNP) granules as well as many organelles in prokaryotic cells.

# 504651 Accessing the Inaccessible: Studying the Liquid-to-Solid Transition in Molecular Simulations


**Abstract:**

Nucleation â€“ that is the onset of a new phase from a metastable phase â€“ is a difficult phenomenon to study both experimentally and computationally. Nucleation relevant spatial and temporal resolution is difficult to access in experiments. In contrast, while computer simulations are perfectly suited to probe these length and timescales, sampling statistical number of nucleation events is computationally expensive. In our research, we use advanced sampling techniques in conjunction with molecular dynamics simulations to sample hundreds and thousands of nucleation events. We apply these methods to study liquid-to-solid phase transitions in Lennard-Jones systems, water and aqueous solutions. In my talk, I will discuss key results obtained for homogeneous nucleation in Lennard-Jones-like liquids and in aqueous solutions. The latter specifically focuses on nucleation of clathrate hydrates of water soluble guest molecules. I will also discuss the limitations of the current methodologies and our approaches to improve the efficiency of the advanced sampling techniques. We have developed novel sampling methods that have the potential to be more efficient in sampling rare events thereby, making studies of phenomena such as nucleation more viable and accessible.

# 504652 A Brief History of Exploring Hypothetical Crystal Structures

**Abstract:**

The ability to design a crystal, both in terms of its structure and function, is the defining quest of the field of 'crystal engineering.' Despite impressive progress over the past several decades, particularly among the class of crystals known as metal-organic frameworks (MOFs), it is by no means a solved problem. There are many distinct aspects to this challenge, one of which relates to the vastness of the sea of possible crystal structures that could plausibly be synthesized. Namely, if one wants to design "the best" crystal for a target application, how does one find it within that vast and deep sea? An even simpler question is troublingly difficult to address, just how vast is this sea exactly? Does the sea have a well-defined basin and coastline whose shape we are able to determine? Exploring the space of possible, or hypothetical, crystals is a task that many research groups have engaged in, each contributing their own insights toward these broad questions. In this talk I will (try to) briefly summarize the large body of past work on hypothetical crystal structure generation, along with my own drop in that ocean.

# 504653 Molecular Modeling of Adsorption of $CO_2$ and Water in Hydrophobic Metal-Organic Frameworks

**Abstract:**

Materials having hydrophobic nanopores are potentially useful for selectively adsorbing molecules of interest out of aqueous solution or from the humid atmosphere. This talk will focus on recent molecular-level modeling of $CO_2$ and water in hydrophobic metal-organic frameworks (MOFs). First, we focus on adsorption of $CO_2$, water, and their mixtures in the flexible, hydrophobic MOF known as FMOF-1. A combination of simulation and various experimental measurements provide a unified picture of unusual isotherm shapes in this material, resulting from framework flexibility and progressive filling of adsorption sites with increasing pressure. We also report a case study of the difficulties in simulating adsorption isotherms for water in ZIF-8, another hydrophobic MOF, where the simulations take a very long time to equilibrate. Based on insights obtained about the adsorption mechanism, we investigated several advanced Monte Carlo algorithms including energy-bias moves and continuous fractional component Monte Carlo (CFC MC) and were able to accelerate the simulation speed by a factor of 6.7 over the conventional grand canonical Monte Carlo algorithm. The insights obtained from this work may also help improve the molecular simulation efficiency for studies of water adsorption in other hydrophobic materials.

# 489803 Progress Towards Ultra-Fast Screening of Porous Sorbents for Chemical Separations

**Abstract:**

Crystalline nanoporous materials such as zeolites and metal-organic frameworks have potential applications in a wide range of chemical separations, including adsorption-based and membrane-based processes. We will discuss progress in methods to screen large libraries of materials for chemical separations. A particular problem of interest is whether computational methods can be used to efficiently characterize not only a large number of adsorbent materials but also large collections of potential adsorbing molecules. Initial results from considering this question allow us to discuss whether â€œprivilegedâ€ adsorbents exist that have high performance for a large number of distinct chemical separations.

# 498533 Identifying New Descriptors for Gas Storage in Nanoporous Materials

**Abstract:**

The low volumetric density of hydrogen and methane is a major limitation to their use as low-emission transportation fuels. Filling a fuel tank with nanoporous materials, such as metal-organic frameworks (MOFs), could greatly improve the deliverable capacity of these tanks if appropriate materials could be found. However, the large number of possible structures makes it challenging to select the best MOF for a given storage application. We have developed new descriptors for gas adsorption in MOFs to characterize the strength of MOF-guest interactions to rapidly predict the storage capacity and efficiently filter large materials databases. By applying this method to a series of simple guest molecules, we will comment on the method's generalizability and applications to identify key structure/property relationships in materials design.

# 499869 Discovery of High-Performing MOFs Via Machine Learning

**Abstract:**

We describe a data-driven approach to developing metal-organic frameworks (MOFs) with high hydrogen storage capacities. MOFs are crystalline nanoporous materials containing a metal cluster bonded to organic linkers. High-throughput (HT) grand canonical Monte Carlo simulations and a Voronoi networks-based technique are used to compute the capacities and properties of nearly 500,000 compounds. The large properties-performance database resulting from these calculations presents a unique opportunity for machine learning (ML). Using ML, we predict the necessary structural properties of a MOF needed to achieve specific capacity targets. We refer to this approach as â€˜reverse crystal engineering.â€™ This method illustrates the possibility of developing *purpose-built* materials with specified functionalities.

# 495533 Finding Truth in Fiction: Efficiently Discovering Physical Structure-Property Relationships By Screening Unphysical Porous Materials

**Abstract:**

The range of physically accessible physical adsorption (i..e, physisorption) behavior in porous materials is not yet well understood. This lack of understanding has implications on our knowledge of the theoretical efficiency limits on important adsorption-related processes, such as post-combustion carbon capture. In this talk we will use classical molecular modeling to simulate adsorption of various gases in unphysical "pseudomaterials," which are just random periodic arrangements Lennard-Jones (LJ) sphere systems. Importantly, within the class of pseudomaterials are all of the real materials. Therefore, by determining the limits of adsorption in pseudomaterials we necessarily also determine the limiting behavior of real materials. In this talk we will describe the methodology developed to explore this high-dimensional phase space and some results.

# 495326 Mapping Transition Metal Chemical Space for Machine Learning Models

**Abstract:**

The unique and tunable electronic properties of transition metal complexes make them ideal targets for molecular design. However, the high complexity and dimensionality of transition metal chemical space both presents challenges for and necessitates new approaches to virtual screening. Data-driven models from machine learning can circumvent the high computational cost of first-principles simulation. However, predictive machine learning models require knowledge of how to optimally map heuristic chemical and topological properties of transition metal complexes to energetic outputs. We have recently trained the first artificial neural network (ANN) to predict electronic structure properties of transition metal complexes to 3 kcal/mol accuracy, outperforming previously developed descriptors (i.e., for organic molecules) by an order of magnitude[1]. We have also implemented this ANN in our virtual high-throughput screening toolkit, molSimplify[2], to enable both prediction of structure and electronic properties prior to first-principles simulation. We will discuss this set and recent modifications to the descriptors through our development of fully continuous variable representations. We will describe how we have used established feature selection techniques from this widened space of candidate descriptors to identify the most important subsets of variables for predictive models. We analyze selected feature sets through how well they resolve differences and similarities among representative transition metal complexes using self-organizing maps and principal component analysis. Finally, we conclude with our approaches for decoding continuous variable representations to lead candidate transition molecules for integration of our predictive machine learning models into multi-level molecular design workflows.

[1]J. P. Janet and H. J. Kulik "Predicting Electronic Structure Properties of Transition Metal Complexes with Neural Networks" arXiv preprint arXiv:1702.05771 (2017)

[2]E. I. Ioannidis, T. Z. H. Gani, and H. J. Kulik "molSimplify: A toolkit for automating discovery in inorganic chemistry" J. Comput. Chem. 37, 2106-2117 (2016).

# 491811 Resolving 3D Structures of Metallic Nanoparticles from X-Ray Absorption Data Using Artificial Neural Network

**Abstract:**

Many applications of metallic clusters and nanoparticles (NPs) rely on the observation that properties (for example, magnetic, optical, chemical, catalytic properties) of such materials are different from those of bulk metals. These unique properties are a result of interplay between NPs core sites and undercoordinated surface sites, the quantum confinement effects that modify the electronic properties of metal atoms, and the interactions between metal sites and support and/or adsorbates. Understanding NPs atomic structure and its relation to the NPs properties is thus important to enable new potential applications of nanostructured materials. More importantly, the atomic structure of NPs can change *in-situ*, e.g., during the chemical reactions, where NPs are used as catalysts.

One of the factors that hinder such understanding is that structural information, provided by experimental techniques, is scarce in case of small (with sizes ca 1 â€“ 2 nm) NPs, especially under *operando* conditions. Conventional crystallographical approaches that commonly are used to learn the structure of bulk materials are not applicable in this case, and the sensitivity of high resolution electron microscopy methods is not sufficient to resolve in details the atomic structure of such small NPs. X-ray absorption spectroscopy (XAS) is one of a few methods that are useful in this case, due to its sensitivity to the chemical state of absorbing atoms and to the types and arrangements of atoms around the absorber. Importantly, XAS method also allows *in-situ* and *in-operando* investigations, thus allowing one to directly correlate NPs structure and properties.

While the extended X-ray absorption fine structure (EXAFS) is widely used in NPs structure studies, much less attention has been paid to the information encoded in X-ray absorption near edge structure (XANES). At the same time, there are number of reasons, why analysis of XANES data can be more beneficial for studies of small NPs. First of all, XANES is less sensitive to disorder effects, which affect severely EXAFS spectra quality and complicate the interpretation of EXAFS data. Secondly, XANES is more sensitive to the 3D geometry of the environment around absorbing atoms. The main challenge that hinders the usage of XANES for quantitative analysis is the lack of methodology that would allow one to extract such 3D structural information from experimental data.

Recent advances in data-enabled discovery science methods in chemical research provide a key to this problem. In this study for the first time we employ machine learning methods and ab-initio XANES calculations to correlate XANES features with the descriptors of 3D local structure of metallic nanoparticles. Here we demonstrate the potentiality of this method on the example of XANES study of small supported platinum NPs. We use theoretical site-specific XANES spectra, calculated by ab-initio codes, to train an artificial neural network (NN). The trained NN is then used to extract structural information from the experimental XANES data, and to determine the size and shape of investigated metallic NPs.

# 494998 Identifying Descriptors for Materials Science Via Genetic Programming: A Case Study for Dielectric Breakdown Strength

**Abstract:**

The development of simple phenomenological theories via descriptor identification makes possible rapid screening large numbers of materials and ultimately facilitates the design of new materials. However, determining how to utilize large material data sets to identify the best descriptors for a given property remains one of the biggest challenges in materials science. In this talk, we demonstrate the use of genetic programming to identify descriptors of dielectric breakdown based on 82 representative crystalline materials. Band gap $E_g$ and phonon cut-off frequency $w_{max}$ were found to be the two most relevant properties, and new classes of phenomenological models featuring functions of $E_g$ and $w_{max}$ were discovered. The genetic programming model was found to outperform other machine learning models in descriptor identification, and we discussed how this approach may be extended to other problems in materials science.

# 490196 Data-Driven Prediction of Materials Properties in an Automated Fashion

**Abstract:**

There is pressing need for the use of rapid and reliable data-driven prediction scheme for materials development and optimization. It can drastically speed up the process of assessing key control variables for materials properties, avoiding the needs of scanning the entire design space with costly experimental measurements and computationally intensive simulations. However, complexity of data generation, model building, and validation procedures for the learned-model approaches could pose as a major obstacle, making them less accessible from the materials science and engineering community. In this work, we showcase the latest development in computerized algorithms for automated generation and ranking of predictive regression models, which is readily available for the design of new chemistry in molecular space. The methodology is demonstrated with large-scale virtual screening of organometallic phosphors, identification of design space for thermally activated delayed fluorescence (TADF) materials, turnover frequency prediction for Ziegler-Natta catalysts, and design of organic electronic compounds with high glass transition temperature ($T_g$). The automated data-driven predictive scheme provides unbiased measures to quickly assess the key design rules for a wide variety of applications, which could significantly lower the barrier towards large-scale virtual screening for developing novel materials solutions.

# 489006 Computer-Aided Design of Novel Materials with Desired Electronic and Physical Properties


**Abstract:**

Historically, materials discovery is driven by a laborious trial-and-error process. The growth of materials databases and emerging informatics approaches finally offer the opportunity to transform this practice into data- and knowledge-driven rational designâ€"accelerating discovery of novel materials exhibiting desired properties.

By using data from the AFLOW repository for high-throughput ab-initio calculations, we have generated Quantitative Materials Structure-Property Relationship (QMSPR) models to predict three critical material properties, namely the metal/insulator classification, bulk modulus, Fermi energy, and band gap energy. To enable these calculation, we have developed novel materials descriptors such as universal property-labelled fragments (PLMF).[1] We have established that the accuracy of predictions obtained with machine learning models approaches that of GGA DFT functionals yet model development requires a minute fraction of computational time as compared to ab initio calculations. Notably, due to the representation of materials with PLMF the QMSPR models are broadly applicable to virtually any stoichiometric inorganic materials. This representation also affords straightforward model interpretation in terms of simple heuristic design rules that could guide rational design of novel materials.

As a proof-of-concept study we used materials informatics approach for solar energy conversion, specifically to find a novel photocathode material for dye-sensitized solar cells (DSSCs). By conducting a virtual screening of 50,000 known inorganic compounds, we have identified lead titanate ($PbTiO_3$), a perovskite, as the most promising photocathode material, far from the traditionally used base elements or crystal structures. Subsequently this finding was confirmed experimentally.



1. O. Isayev*, C. Oses, S. Curtarolo, A. Tropsha. Universal Fragment Descriptors for Predicting Electronic Properties of Inorganic Crystals. *Nature Comm*. 2017, Accepted. Preprint: https://arxiv.org/abs/1608.04782

# 502489 Stability Prediction of Hypervalent Compaounds Based on Data-Centric Modelling

**Abstract:**

Hypervalent iodine compounds have become widely used reagents for the transfer of electrophilic substituents to donors. However, being very reactive by definition, many of these compounds are stable only because of a high barrier protecting them from transformation to chemically inactive forms.

In this study we explore hundreds of reagents of the general type shown in Figure 1 in view of their stability, i.e. the possibility of their use for the transfer of the substituent marked L.

For that matter, a support vector machine (SVM) was trained using simple descriptors based on the molecular structure and the atomic charges observed in the reactive center. The SVM, a popular machine learning tool for yes/no-type classification, one trained properly, was shown to successfully predict (in)stability of these compounds based on the descriptors selected.

In this talk we will focus on the selection of descriptors, the training of the SVM and its application to a large array of know and (still) unknown compounds. Relative to the explicit determination of the kinetic and thermodynamic stability, this approach allows a prediction at a small fraction of the cost since only the molecular structures and the atomic charges need to be available.

We will also show that in machine learning the availability of â€œnegativeâ€ information, i.e. information on non-existing compounds, is essential. The fact that negative information is much more difficult to find, presents a challenge and calls for rethinking of the existing, success-based publication culture.

Figure 1:

# 495439 Human-Interpretable Reaction Informatics

**Abstract:**

Cognitive technologies in the areas dominated by expert knowledge, such as chemical and material sciences, require cheminformatic models that are human-interpretable and can be mapped onto the existing structure of the domain knowledge. Advanced analysis of chemical data associated with molecules and reactions relies on fingerprinting methods that enable direct comparison of chemical entities with each other. We will present a new strategy of reaction fingerprinting that meets requirements of human readability and relates to the chemical intuition better than the strategies explored so far. Its application to different classes of reactions will be discussed.

# 494505 Understanding the Removal of Pharmaceuticals from Water Effluents By Adsorption in Activated Carbons – a Molecular Simulation Approach

**Abstract:**

There is an increased concern in recent years on the removal of pharmaceuticals from water effluents. Many of these compounds are at concentrations of mg/L or below, are hardly biodegradable even at the outlet of wastewater treatment plants and much on their potential impact on environment and human health are still not completely known. Among the main removal technologies reported in the literature, removal by adsorption is one of the most promising techniques because of its convenience when applied in current water treatment processes. Adsorption using activated carbon is ideal for removing organic compounds due to the availability of high surface area and combination of well-developed pore structure and surface functional group properties [1]. Hence, the identification of the mechanisms governing the adsorption phenomena of pharmaceuticals in aqueous solutions is of fundamental importance.

The main objectives of this study was to evaluate, by molecular simulations, the competitive adsorption of pharmaceutical compounds as pollutants contented in water on activated carbons of varied characteristics, and to compare the removal capacities from equilibrium simulations on single component and aqueous mixtures. Molecular simulations using the Grand Canonical Monte Carlo (GCMC) method have been performed in order to understand the interaction of pharmaceutical molecules (ibuprofen, diclofenac, naproxen, paracetamol and amoxicillin, all with similar dimensions, except amoxicillin) on activated carbons in both, pure and aqueous mixtures. "Realistic" nanoporous carbon models based on units of polyaromatic molecules with different number of rings, defects and polar-oxygenated sites have been employed for this purpose [2].

The modeling approach we have used for the activated carbons, inspired by the early work of Gubbins and co-authors (see, for instance, reference [3]) represents a middle ground between the slit pore model and the fully atomistic models of carbon; it is an extension of the model packing idealized structures first developed by Segarra and Glandt (SG) [4]. We have modeled graphene planes as platelets, but including polar groups and defects, not explicitly considered in the SG model, and in a similar fashion of the extended Liu-Monson model [5]. Furthermore, in order to create imperfections on the surface, increased surface roughness were created by deleting up to 10–20% of the atoms in the inner carbon layers. Details on the calculations can be found in [2].

Our methodology allowed us to obtain a successful prediction of the drug adsorption properties. Comparison of the simulated capacities with experimental reported values is highly consistent. The model is sufficiently flexible to tune its parameters in a systematic way and introduce additional features to reflect specific properties of the material of interest, *i.e.*, effects of platelet sizes and geometries, surface area, concentration and nature of surface groups, among others. Results confirm the role of surface sites in the enhancement of water adsorption and reduction of drug adsorption as the oxygenated sites exceeds 10%wt. Reduction of the surface functionality gives a higher adsorption capacity of pharmaceuticals in aqueous mixture, since there is not competition with water molecules.

This work is a further step in order predict macroscopic performance of activated carbons in drug removal applications, providing useful molecular-level insights into the drug/water adsorption mechanism and allowing finding the most favorable material.

References:

1. L. Nielsen, T. J. Bandosz, Analysis of the competitive adsorption of pharmaceuticals on waste

derived materials, Chemical Engineering Journal 287 (2016) 139–147.

2. D. Bahamon, L. Carro, S. Guri, L.F. Vega, Computational study of ibuprofen removal from water by adsorption in realistic activated carbons, Journal of Colloid and Interface Science, 498 (2017), 323–334.

3. J. Pikunic, C. Clinard, N. Cohaut, K.E. Gubbins, J.M. Guet, R.J.M. Pellenq, I. Rannou, J.-N. Rouzaud, Structural modeling of porous carbons: constrained reverse Monte Carlo method, Langmuir 19 (2003) 8565–8582.

4. E.I. Segarra, E.D. Glandt, Model microporous carbons: microstructure, surface polarity and gas adsorption, Chem. Eng. Sci. 49 (1994) 2953–2965.

5. J.C. Liu, P.A. Monson, Monte Carlo simulation study of water simulation study of water adsorption in activated carbon, Ind. Eng. Chem. Res. 45 (2006) 5649–5656.

# 504009 Insights into the Adsorption and Phase Behaviour of Fluids in Nanoporous Materials with Hierarchical Pore Structure:Â Towards an Advanced Textural Characterization

**Abstract:**

In recent years, major synthetic efforts have been focused on the introduction of hierarchical pore structures into many different materials such as zeolites, carbons, silicas, MOFs, hybrid layered and pillared structures. Nanoporous materials which consist of hierarchical pore networks with pore widths spanning the entire micro-mesopore range (i.e. up to 50 nm) allow one to overcome accessibility and diffusion limitations associated with classical microporous materials for catalysis and separations.

In this talk recent developments in the structural characterization of these materials will be described. An accurate textural characterization is crucial not only within the discovery process of novel hierarchically ordered nanoporous materials but also contributes to advance their application in a variety of areas including heterogenous catalysis, separation, battery devices, drug delivery. Among a variety of techniques, gas adsorption is widely applied because it assesses a wide range of pore sizes, spanning the entire micro- and mesopore range. Within the last two decades major progress has been achieved in physical adsorption characterization also because of the development of advanced approaches based on statistical mechanics such as molecular simulation and density functional theory (DFT). This progress, coupled with the availability of high resolution experimental methodologies for the adsorption of various subcritical fluids (also allowing one to couple adsorption techniques with complimentary techniques such as small angle scattering), has led to major advances in the structural characterization of nanoporous materials [1]. However, major challenges still exist concerning an in-*depth* characterization of complex pore networks. Application of characterization techniques to hierarchical materials has revealed new phenomena, such as cavitation processes during desorption, and led to significant refinement of these techniques [2].

[1] M. Thommes, K. Kaneko, A.V. Neimark, J.P. Olivier, F. Rodriguez Reinoso, J. Rouquerol and K.S.W Sing, (IUPAC Technical Report), *Pure Appl. Chem.* **87**, 1051 (2015)

[2] K. A. Cychosz, R. Guillet-Nicolas, J. Garcia-Martinez, J., M. Thommes, M. *Chem. Soc. Rev.* **46**, 389, (2017)

# 504010 Molecular Studies of Supercapacitors: Ionic Liquids Adsorbed into Porous Carbon Electrodes

**Abstract:**

Energy storage continues to attract the attention of the scientific and engineering research communities. Renewable energy sources, such as solar and wind, require so-called grid-based energy storage so that energy generated during periods of high energy production can be made available during periods of low or zero energy production. Mobile energy storage (for everything from mobile phones to automobiles and beyond) is clearly increasing exponentially in importance. Among the many choices for energy storage devices, electrical double layer capacitors (EDLCs), also called supercapacitors, are attracting considerable attention. Supercapacitors store electrical energy via ion electrosorption directly in the EDLs at the electrolyte-electrode interface, suggesting that such liquid-solid interfaces play a dominant role in the underlying energy storage mechanism and the resulting device performance. Because electrical energy in supercapacitors is stored based on physical phenomena rather than chemical reaction (as in batteries), supercapacitors have fast rates of charge/discharge and a virtually limitless number of charge cycles (unlike batteries, which are often limited to $10^4$or less cycles). Much of the goal of supercapacitor research is aimed at increasing the amount of energy stored (energy density is the strong point in favor of batteries), which in turn focuses attention on the electrolyte, the nature of the electrode, and the electrode-electrolyte interactions.

To date, ionic liquids (ILs) have become emerging candidates for electrolytes used in supercapacitors, due to their exceptionally wide electrochemical window, excellent thermal stability, nonvolatility, and relatively inert nature; meanwhile carbons are the most widely used electrode materials in supercapacitors, due to their high specific surface area, good electrical conductivity, chemical stability in a variety of electrolytes, and relatively low cost. To improve the energy density and the transport properties of the charge carriers in supercapacitors, carbons have been developed in diverse forms such as activated carbons, carbon nanotubes (CNTs), onion-like carbons (OLCs), carbode-derived carbons and graphene. Using molecular modeling combined with molecular experimental probes, such as SAXS, SANS, NMR, and AFM, we report on our investigations into the interfacial phenomena occurring between the IL electrolytes and electrodes of varying geometries to understand the energy storage mechanism of supercapacitors that rely on EDLs established at IL-electrode interfaces.

# 504011 Structure of Ice in Confinement; Water in Mesopores

**Abstract:**

We report X â€“rays diffraction studies of water adsorbed in nanoporous activated carbon fibres (ACFs) and CMK-3 and CMK-8 carbon mesopores of different pore sizes. The fibres are built of turbostratic nanoparticles separated by quasi two-dimensional voids, forming narrow slit-shaped pores; CMK-3 and CMK-8 are the reverse carbon replica of silica SBA-15 and KIT-6 porous matrices. In order to determine the structure of water within the pores and its influence on the fibresâ€™ structure, mean interatomic and intermolecular distances have been estimated from the positions of the maxima of the normalized angular distribution functions obtained by X-ray diffraction [1]. We observe significant changes in the interlayer distance of the carbon nanoparticles; the results suggest that very high pressures arise within the pores,of the order a few hundreds MPa, as has been observed in molecular simulations [1,2]. Such results are confirmed by observation of high pressure forms of ice in cylindrical nanocarbons , using neutron and X â €“ray diffraction [3]. For water in multi-walled carbon nanotubes (MWCN), below the pore melting point cubic ice was observed. This structure of ice can be obtained during re-crystallisation from high pressure phases at low temperature for bulk water. For CMK-3 and CMK-8 mesopores we observed the existence of a stacking-disordered ice, $I_{sd}$ [4,5]. This metastable ice is neither cubic nor hexagonal, and is not a simple mixture of the two, but a combination of cubic sequences intertwined with hexagonal sequences, which was identified as having the space group p3m1 [4]. Moreover, the stacking disorder can vary in complexity depending on the way the ice is formed and on the prevailing thermal conditions during this process. An analysis of the kind of $I_{sd}$ formed in CMK-3 and CMK-8 of different pore sizes will be presented. These crystal forms, which occur in bulk water only at temperatures below 180 K in the case of cubic ice, and at pressures of hundreds or thousands of MPa are stabilized by the confinement.

**References**

[1] M.Sliwinska-Bartkowiak, M.Drozdowski, M.Jazdzewska, Y.Long, J.Palmer, K.E.Gubbins, Phys. Chem. Chem. Phys., **14**, 71454, 2012.

[2] Y. Long, J.Palmer, B.Coasne, M.Sliwinska-Bartkowiak, K.E.Gubbins, Phys. Chem. Chem. Phy*s* , **14** 17163, 2011 .

[3] M.Jazdzewska , M. Sliwinska-Bartkowiak, A.I. Beskrovnyy, S.G. Vasilovskiy, S.W. Ting, K.Y. Chan, L.L. Huang, K.E. Gubbins, Phys. Chem. Chem. Phys., **13** ,9008 ,2011

[4] Malkin, T. L., Murray, B. J., Salzmann, C., Molinero, V., Pickering, S. J., & Whale, T. F. Phys. Chem. Chem. Phys., **17(1)**, 60-76, 2014; A.Haji-Akbari, P.G.Debenedetti, PNAS, **112,** 10582, 2015

[5]K.Domin,K.Y.Chan,H.Yung,A.Sterczynska,M.Jurek,K.E.Gubbins, M.Sliwinska-Bartkowiak, J.Chem..Eng. Data, 61, 4252-4260, 2016

# 504012 Towards Understanding the Role of Microstructure in Energetic Material Response: Coarse-Grain Modeling and Simulation

**Abstract:**

Mechanical and thermal loading of energetic material (EM) composites can incite responses over a wide range of spatial and temporal scales due to inherent microscale features. Many energy transfer processes within these materials are atomistically governed, yet the overall material response is manifested at the micro- and mesoscale â€“ scales beyond those amenable to atomistic simulation techniques. Moreover, continuum level approaches rely on field-based formulations that are empirically based, which lack sufficient fidelity to capture microstructure effects.

Particle-based microscale simulation methods that utilize coarse-grain models offer a promising route for extending the attributes of atomistic modeling toward the mesoscale. We have developed such microscale methods based upon the constant-energy dissipative particle dynamics method that includes chemical reactions. Coarse-grain models of EMs have been built using the Multiscale Coarse-Graining approach. Time-resolved depictions of an EM responding to insult has been simulated for a variety of microstructures, including samples with intra- and inter-granular voids of varying shape, size and relative orientation, varying grain shapes and sizes, and polymer binder. A sampling of these results will be presented.

# 504014 Adsorption and Transport in Multiscale Porous Media

**Abstract:**

Hierarchical porous materials such as hierarchical zeolites, which combine several porosity scales, are widely used in industry (adsorption, separation, catalysis) to overcome slow diffusion in microporous solids (< 2 nm) and enhance access to their large surface area. Available modeling approaches for adsorption and transport in such multiscale porous media are limited to empirical parameters which cannot be derived from molecular coefficients. In particular, existing approaches do not offer the ground for a bottom up model of adsorption/transport in multiscale materials as (1) they describe empirically the adsorption/transport interplay and (2) they do not account for the breakdown of hydrodynamics at the nm scale.

In this talk, I will present a multiscale model of adsorption and transport in hierarchical materials obtained by adding mesopores (~ few nm) and macropores (> 10 nm) to existing microporous crystals [1]. I will first show how adsorption, permeance, and transport in such media can be described without having to rely on macroscopic concepts such as hydrodynamics [2,3]. Using fundamental parameters and coefficients available to simple experiments, we will see how transport coefficients can be rigorously obtained from simple models in the framework of Statistical Mechanics. Then, I will present a multiscale model of adsorption and transport in hierarchical materials [4]. This approach consists of upscaling accurate molecular simulations in a lattice model. Thanks to the use of atom-scale simulations, which capture the different adsorption and transport regimes upon varying the temperature, pore size, pressure, etc. this bottom-up model does not rely on hydrodynamics and, hence, does not require assuming a given adsorption or flow type. I will also discuss NMR experimental results on transport in hierarchical zeolites [5].

[1] B. Coasne et al., Langmuir 29, 7864 (2013).

[2] K. Falk et al. Nature Comm. 6, 6949 (2015).

[3] T. Lee et al. Nature Comm. 7, 11890 (2016).

[4] A. Botan et al. Phys. Rev. E 91, 032133 (2015).

[5] A. Galarneau et al., J. Phys. Chem. C 120, 1562 (2016).

# 495752 Developing Multi-Scale Models of Bimetallic Catalysts for the Hydrodeoxygenation of Bio-Oil Compounds

**Abstract:**

With the ever-increasing need to find a sustainable source of renewable energy, bio-oil produced via the fast pyrolysis of biomass is a promising source of liquid fuel; however, the resulting bio-oil contains oxygenated products that contribute to poor fuel quality [1]. Hydrodeoxygenation (HDO) is used to refine the bio-oil by reducing the oxygen content, ideally using a minimal amount of $H_2$ in order to form $H_2O$. Bimetallic catalysts such as Pd on Fe have demonstrated synergistic behavior that contributes to a more cost-effective and longer-lasting catalyst [2-4]. Similarly, Pt/Sn catalysts have also exhibited traits favorable for the HDO of phenolic compounds [5], namely a propensity to promote lower temperature desorption of aromatic compounds like benzene [6]. Identifying the cause of the different synergetic effects of these systems can provide insight into the synthesis of a superior HDO bimetallic catalyst. However, the behavior of bimetallic catalysts is complex and necessitates a thorough understanding of the nanoscale behavior of these systems in order to develop a truly predictive model on relevant time scales. In particular, we present a DFT study that examines the importance of including coverage effects into theoretical models in order to accurately predict experimental behavior.

Quantifying the lateral interactions between adspecies is necessary in the development of truly predictive, theoretical models for bimetallic catalysts, as they will significantly affect the coverage and diffusion of adspecies and therefore the heterogeneous catalytic environment. However, such a characterization of the lateral interactions between adspecies for the highly complex HDO reaction has not yet been attempted. To that end, we have characterized the benzene-benzene lateral interactions on a Pt (111) surface and a $Pt_3Sn$ (111) surface alloy. This coverage-dependent adsorption behavior was incorporated into a kinetic model for each surface and used to simulate the temperature-programmed desorption (TPD) spectra. We benchmarked our simulated spectra by comparing it to experimental TPD [6]. We found that it is imperative to incorporate the effect of surface coverage in these theoretical models if the desorption behavior is to be accurately predicted (see Figure 1). Furthermore, we found that a mean-field model is sufficient in describing the benzene-benzene lateral interactions. Using our models, we were able to interpret the desorption behavior at an atomistic level and provide a deeper level of insight into these systems. While it was previously speculated that the broad desorption peak for benzene on Pt (111) was the result of desorption from two adsorption sites of different binding strengths [6], our results indicate that the behavior is due to benzene desorbing from a single site but experiencing strong lateral interactions on the surface. We also found that the adsorption of benzene was significantly weaker on the $Pt_3Sn$ (111) than on Pt (111), and as such, we were able to assign the monolayer desorption of benzene to a TPD peak previously attributed entirely to multilayer benzene desorption [6]. In this regard, we were also able to attribute the higher temperature desorption peaks observed on $Pt_3Sn$ (111) to benzene desorbing from defects in the alloy surface, where Pt atoms were not substituted for Sn atoms in the alloying process.

Our investigation into modeling the lateral interactions of an aromatic molecule on both the Pt (111) and the $Pt_3Sn$ (111) surfaces warrants the application of our approach onto systems less experimentally characterized, such as Fe surfaces doped with different noble metals.

**Figure 1.** TPD spectra for benzene on Pt (111) simulated with coverage effects (blue) and without coverage effects (light green) compared to experiment (black) [6].

**References**

[1] H. Wang, J. Male, Y. Wang, ACS Catal. 3 (2013) 1047-1070.

[2] Y. Hong, H. Zhang, J. Sun, K.M. Ayman, A.J.R. Hensley, M. Gu, M.H. Engelhard, J.-S. McEwen, Y. Wang, ACS Catal. 4 (2014) 3335-3345.

[3] J. Sun, A.M. Karim, H. Zhang, L. Kovarik, X.S. Li, A.J. Hensley, J.-S. McEwen, Y. Wang, J. Catal. 306 (2013) 47-57.

[4] A.J.R. Hensley, Y. Hong, R. Zhang, H. Zhang, J. Sun, Y. Wang, J.-S. McEwen, ACS Catal. 4 (2014) 3381-3392.

[6] C. Xu, Y.L. Tsai, B.E. Koel, J. Phys. Chem. 98 (1994) 585-593.

# 496761 Mechanistic Insights into Hydroeoxygenation of Phenol on Bimetallic Phosphide Catalyst

**Abstract:**

Bimetallic transition metal catalysts provide immense opportunity to tailor catalytic function via composition modulation to promote desired chemical transformation. Addition of non-metal components such as oxygen, nitrogen, and phosphorous can provide for better charge density distribution to enhance Lewis acidic character. Recent work by Rensel et al.[1] has shown that FeMoP catalyst is highly selective for Câ€"O bond cleavage for aryl ethers. Furthermore, the catalytic function of FeMoP can be tuned by altering the Fe/Mo ratio to improve benzene selectivity upon hydrodeoxygenation of phenol.[2] In this presentation, we will share density functional theory calculations that provide mechanistic insights into hydrodeoxygenation of phenol on $Fe_XMo_{2-X}P$ catalyst. We find that optimal Fe/Mo composition has smallest activation energy barrier for Câ€"O bond cleavage which correlates well with experimental TOFs. Bader charges and density of states analysis provide pointers to the change in the electronic structure of the catalyst upon changing Fe/Mo composition.

# 495987 Catalytic Conversion of Furfural to Methylfuran: Investigating Reaction Mechanisms on Ni and the Effect of Boron Doping on the Activity and Selectivity of the Catalyst

**Abstract:**

Fast pyrolysis of lignocellulosic biomass is emerging as a promising route for the production of liquid transportation fuels. However, the bio-oil produced from pyrolysis has a high oxygen content (~45%) which results in a poor heating value, thus limiting its direct usability as a fuel. The catalytic hydrodeoxygenation (HDO) of bio-oil in the presence of hydrogen is an effective technique for the removal of oxygen groups from bio-oil. Furfural is one of the major bio-oil components and is a potential starting material for the production of compounds like 2-methylfuran (MF), which can be blended into transportation fuels. The challenge is to develop catalysts which can selectively break C – O bonds, while preserving the C – C bonds. In this context, hydrogenolysis of furfural ($C_{formyl}$ – O cleavage) to produce MF is desirable, as the high octane number of MF allows easy blending into gasoline/diesel pools. Vapor phase HDO of furfural over Ni catalysts leads to furfuryl alcohol (via hydrogenation of formyl group) as the dominant product at low temperatures, while furans (via decarbonylation by C – C cleavage) and C4 products (via furan ring opening) dominate at higher temperatures[1-2]. We investigated the potential competitive pathways for the catalytic conversion of furfural on model Ni(111) surfaces using Density Functional Theory. Our calculations show that decarbonylation (C – C cleavage) and ring opening ($C_{ring}$ – O cleavage) of furfural to form furan and ring opened C4 products are kinetically favored pathways on Ni(111) surface, in agreement with experimental studies. Recent studies have suggested that the activity and selectivity of catalyst can be improved by doping boron[3-6]. On Ni(111) surfaces boron binds strongly at the subsurface positions, resulting in a corrugated Ni surface, similar to a stepped surface. The presence of the subsurface, interstitial boron expands the lattice of Ni and also alters its electronic properties by donating electrons. The incorporation of B in the Ni surface was found to increase the C – C cleavage and ring opening barriers and drastically lower the $C_{formyl}$ – O cleavage barriers, relative to the Ni(111) surfaces. We propose the hydrogenolysis of furfuryl alcohol to form MF to be kinetically favorable over C – C cleavage (furan formation) and ring opening (C4 products), leading to significantly enhanced MF selectivity in the vapor phase hydrogenation of furfural on boron doped Ni surfaces.

**References**

1. S. Sitthisa, D. E. Resasco, Catalysis Letters 2011, 141, 784-791.
2. S. Sitthisa, W. An, D. E. Resasco, Journal of Catalysis 2011, 284, 90-101.
3. J. Xu, M. Saeys, Journal of Physical Chemistry C 2009, 113, 4099-4106.
4. T. Q. Thang, A. V. Nguyen, D. C. Huynh, T. H. Pham, S. H. Mushrif, Catalysis Science and Technology 2016, 6, 5871-5883.
5. B. Yang, R. Burch, C. Hardacre, P. Hu and P. Hughes, Journal of Physical Chemistry C, 2014, 118, 3664-3671.

6. J. Xu, L. Chen, K. F. Tan, A. Borgna and M. Saeys, Journal of Catalysis, 2009, 261, 158-165.

# 494725 First-Principles Insights into the Mechanisms and Sites for Base Catalyzed Aldol Condensation and Esterification over Copper

**Abstract:**

Base-catalyzed aldol condensation and esterification are important reactions in the synthesis of large organic molecules from biomass-derived precursors.[1] These molecules can then subsequently be converted to fuels and useful chemicals. Such reactions are typically catalyzed on homogenous catalysts in the presence of a basic promoter or on basic metal oxides, that catalyze C-C and C-O bond formation.[1-3] Recent studies have shown that non-basic materials such as Cu supported on $SiO_2$ can actively catalyze both condensation and esterification, when equilibrated mixtures of propanol, propanal and $H_2$ are reacted.[4] These supported Cu systems result in the formation of 47% 3-pentanone (C-C formation via condensation) as well as 34% propyl propionate (C-O formation via esterification).[4] The results suggest that mixtures of propanol, propanal and $H_2$ readily equilibrated via the formation of surface propoxide intermediates, that can act as a base, thus, catalyzing both condensation and esterification. Ab-initio density functional theoretical calculations were carried along with detailed kinetic analyses to elucidate elementary steps and sites involved in propanal conversion to propanol and the subsequent C-C and C-O bond formation steps. Detailed charge analyses show that alkoxides which readily form on Cu as well as other group 11 metals readily extract electron density from the metal surface and behave as basic intermediates. For the aldol condensation path, the surface bound propoxide abstracts the weakly acidic hydrogen from a bound propanal intermediate to generate an enolate species in a rate-determining step having a Gibbs free energy barrier of 69 kJ/mol. The enolate subsequently undergoes rapid C-C coupling with a second vicinally-bound propanal to generate a Î$^{2}$-alkoxide intemediate. The Î$^{2}$-alkoxide can either eliminate hydrogen to the Cu or react with a surface propanal to form a 2-formyl-3-pentanone intermediate that subsequently decarbonylate to form 3-pentanone. The same alkoxide intermediates can also catalyze esterification. The calculations show that the bound propoxide can readily carry out a nucleophilic attack on the carbonyl group of a surface bound propanal in a rate-determining step to form a hemiacetalate intermediate with a Gibbs free energy barrier of 66 kJ/mol. The hemiacetalate further reacts on Cu to form the propyl propionate. The theoretical results indicate that the rate limiting steps for both C-C and C-O formation essentially involve the same propoxide and propanal surface intermediates, which is fully consistent with experimental findings. We predict similar activation barriers for both C-C and C-O formation with C-O formation being more favorable by 3 kJ/mol, while experiments report that C-C formation is slightly more favorable than C-O formation. In addition, we show that esterification and condensation preferentially proceed over coordinatively-saturated Cu 111 terrace sites. The alkoxide as well as other critical intermediates bind too strongly to the edge and corner sites and as such are blocked and much less active than the terrace sites. The active alkoxide intermediates bind much more weakly to the terrace sites and thus catalyze condensation and esterification. This is is fully consistent with the experimental results which show that larger particles are more active than the smaller Cu particles.

**References**

[1]Chheda, J.N., and Dumesic, J. A. Catal. Today (2007), 123, 59.

[2]Hamilton, C. A., Jackson, S.D., and Kelly, G.J. Appl. Catal. A (2004), 263, 63.

[3]West, R.M., Liu, Z.Y., Peter, M., Gartner, C.A., and Dumesic, J.A. J. Mol. Catal. A: Chem. (2008), 296, 18.

[4]Sad, M.E., Neurock, M., and Iglesia, E. J.Am.Chem.Soc. (2011), 133, 20384.

# 500107 Advancing the Selective Oxidation of Ethylene Glycol Via Combining Novel Catalyst Design and Density Functional Theory (DFT) Calculations

**Abstract:**

Ethylene glycol (EG) is the simplest diol that can be produced from hydration of ethylene oxide or hydrogenolysis of biomass (cellulose) [1]. In recent work, significant research work is focused on EG electroxidation for fuel cells [2]. This status is attributed to the fact that EG has higher boiling point (198 $^{o}$C) making it a safer liquid fuel and has higher energy density compared to ethanol and methanol. However, in order to harvest the total amount of energy during this process, EG is fully converted into $CO_2$ that aggravates global warming. In this context, conversion of EG into value-added products chemically could be more attractive. A number of papers are available on EG oxidation in aqueous alkali medium using a non-hazardous oxidizing agent – oxygen gas [3-6]. However, the data reported on this reaction system are mainly on EG conversion and little is known on the reaction mechanism or the viability of applying different catalysts. Moreover, glycolic acid is usually the main product, while formic acid and $CO_2$ are believed as the side products of over oxidation rather than desirable oxalic acid (dicarboxylic acid) [6]. At the same time, current oxalic acid production suffers from the problems of using corrosive and stoichiometric reagents (nitric acid and sulfuric acid), with low yield and selectivity [7,8]. Suitable catalysts for a single-step conversion of EG to oxalic acid are highly desirable, that would be important to both industrial application and fundamental understanding of alcoholic group oxidation in polyols.

In this study, a series of monometallic catalysts (including Ag, Pd, Ru, Rd, Fe, Pt) supported on $CeO_2$ were synthesized by using a novel solvothermal method. The oxidation tests were conducted at a mild condition (70 $^{o}$C, ambient pressure $O_2$) in aqueous alkali solution. The catalysts were characterized by $H_2$ chemisorption, inductively coupled plasma (ICP), transmission electron microscopy (TEM) for the metal content, dispersion and morphology information. Volcano plots of normalized TOF values vs. density functional theory (DFT) calculated Gibbs Free energy was established for EG oxidation. Furthermore, a bimetallic Pt-Fe catalyst showed a significant rate enhancement and selectivity in one-step conversion of EG-to-oxalic acid. The reaction mechanism for EG catalytic oxidation by $O_2$ on (100) and (111) surfaces was studied using DFT calculations which rationalizes the experimental observations that the bimetallic Pt-Fe catalyst has a significantly higher activity (TOF) compared to the monometallic catalysts (Pt and Fe). Fundamental concepts in the computational chemistry and its relation to experimentally observed performance of bimetallic catalysts will be presented in this work.

**References**

1. Yue, H., Zhao, Y., Ma, X., & Gong, J. (2012). Ethylene glycol: properties, synthesis, and applications. *Chemical Society Reviews*, *41*(11), 4218-4244.

2. Serov, A., & Kwak, C. (2009). Review of non-platinum anode catalysts for DMFC and PEMFC application. *Applied Catalysis B: Environmental*, *90*(3), 313-320.

3. Sakharov, A. M., Sakharov, P. A., & Zaikov, G. E. (2012). Catalytic Oxidation of Ethylene Glycol by Dioxygen in Alkaline Medium. The New Example of One-Stage Oxidative Cleavage of C–C Bond. *Molecular Crystals and Liquid Crystals*, *555*(1), 168-176.

4. Berndt, H., Pitsch, I., Evert, S., Struve, K., Pohl, M. M., Radnik, J., & Martin, A. (2003). Oxygen adsorption on Au/Al 2 O 3 catalysts and relation to the catalytic oxidation of ethylene glycol to glycolic acid. *Applied Catalysis A: General*, *244*(1), 169-179.

5. Case, J. M. (2009). *Gold catalysts prepared by ion exchange for use in ethylene glycol oxidation: An exploratory study* (Doctoral dissertation, University of Cape Town).

6. Kopeykina, L. Y., Vodyankina, O. V., Kurina, L. N., & Golovko, A. K. Selective ethylene glycol oxidation on the oxide catalysts. In *Science and Technology, 2001. KORUS'01. Proceedings. The Fifth Russian-Korean International Symposium on* (Vol. 2, pp. 177-179). IEEE.

7. Isshiki, T., Suzuki, T., Yashima, Y., & Yonemitsu, E. (1972). *U.S. Patent No. 3,678,107.* Washington, DC: U.S. Patent and Trademark Office.

8. Knepper, W., & Rohl, H. (1975). *U.S. Patent No. 3,864,393.* Washington, DC: U.S. Patent and Trademark Office.

# 488083 Quantum Chemical Characterization of Catalytic Ester Decarbonylation: Olefins from Biomass

**Abstract:**

**Quantum Chemical Characterization of Catalytic Ester Decarbonylation: Olefins from Biomass**

Synthetic transformations of carboxylic acids into olefins are of paramount importance since bio-mass derived carboxylic acids could then be a renewable source for polyolefin syntheses. An important goal, then, is to develop catalytic transformations that sustainably transform these structurally diverse acid educts into olefin products with high selectivity. Production of olefins via transition-metal-catalyzed decarbonylation of bio-derived carboxylic acids is one approach that has been explored and demonstrated to be possible with low toxicity, and tolerant of diversity.[1-2]. As a subclass, linear a-olefins (LAOs) are important commodity chemicals that find utility in a wide range of industrial applications. LAOs are usually generated from petroleum, which is a nonrenewable and depleting resource. A green and sustainable approach would be to produce LAOs by decarbonylative dehydration of fatty acids, which are inexpensive and renewable.

Direct routes for the conversion of carboxylic acids to olefins typically proceed with high activation energies. However, carboxylic acid derivatives such as esters are more reactive and can serve as key precursors in carboxylic acid transformations, and in particular transition-metal-based catalysts have been shown capable of transforming them to high-value olefins. However, controlling selectivity for terminal alkene formation is a major challenge. A highly selective, palladium-catalyzed decarbonylation of *p*-nitrophenyl esters of fatty acids to their corresponding LAOs has been very recently developed.[3] A cooperative ligand system including phosphines and N-heterocyclic carbenes is reported to exhibit enhanced selectivity to terminal olefins. Herein, we investigate the decarbonylation mechanism of *p*-nitrophenylbutyrate to rationalize the switching reactivity and selectivity trends based on either individual or cooperative roles of phosphine and heterocyclic carbene ligands. The primary goal of the project is to elucidate the origin of terminal alkene **A** selectivity vs the thermodynamically more stable internal alkene **B** which can be generated as an undesirable alternative product (Scheme 1).

**Scheme 1.** Pd-catalyzed decarbonylation of *p*-nitrophenylester to a-olefins. XantPhos and IPr ligands comprise the dual ligand system.

To gain mechanistic insight at the atomic level of detail, we performed density functional theory (DFT) modeling. We focused on the Pd-catalyzed decarbonylation of *p*-nitrophenylbutyrate in order to fully characterize reaction coordinates for the formation of terminal **A** and internal **B** alkene products and to explain the roles of the ligands in promoting a-selectivity. To assess selectivity, we have thoroughly compared the competing reaction mechanisms using three ligand systems: IPr, XantPhos, and a mixture of IPr and XantPhos. Calculations suggest that the bulkiness of IPr combined with the bidentate nature of XantPhos prompt a rapid decoordination of the terminal alkene, thus precluding the alkene isomerization pathway.

**References**

1. Rodriguez, N.; Goossen, L. J., Decarboxylative coupling reactions: a modern strategy for C-C-bond formation. *Chemical Society Reviews* 2011, *40* (10), 5030-5048.
2. Le Nôtre, J.; Scott, E. L.; Franssen, M. C. R.; Sanders, J. P. M., Selective preparation of terminal alkenes from aliphatic carboxylic acids by a palladium-catalyzed decarbonylation–elimination reaction. *Tetrahedron Letters* 2010, *51* (29), 3712-3715.
3. Alex John, Levi T. Hogan, Marc A. Hillmyer and William B. Tolman, Olefins from biomass feedstocks: catalytic ester decarbonylation and tandem Heck-type coupling, Chem. Commun., 2015, 51, 2731.

# Elucidating the Effect of External Electric Fields and Surface Dopants

**Abstract:**

Fe catalysts are of great interest as they have been shown to be highly selective for the catalytic upgrading, via hydrodeoxygenation (HDO), of bio-oil compounds. Unfortunately, these catalysts alone are unacceptably prone to oxidative deactivation. One method of improving the oxidation resistance of Fe catalysts has been to dope the Fe with a second metal (e.g. Pd, Pt, etc.). However, the addition of the second metal increases the catalyst cost and complexity, thereby making it harder to experimentally optimize such systems to both increase activity and prevent oxidative deactivation. Previous work has shown that externally applied electric fields can dramatically improve the coke resistance of a Ni catalyst used in methane steam reforming (MSR).[6-9] These results suggest that external electric fields are another adjustable parameter that can be used to increase activity, tune selectivity, or prevent catalyst deactivation. Due to the highly polar nature of oxygen-containing compounds, it is expected that electric fields will have a large effect on HDO and could potentially be used in conjunction with doping to create a highly stable and selective Fe-based HDO catalyst. To this end, we must develop multi-scale models that can guide the selection of dopants in a rational manner—with both surface coverage and electric fields explicitly accounted for.

Here, we develop density functional theory (DFT) parameterized lattice gas models of oxygen on both undoped and doped Fe(100) surfaces to show the direct effect of an applied electric field on oxygen adsorption. Typical DFT models consider a uniform coverage distribution of adsorbates on the surface, when in reality, this is not the case.[2-4] Our cluster expansions of oxygen on Fe(100) surfaces both with and without an externally applied electric field accounts for the more realistic non-uniform adsorbate distributions and characterizes the lateral interactions between oxygen adatoms.[5] The apparent sensitivity of the lateral interactions on an applied external electric field reveals their influence on the oxidative deactivation of Fe catalysts.

Our preliminary results for O/Fe(100) indicate that an applied electric field has a significant effect on the attractive/repulsive interactions between adsorbed oxygen atoms. As seen in Figure 1, the clusters that are most important to the predictiveness of the model also change after the application of an electric field. These changes suggests that electric fields will have major effects on the Fe oxidation process as the electric field has a significant effect on the dynamic surface behavior of oxygen. Using such a lattice gas model for oxygen on both undoped and doped Fe(100), we will characterize the synergy between surface dopants and external electric fields in order to minimize the oxidation potential of the Fe surface. Our preliminary results suggest that one can expect significant deviations in catalytic activity after the application of external electric fields, but also, they suggest that mean-field models will not be sufficiently predictive for the work needed moving forward. Overall, the work presented here provides the first ingredients necessary for the selection of potential dopants for optimizing Fe-based HDO catalysts in the presence of external electric fields.

Figure 1: Effect of an electric field on oxygen cluster lateral interactions on the Fe(100) surface.

References:

1. Wang, H., Male, J. & Wang, Y. Recent Advances in Hydrotreating of Pyrolysis Bio-Oil and Its Oxygen-Containing Model Compounds. *ACS Catal.* **3,** 1047–1070 (2013).

2. Govender, A., Curulla Ferré, D. & Niemantsverdriet, J. W. The Surface Chemistry of Water on Fe(100): A Density Functional Theory Study. *ChemPhysChem* **13,** 1583–1590 (2012).

3. Blonski, P., Kiejna, A. & Hafner, J. Theoretical study of oxygen adsorption at the Fe (110) and (100) surfaces. *Surface Science* **590,** 88–100 (2005).

4. Freitas, R. R. Q., Rivelino, R., de Brito Mota, F. & de Castilho, C. M. C. Dissociative Adsorption and Aggregation of Water on the Fe(100) Surface: A DFT Study. *J. Phys. Chem. C* **116,** 20306–20314 (2012).

5. Frey, K., Schmidt, D. J. & Wolverton, C. Implications of coverage-dependent O adsorption for catalytic NO oxidation on the late transition metals. *Catalysis Science & ...* (2014). doi:10.1039/C4CY00763H

6. Che, F., Gray, J. T., Ha, S. & McEwen, J.-S. Improving Ni Catalysts Using Electric Fields: A DFT and Experimental Study of the Methane Steam Reforming Reaction. *ACS Catal.* (2016). doi:10.1021/acscatal.6b02318

7. Che, F., Ha, S., McEwen, J. S., Hensley, A. J. & Zhang, R. Elucidating the Role of the Electric Field at the Ni/YSZ Electrode: A DFT Study. *J. Phys. Chem. C* (2016). doi:10.1021/acs.jpcc.6b01292

8. Che, F., Ha, S. & McEwen, J. S. Hydrogen Oxidation and Water Dissociation over an Oxygen-Enriched Ni/YSZ Electrode in the Presence of an Electric Field: A First Principles-Based Microkinetic .... *Industrial & Engineering Chemistry ...* (2017). doi:10.1021/acs.iecr.6b04028

9. Che, F., Gray, J. T., Ha, S. & McEwen, J. S. Catalytic water dehydrogenation and formation on nickel: Dual path mechanism in high electric fields. *Journal of Catalysis* (2015). doi:10.1016/j.jcat.2015.09.010

# 495389 Uncovering Heterogeneous Ice Nucleation Using Advanced Molecular Simulations [Invited Talk]

**Abstract:**

Heterogeneous ice nucleation is relevant to a broad range of fields from atmospheric chemistry to food and transportation industries. In principle, because molecular simulations can access molecular lengthscales they are ideal to study nucleation. However, since nucleation is a rare event in molecular simulations straightforward simulation techniques cannot be used to sample statistically significant nucleation events. In our research, we have addressed this challenge by using the advanced sampling technique called forward flux sampling (FFS). Further, we have developed a software platform called SAFFIRE to enable largescale FFS calculations in high performance computing environment with computational and user efficiency.

In this talk, we will discuss the results obtained from largescale FFS studies of heterogeneous ice nucleation performed using SAFFIRE. We studied ice nucleation on silver iodide (AgI) and kaolinite-like surface. Both these surfaces are considered to be good ice nucleating agents and have a good lattice match with ice lattice. Traditionally, lattice match has been used a governing parameter to determine the ice nucleating efficiency of surface. Our extensive calculations show that the rate of ice nucleation on AgI and the modified kaolinite surface is different. Our results clearly demonstrate that lattice spacing is not a sufficient measure of ice nucleating efficiency. Exhaustive calculations of committor probability further enables us to comment on the difference in the nucleation mechanisms and role of surface properties in determining ice nucleating efficiency. To the best of our knowledge, these studies are the largest scale FFS calculations of heterogeneous ice nucleation using all-atom water model.

# 497108 Investigation of the Chromatographic Separation of Chiral Drugs By Molecular Dynamics Simulation

**Abstract:**

Many drugs have a chiral center as a key constituent. Chiral molecules have two enantiomers, which have the same molecular composition but different three-dimensional structures. The two enantiomers could have different therapeutic results; one enantiomer could be beneficial while the other could be toxic. Consequently, over the past decades pharmaceutical companies have made significant investments in purifying drugs to create an enantio-pure product; however, the design of these purification systems is challenging. One of the most effective methods for chiral separation is chromatography; but method development is still slow owing to a large number of combinations of chiral stationary phases (CSPs), mobile phases, pH, and temperatures, in the screening process to purify the racemate.

To help speed up the process of method development and to elucidate the chiral recognition mechanism, we have used molecular dynamics (MD) simulations to investigate the interaction between the CSP [in this case Amylose tris-(3,5-dimethyl-phenylcarbamate)] and various drug molecules. Our aim is to successfully relate the calculations from MD to the experimental observations. We first constructed and validated the model of the CSP by comparing the configuration generated from simulations with that observed in the NMR study (left-handed helical backbone structure). We then identified properties such as the lifetime of the hydrogen bonds between the CSP and the drug, which correlated well with the elution order and the selectivity of the drug racemates observed in the lab. Finally, the effect of the mobile phases on the chiral separation and on the conformation of the CSPs was investigated. The mobile phases were found to have a significant effect on the conformation of the CSP, thus affecting the chiral separation.

Our ultimate goal is to develop a model that will successfully predict the chiral stationary phase and mobile phase combination required for enantiomeric separations. The successful development of our predictive model will accelerate drug discovery in the pharmaceutical industry by reducing R&D time and costs.

# 486309 Molecular Simulation of $CO_2$ Absorption in Sorbent-Solvent Â Suspension and Interface Regions

**Abstract:**

$CO_2$ absorption by using solvents and adsorption by using sorbents have their respective advantages and disadvantages in carbon capture applications. Recently, Smit and coworkers [1] have investigated the slurry concept for $CO_2$ capture. By suspending a ZIF8 nano-prorous sorbent in the 2-methylimidazole glycol mixture solvent, the slurry system combines the low cost and efficiency of a nano-porous sorbent and the processing ease of a solvent-based separation process. Additionally, Azizi *et al.* have found that propane and propylene gas solubilities and absorption rates in the N-methyl pyrolidone solvent are increased when dispersing only 0.3 vol% $TiO_2$ nanoparticles in the solvent [2], which is partly due to the solvent structure perturbation in the presence of the nano-particle [3]. Even though all these experimental results indicate that sorbent-solvent suspension/slurry could be useful in $CO_2$ capture, there are still many important questions to be addressed, such as the suspension thermo-physical properties, $CO_2$ solubility and diffusivity in the bulk solution and in the interface region, $CO_2$ selectivity over other gases, and the effect of water and other gas impurities on the suspension performance. In this work, we address some of these important questions, which will be very useful to develop more efficient suspension systems for $CO_2$ capture. Our emphasis in particular is on the properties of the interface region, which will significantly affect $CO_2$ absorption in the suspension but cannot be determined using common experimental characterization procedures.

In this work, we have modeled a suspension system by dispensing nano-sized ZIF8 porous material into a polyethylene glycol siloxane (PEGS) solvent. It was found that the PEGS density in the ZIF8-PEGS interface region is about 10% larger compared with the bulk PEGS solvent density. Due to this increased PEGS density in the interface region, the $CO_2$ loading, diffusivity, in the PEGS interface region are significantly smaller compared with the corresponding values in the bulk PEGS solvent. Additionally, we present the viscosity, surface tension, and $CO_2$ absorption for the suspension system.

1. Liu, H.; et al. *Nature Communications* **2014**, *5*, 1-7.

1. Azizi, S.; et al. *Heat Mass Transfer* **2014**, *50*, 1699-1706 .

1. Shi, w.; et al. *J. Phys. Chem. B* **2010**, *114*, 15029-15041.

# 489885 Acid Gas Adsorption on Metal–Organic Framework Nanosheets As a Model of an "All-Surface" Material

**Abstract:**

To establish a model of metal–organic framework (MOF) surfaces and build an understanding of surface-specific ligand adsorption phenomena in MOFs, we present a computational study exploring multiple models of a series of MOF-2 nanosheet materials, "M-BDCs", with M = Zn, Cu, and Co and BDC = benzene-1,4-dicarboxylate. We study and assess the appropriateness of a series of models ranging from small clusters (18 atoms) to fully periodic sheet models. We additionally study the interactions of these models with acid gases and energy-relevant small molecules (CO, $CO_2$, $H_2O$, $SO_2$, $NO_2$, and $H_2S$). We employ computational methods ranging from DFT with various exchange–correlation functionals to perturbative and coupled-cluster methods. For these systems, we present binding energies and enthalpies with the various ligands studied as well as IR frequency shifts for the normal modes of these ligands upon complexation with the open-metal sites of these materials. Our calculations lead to an understanding of phenomena unique to MOF surfaces and the importance of the periodicity in these materials in capturing surface-specific adsorption behaviors.

Furthermore, we study adsorption of a set of acid gases to the exposed OMSs (open metal sites) existing at the surface of these materials in the presence of water, including competing co-adsorption motif and cooperative co-adsorption motif. Our computational results suggest that preadsorbed acid gases ($CO_2$, $SO_2$, $NO_2$, and $H_2S$) on Cu-BDC nanosheets would be replaced by water in the presence of water, leading to acid gas adsorbed on water coordinated OMSs instead of pure OMSs. For $CO_2$ and $H_2S$, the binding energies on water coordinated OMSs are lower than vacant OMSs, while for $SO_2$ and $NO_2$, the binding energies on water coordinated OMSs are higher. The strong double hydrogen bonds established by two oxygen atoms in $SO_2$ and $NO_2$ and two hydrogen atoms in water contribute to the enhanced binding energies. Even though water has a high probability to compete with acid gas for the same adsorption site, and even replace preadsorbed acid gases, in $SO_2$ and $NO_2$ cases, preadsorbing a small amount of water can increase the $SO_2$ and $NO_2$ binding strength and attack on Cu-BDC nanosheets.

# 498901 Contaminant Adsorption on Î±-Alumina Surface As Predicted By the Plane-Wave Density Functional Theory

**Abstract:**

Alumina is one of the most important ceramic materials and possesses highly insulation property. It is characterized by the combination of both ionic and covalent bonding. The Î±-phase of the alumina (Î±-$Al_2O_3$) which is also called corundum is the most stable phase. The Al-terminated layer of the (0001) surface of Î±-$Al_2O_3$ has been suggested to be more stable than the oxygen-terminated surface. Water binding on the metal-oxide surfaces is a complex phenomenon. Such binding is affected by several factors such as surface structure, and acidity and basicity of surface metal and oxygen sites respectively. The Al-terminated Î±-alumina surface is hydroxylated in the presence of water. Adsorption of high energy military contaminants on different surfaces is an important area of research since such studies can provide significant information on fate and effects of these contaminants in the environments. In the present work, we discuss the adsorption of some explosive compounds on the alumina surface at the plane-wave Density Functional Theory level using the PBE and van der Waals functionals. Vanderbilt ultrasoft pseudopotentials were used to model interaction of electron with core. Both parallel and perpendicular orientations of adsorbates with respect to alumina surface were considered. The electronic structures and the nature of interaction of adsorbates on alumina surface with and without the presence of water molecules will be discussed.

# 499454 Conformal Sites Model for Adsorbed Films on Energetically Heterogeneous Surface

**Abstract:**

Molecular simulations and theoretical analyses of thin films adsorbed onto solid surfaces or within narrow pores have usually treated the solid surfaces as atomically smooth and energetically homogeneous. Most surfaces, however, exhibit roughness through geometric curvature, variations of surface composition and chemical groups attached to the surfaces. Such oversimplification of the surfaces thus misses important features of adsorption, especially at low bulk pressure. In this research, a flexible conformal sites model, mapping a real interfacial system with an energetically heterogeneous surface onto a reference system with an energetically homogeneous surface, is developed. Grand Canonical Monte Carlo simulation is used to explore the limits of applicability of this model to the prediction of adsorption isotherms. Preliminary simulation results show that our conformal sites theory is quite successful in reproducing the adsorption behavior in non-polar systems and in slightly heterogeneous polar systems. Due to the simplicity and accuracy of the theory, we anticipate this new method will be incorporated in the software of experimental instrument to study porous materials by taking advantage of the corresponding states theory.

# 500599 Identifying Relationships between Terminal Group Chemistry and Interfacial Friction in Monolayer-Based Lubrication through a Molecular Dynamics Screening Approach

**Abstract:**

Nanoscale devices featuring surfaces in sliding contact are subject to failure resulting from high surface area-to-volume ratios and high-energy surfaces that promote stiction and surface wear. Alkane-based monolayers have been proposed as a potential solution to these tribological issues, having been shown to reduce stiction while yielding low coefficients of friction. The tribological performance of monolayer lubricants is closely linked to their terminal group functionality, with, for example, group size, hydrophobicity, and hydrogen bonding shown to alter adhesion and frictional performance of monolayer films [1], [2]. Monolayer chemistry represents a vast parameter space, and thus a wide screening of various terminal group functionalities is desirable to obtain a thorough understanding of the links between chemistry and optimal tribological properties.

To this end, several tools recently developed in our group as part of the MoSDeF software suite [3], including the open-source mBuild [4], [5] and Foyer [6] packages, in conjunction with the Signac-flow workflow manager [7], have been designed to facilitate screening over such chemical/structural parameter spaces. In this work, we demonstrate the use of these tools to screen a pool of terminal group functionalities, spanning a wide chemical space, for optimal tribological properties. The effects on friction of various properties of the terminal group, such as hydrophobicity, aromaticity, and chemical family are analyzed, with relation to their influence on the underlying monolayer structure. Our work helps demonstrate how molecular screening can provide a framework for the design of better-performing monolayer-based lubricants.

[1] B. Park, C. D. Lorenz, M. Chandross, M. J. Stevens, G. S. Grest, and O. A. Borodin, "Frictional Dynamics of Fluorine-Terminated Alkanethiol Self-Assembled Monolayers.," Langmuir, vol. 20, no. 23, pp. 10007–14, Nov. 2004.

[2] J. L. Rivera, G. K. Jennings, and C. McCabe, "Examining the frictional forces between mixed hydrophobic-hydrophilic alkylsilane monolayers.," J. Chem. Phys., vol. 136, no. 24, p. 244701, Jun. 2012.

[3] "MoSDeF" [Online]. Available: https://github.com/mosdef-hub.

[4] C. Klein, J. Sallai, T. J. Jones, C. R. Iacovella, C. McCabe, and P. T. Cummings, "A Hierarchical, Component Based Approach to Screening Properties of Soft Matter," in Foundations of Molecular Modeling and Simulation, 2016, pp. 79–92.

[5] C. Klein, A. Z. Summers, T. Ma, C. R. Iacovella, and J. Sallai, "mBuild." [Online]. Available: https://github.com/mosdef-hub/mbuild.

[6] C. Klein, A. Z. Summers, T. Ma, C. R. Iacovella, and J. Sallai, "Foyer." [Online]. Available: https://github.com/mosdef-hub/foyer.

[7] C. S. Adorf and P. Dodd, "Signac-Flow." [Online]. Available: https://bitbucket.org/glotzer/signac-flow.

# 498058 Molecular Simulation Study of Aluminum – Noble Gas Interfacial Thermal Accommodation Coefficients

**Abstract:**

Thermal accommodation coefficients (TAC) between aluminum and noble gases were studied with molecular dynamics (MD) simulations. Gases interacting with aluminum substrates were modeled by MD with gas velocities sampled from the Maxwell – Boltzmann distribution to give accumulated TAC results. Different implementations of the equation to calculations TAC, which differ in the amount of information gleaned from MD and the corresponding simulation results, were carefully discussed. The best formula for molecular dynamics modeling in finite simulation time was determined. In addition, the influence of the combining rules applied on aluminum – noble gas interatomic potential for was characterized with the well-known Lennard – Jones 12 – 6 potential combined with Lorentz – Berthelot and Fender – Halsey mixing rules. The results were compared with experimental values and previous analytical model. TACs simulated with Fender – Halsey rules present excellent agreement with the experimental values. Detailed TAC distributions and accumulated TAC convergence with a TAC cutoff value are also included in the discussion.

# 485598 The Water Flow through Graphene Slit Pores: Insights from Non-Equilibrium Molecular Dynamics Simulations

**Abstract:**

Water flowing through graphene membranes draws much attention of scientists since the excellent performance of water flux. The efforts at preparing graphene membranes have been made for years, but there are few cases of stable graphene membranes that could be applied into industrial applications. Some researchers tried to figure out the mechanism of high water flux in graphene pores, so as to design some polymer or inorganic membranes with excellent mass transport properties. In this work, we investigate the phenomenon of water flow through graphene slit pores by non-equilibrium molecular dynamics (NEMD) simulations. As was expected, the average velocities of water molecules get boosted with the additional force being enhanced. Besides, we tuned the slit width of graphene pores to figure out the effect of geometric size on the water flow. The slit pore size was chosen from 0.65 nm to 1.40 nm, so the water structure inside the graphene pore is switching from single layer to double layers. Since the microstructure of water confined in graphene is different, the mechanism of water transport is various. Unexpectedly, the water flux is not promoted with the slit width monotonically. The flux of water in 0.85 nm graphene slit pores exhibits minimum, probably due to the single â€“ double layer switching.

# 501528 Enhanced Sampling Methods for Modulating Density Fields

**Abstract:**

Driven by hydrophobic effects, aqueous surfactant solutions self-assemble into mesostructures with distinct symmetries, such as micelles, vesicles and bilayers. Molecular simulations play an important role in connecting the molecular characteristics of the surfactants to the properties of the mesostructures, the thermodynamic stability of the mesophases, and transitions between them. However, studying these mesostructures using straightforward simulations can be challenging, due to the large separation in timescales between the relaxation of individual molecules and that of the mesostructures. Here, we present an enhanced sampling simulation technique to address this challenge. Our approach involves organizing the molecules of interest into structures with well-defined symmetries, through the application of an external potential, which biases the molecular density field. The response of the system to the biasing potentials allows us to estimate mesostructure properties, as well as the thermodynamic stability of a particular mesophase with respect to another.

# 496866 Mapped Averaging Methods for Accurate and Precise Evaluation of Free Energies and Other Properties By Molecular Simulation

**Abstract:**

"Mapped averaging" is a recently published scheme that provides alternative, rigorous expressions for the ensemble averages that underlie molecular simulation.[1-3] The approach introduces knowledge from approximate theories while relying on the molecular simulation to measure the deviation from the theory. The approach also exploits the knowledge of the forces and the Hessian of the potential. In this manner, calculation of the mapped averages by molecular simulation can proceed without contamination by noise produced by behavior that has already been captured exactly by the theory, thereby yielding results of unprecedented precision with minimal computational effort. The approach can be applied in principle to any property that can be expressed as a derivative of the free energy. It does not affect sampling, so it can be used for many properties in the same simulation. We describe our recent advances in formulating and applying these methods, focusing in particular on two broad cases: (1) the free energy, which can be obtained with great efficiency via thermodynamic integration of first-derivative properties; and (2) second-derivative properties such as the dielectric constant, which are of great practical interest, but which suffer particularly from stochastic uncertainty because their evaluation is based on averaging ensemble fluctuations when performed using conventional methods.

(1) Schultz, A. J.; Moustafa, S. G.; Lin, W.; Weinstein, S. J.; and Kofke, D. A. Reformulation of Ensemble Averages via Coordinate Mapping, *Journal of Chemical Theory and Computation* 2016.

(2) Moustafa, S. G.; Schultz, A. J.; and Kofke, D. A. Very fast averaging of thermal properties of crystals by molecular simulation, *Phys. Rev. E* 2015, *92*, 043303.

(3) Moustafa, S. G.; Schultz, A. J.; and Kofke, D. A. Harmonically Assisted Methods for Computing the Free Energy of Classical Crystals by Molecular Simulation: A Comparative Study, *Journal of Chemical Theory and Computation* 2017, *13*, 825-834.

# 489235 Predicting the Free Energy Landscape of Multicomponent Fluids

**Abstract:**

We derive a statistical mechanical framework for predicting the grand canonical free energy landscape of multicomponent fluids obtained from flat-histogram Monte Carlo simulations at chemical potentials and temperatures other than those used during a simulation itself. This is accomplished by expanding the landscape in a multidimensional Taylor series at each macrostate on the surface defined by the sampling order parameter; we illustrate how all equations necessary to precisely obtain the coefficients in these series may be obtained using classical statistical mechanics, and may be computed using data trivially available during the simulation. While each individual series may be relatively simplistic, when used in concert they are capable of producing feature-rich landscapes out of relatively simple ones. This approach allows us to obtain the thermodynamic behavior of multicomponent fluid systems, for instance, at low temperatures using only a single simulation at a much higher one. For multicomponent systems, we illustrate how a small number of different simulations may be rationally combined in a self-consistent fashion that yields a continuous thermodynamic phase phase for the fluid. This enables the equation of state for multicomponent fluid systems to be easily computed, and phase diagrams to be obtained over a broad range of conditions. After surveying a variety of different fluid systems, we identify a range of temperature and chemical potential differences over which the extrapolation of simulations tends to quantitatively predict various thermodynamic properties accurately. Beyond this range, extrapolation still provides a reasonable estimate of the free energy landscape; this requires less computational effort to refine via additional simulation(s) at the desired conditions than construction of the surface without any initially informed estimate. In either case, this method significantly increases the computational efficiency of flat-histogram Monte Carlo simulations when investigating thermodynamic properties of multicomponent fluids over a wide range of conditions.

# 498180 A Novel Molecular Simulation Method for Liquid-Liquid Equilibria Predictions and *in silico* Screening of Desalination Solvents

**Abstract:**

Conventional desalination technologies include energy-intensive distillation or pressure driven membrane processes. Recently, directional solvent extraction (DSE) has emerged as a viable low-temperature and low-energy alternative which utilizes a solvent capable of selectively dissolving water by rejecting the salt ions and at the same time having negligible solubility in water phase. The dissolved water is then extracted either by heating or cooling the water-solvent phase. Thus far, the proof-of-concept has been demonstrated with decanoic acid for which the extraction efficiency, defined as amount of pure water produced per 100 kg of solvent, is only 0.5% suggesting that improvements are needed. This can be achieved by *in-silico* screening of a large number of molecular structures that can extract substantially higher amounts of water than decanoic acid. Success of such a screening critically depends on the ability to calculate solubility of water in solvents. As is widely known, liquid-liquid equilibria (LLE) calculations from molecular simulations continue to remain challenging due to the challenges associated with the transfer of molecules between two dense phases. In this presentation, we will describe the development of a novel molecular simulation methodology based on computing the fugacity of water as a function of concentration to caclulate the extent of water dissolution in a range of solvents. First, the applicability of the LLE method is shown considering that the solubility of the solvent in water is negligible. The calculated solubilities are shown to give good agreement with experimental data for carboxylic acids with six to ten alkyl chain lengths over a range of temperatures. This method is also extended and demonstrated for systems exhibiting substantial mutual solubility unlike directional solvents and water. In addition, the structural properties elucidating the role of the hydrogen bonding interactions between water and directional solvents governing the solubility will be presented.

# 492659 Ssages: A Comprehensive Platform for Enhanced Sampling Simulations

**Abstract:**

Molecular simulations are an essential tool in modern research, guiding experimental efforts and providing fundamental insights into underlying phenomena. Well-designed simulations offer a molecule-scale microscope with exquisite control over the physical phenomena of interest. While immensely powerful, the simulation toolbox has some important limitationsâ€"the accuracy of computations requires proper sampling of the system. While this may be achieved by simulating long trajectories or multiple independent copies of a system, the size of the systems and the time-scales that are accessible are still restricted by hardware. Importantly, systems that have rough energetic landscapes can become â€œstuckâ€, impeding proper exploration of configurations.

A palette of techniques has evolved to overcome these difficulties through static and dynamic modifications of the statistical ensemble underlying the sampling; these may be restriction based (as in umbrella sampling and replica exchange techniques) or on-the-fly adaptations (as in the powerful adaptive biasing force and metadynamics methods). Regrettably, the implementation of these techniques is not always straightforward and can be unattainable to users. This has often constrained newly developed, ostensibly general methods to problem or software specific applications, preventing the use of these techniques across platforms and research.

With these concerns in mind, we have developed SSAGES: Software Suite for Advanced General Ensemble Simulations. SSAGES is a multi-platform software that allows the use of enhanced sampling techniques on molecular simulations, which allows users to seamlessly apply enhanced sampling techniques to a variety of systems using widely used simulation engines such as LAMMPS, GROMACS, QBox, and OpenMD. We leverage SSAGES to study a variety of systems using multiple methods and discuss relative merits of flat histogram and string methods. In this talk, we will demonstrate the powerful methods, such as metadynamics, string methods, forward flux sampling and adaptive biasing force to several example systems, including dissociation of NaCl, isomerization of alanine dipeptide, mechanical properties of liquid crystals, self-assembly of porous materials, and discuss applications to related classes of problems.

# 501120 Multiscale Modeling of Multicompartment Micelle Nanoreactors

**Abstract:**

In recent years, research in industrial applications of polymeric materials has begun to explore the field of immobilized catalysis. In particular, the idea of catalysts bound to a micelle backbone, creating a nanoscale molecular reactor (commonly referred to as nanoreactor), has become an area of great interest. From a computational perspective, investigating the potential of micelles as nanoreactors requires analyzing the miscibility of block copolymers, both on a fully atomistic and on a mesoscale basis.

Extensive work has already been completed in this group to establish a robust method of estimating the Flory-Huggins Ï‡ interaction parameter for a given pair of molecules in order to probe polymer-polymer miscibility; this information is necessary for preparing coarse-grained molecular simulations (e.g., micellization simulations). In our present work, we apply miscibility analyses to a relatively nascent technology in immobilized catalysis science, viz. the multicompartment micelle nanoreactor. This technology offers a way to harness both the enhanced reactivity of homogeneous catalysis *and*the ease of separation traditionally enjoyed by heterogeneous catalysis.

Through the use of mesoscale calculations, we will study the feasibility of a three-compartment micelle nanoreactor composed of hydrophilic, lipophilic, and fluorophilic blocks. Both aromatics and alkyls will be explored for the fluorophilic block, in order to find the polymer architecture which allows for maximum phase segregation. Additionally, we will investigate reactant and product diffusion through the micelle compartments in order to explore potential applications of this technology to polymer production. We hope to demonstrate the viability of a micelle nanoreactor capable of reaction compartmentalization and tandem catalysis, two hugely promising capabilities for highly selective multistep-catalyzed reactions in polymer science and manufacturing.

# 499342 Reaction Ensemble Monte Carlo: Applications to Ionic Liquids

**Abstract:**

Reaction ensemble Monte Carlo (RxMC) has been successfully used to study equilibrium compositions in many non-ideal environments, e.g. in nanopores or at high temperatures and pressures. In classical simulations, a reaction move is accomplished by deleting reactant molecules and inserting product molecules. Random insertions and deletions in condensed phases are rarely accepted. In simple liquids, this problem can be overcome using continuous fractional component (CFC) Monte Carlo with Wang-Landau (WL) biasing. But in ionic liquids, the extremely low vapor pressure and heterogeneous nanostructure cause even CFC-WL to fail. We present a simple adaptation of the original reaction move to align both the position and orientation of inserted product molecules to the deleted reactant molecules. We discuss applications to absorption of $CO_2$ in the ionic liquid triethyloctylphosphonium 2-cyanopyrrolide, [P2228][cnp] and the dissociation of the acid-base pair acetic acid and N-methylpyrrolidine.

# 502781 Temperature-Dependent Physicochemical Properties of Nitrotoluenes from Solvation Free Energies


**Abstract:**

The temperature-dependent vapor pressure, solubility, Henryâ€™s law constant, and partition coefficients are important in many industrial applications and in environmental modelling. However, the reliable data for these properties are not always available, especially, for hazardous and exothermic compounds, such as the nitrotoluenes. Here we demonstrate how force field-based solvation and self-solvation free energies as functions of temperature can be used to compute a wide range of physicochemical solvation properties. The temperature-dependent solvation free energies in water, 1-octanol, and in the pure liquid or subcooled phases for a set of nitrotoluenes are computed using expanded ensemble molecular simulation method over the temperature range of 273 K to 330 K. TraPPE force field parameters with CHELPG charges were used to model the nitrotoluenes. TIP3P and TraPPE-UA force field parameters were used to model water and 1-octanol, respectively. Subcooled and liquid-phase vapor pressures, solubilities, Henryâ€™s law constant, and partition coefficients (air-water, air-octanol, and water-octanol) are predicted and compared with available experimental data. The changes in entropy, enthalpy, and heat capacities are estimated both for the solvation of the solute in water and in 1-octanol, and for the transfer of the solute from water to 1-octanol. The effects of functional groups and molecular structure on physicochemical and thermodynamic properties are also discussed. The protocol followed here can be used for compounds for which temperature-dependent data are scarce or not accessible due to hazardous or exothermic nature.

# 489377 Probing Topology and Reactant Effects on Hydride Transfer in Various Zeolites from First Principles

**Abstract:**


**Theoretical Insights into the Effect of the Zeolite Environment and Reactant Identity on Alkane-Alkoxide Hydride Transfer**



Thomas Chen and Matthew Neurock, Deparment of the Chemical Engineering and Materials Science, University of Minnesota, Minneapolis, MN, 55455.



The thermal stability and tunability of the pore topology of zeolites make them attactive solid acid catalysts for carrying out the elementary transformations involved in hydrocarbon chain growth processes important in the production of chemicals and fuels.  Hydride transfer is ubiquitous to hydrocarbon transformation processes. However, detailed mechanistic studies of hydride transfer are limited as it is a secondary reaction that reactive intermediates can undergo, and direct experimental observations of hydride transfer are difficult to obtain. Herein, first principle density functional theory calculations are used to examine light alkane and olefin hydride transfer in TON, MFI, MOR, and BEA zeolites. We examine the effects of the alkane, the alkoxide, and the zeolite and show how transition state selectivity is determined by the zeolite topology. The transition state for hydride transfer are carbenium ions whose stability increase with reactant substitution thus increasing reactivity. However, calculations with TON show the existence of pore size limitations where hydride transfer reactivity decreases when either alkane or alkoxide exceeds a certain size threshold. Geometric and charge analyses are used to describe the competing effects of destabilizing steric interactions and stabilizing solvation effects. These interactions cause the pore size limitation where repulsive interactions outweigh zeolite solvation that delocalizes the positive charge of the carbenium ion transition state. Zeolite comparisons show the relative reactivity of alkane and alkoxide pairs for hydride transfer shifts to bulkier reactants as the pore size and dimensionality increase. Detailed comparison between TON and BEA show differences in energetics and transition state structure. Due to the larger pore size of BEA, no size threshold is observed for the molecules studied indicating a size threshold which is shifted towards bulkier species.

# 499776 Simulations of Ammonia Adsorption for the Characterization ofÂ Acid Sites in Metal-Doped Amorphous Silicates

**Abstract:**

The acidity of amorphous mesoporous silicates is directly related to the catalytic activity of these materials, and metal doping increases the acidic behavior. However, the local structure of the active sites is not well known as it is difficult to probe it experimentally. Additionally, it is not clear what drives the formation of specific acid sites for a specific metal dopant, i.e. Brønsted and/or Lewis, and their relative strength. All of this inhibits the systematic improvement of the catalytic activity of these materials. We use quantum mechanical simulations based on density functional theory to characterize the local structure of Zr-, W-, and Nb-doped amorphous silicates, for which experimental data of acidity is available. We consider a different number of hydroxyl groups directly bonded to the metal centers and of nearby silanol groups, and their effect on the Lewis and Brønsted acidity of the active site. The acidity is studied via adsorption of one or two ammonia molecules on the metal, M-OH, and Si-OH sites, respectively, to model measurements with the $NH_3$ temperature controlled desorption ($NH_3$-TCD) technique. Our calculations reproduce the experimental trends of acidity strength across metals as seen by Ramanathan (2012, 2013, 2014): Zr > W > Nb, where Zr mostly shows Lewis acidity, while Nb and W showing both Brønsted and Lewis sites. Metal atoms that are more grafted into the silica (have more M-O-Si interactions) exhibit the strongest Lewis acidity. Furthermore, silyl oxoniums stabilized by a nearby metal atom exhibit the strongest affinity for ammonia, and represent the most likely source of Brønsted acidity in these materials.

# 501323 First-Principles Grand-Canonical Simulations of Water Adsorption in Proton-Exchanged Zeolites Using a Highly Parallelizable Algorithm

**Abstract:**

Water is a ubiquitous solvent or an unavoidable impurity in many important engineering and biological processes. For reactions catalyzed by microporous materials, quantifying the amount or the structure of adsorbed water clusters at reaction conditions is prerequisite to understanding the relevant reaction mechanisms. Computational studies, such as those based on static electronic structure methods or first-principles molecular dynamics in the canonical ensemble, relies on preloading the framework with an often arbitrary amount of water molecules in order to examine their effects. We present here first-principles calculations in the grand-canonical ensemble of water adsorption onto proton-exchanged zeolites with different Si/Al ratios (Si/Al = 31--95). These simulations use pre-sampling with inexpensive force fields and a new parallelization algorithm to improve simulation efficiency and to bypass the scaling limit of the underlying electronic-structure method. These developments enable us to predict water adsorption under specified thermophysical conditions. Under almost all conditions examined (T = 373 -- 573 K, p = 10 -- 100 kPa), we observe an appreciable amount of water adsorption, and the zeolitic protons are delocalized by participating in the hydrogen-bonding chains formed by adsorbed water molecules.

# 495946 The Role of ABC-6 Zeolite Cavity in Methanol to Olefin Conversion

**Abstract:**

A density functional study (DFT) study is reported to investigate methanol to olefin (MTO) conversion in four ABC-6 zeolite cavities (*cha, avl, aft* and *h1*) differing in shape and size. The reaction barriers and energies of methylations, deprotonations and eliminations based on the side-chain of hydrocarbon pool (HP) mechanism are calculated in these cavities. By comparing their catalytic performance, we find that *cha* with low barriers in methylations and eliminations exhibits a stronger confinement effect due to its elliptical shape and small size. *Avl* and *aft* have low barriers in the first and second methylations, and the elimination barriers similar to *cha* for producing olefins. With a slightly wider dimension, *aft* shows high stability for HP species, especially the charged HP species due to long-range interaction with zeolite framework. These results demonstrate that the activity and selectivity of MTO conversion depend strongly on the size and shape of zeolite cavity, which is highly controlled by the confinement eï¬€ect of zeolite framework. Our calculations also predict a candidate zeolite with *aft* cavity might perform well in MTO conversion. This theoretical study reveals the relationship between zeolite structure and activity performance, and provides an efficient approach to design and screen ideal zeolites for MTO conversion.

# 496463 Understanding the C-H Activation and Dehydrogenation Mechanisms of Alkanes on Metal Oxides

**Abstract:**

Olefins are important chemical building blocks for the production of a wide range of valuable chemicals and plastics. A promising route to produce olefins is the non-oxidative dehydrogenation of alkanes on metal oxides. The Lewis acid-base functionalities of metal oxides play a key role in their catalytic behavior. However, how these functionalities affect the alkane dehydrogenation behavior and how they can be rationalized on multisite surfaces of metal oxides, is still elusive. In this work, we provide fundamental insights into the various mechanisms of alkane dehydrogenation on Î³-$Al_2O_3$ and identify chemically intuitive, structure activity relationships, by using Density Functional Theory calculations. These relationships are able to screen different sites of metal oxides towards their C-H activation and alkane dehydrogenation behavior. The obtained relationships can be utilized to accelerate the discovery of active dehydrogenation metal-oxide catalysts.

# 486831 Prediction and Screening of Product Distribution in Nanoporous Material-Catalyzed Propene Dimerization Via Molecular Simulations

**Abstract:**

The ability to predict the product distribution of a nanoporous catalyst *in silico* would enable better understanding of catalyst structure-function relationships and advance discovery of more selective catalysts for industrially important reactions. One such reaction is the propene dimerization reaction, which yields a mixture of branched and linear $C_6$ olefin isomers, of which the linear isomers are valued as industrial feedstocks for fuels and lubricants[1]. We have developed a conceptual framework based on shape selectivity theory[2] to explain and predict changes in the product distribution of the propene dimerization reaction resulting from the use of different zeolites and metal-organic frameworks (MOFs) as catalysts.

Using our model, experimentally observed changes in the product distribution from the literature[1] could be explained with quantitative agreement in terms of the contribution of the pores to the free energies of formation of the product isomers, which were computed using molecular simulations. We applied our model to screen a database of 118 existing and hypothetical MOF and zeolite structures to study how product distribution could be tuned by changing pore size, shape, and composition of porous materials. The screening identified a few not-yet-synthesized materials with improved linear isomer selectivity compared to experimentally identified catalysts, and determined catalyst properties that enhanced selectivity toward the valued linear olefin isomers, such as presence of open metal sites. This molecular picture of the mechanisms influencing product distribution provides insights for prediction and rational design of improved catalyst materials.

[1] Mlinar, A. N.; Keitz, B. K.; Gygi, D.; Bloch, E. D.; Long, J. R.; Bell, A. T. ACS Catal. 2014, 4, 717â€“721.

[2] Smit, B.; Maesen, T. L. M. Nature 2008, 451, 671â€“678.

# 493406 Catalytic Hydrogenation of Carbon Dioxide in Functionalized Metal Organic Frameworks

**Abstract:**

Metal Organic frameworks (MOFs) are a class of porous materials having potential for both gas adsorption and catalysis. Functionalization of MOFs with catalytic moieties results in heterogeneous catalysts with active sites similar to those of molecular catalysts. In this work, we will present how the capture and conversion of $CO_2$ to C1 products can be enhanced in MOFs by incorporating Lewis pairs (LPs) and chelated metal centers. In contrast to $CO_2$ reduction on metal surfaces, the pathways here involve concerted transfer of a hydride and proton to the adsorbed or co-adsorbed $CO_2$. The density functional theory results presented here were calculated using the CP2K code, with the PBE functional to describe the exchange correlation energy. Girmmeâ€™s D3 dispersion correction was used to account for van der Waals interactions. We computed reaction pathways for $H_2$ dissociation, $CO_2$ chemisorption, and $CO_2$ hydrogenation.

We have shown that inclusion of Lewis pair (LP) moieties into UiO-66 and UiO-67 gives effective catalysts for hydrogenation of $CO_2$ to formic acid and methanol [1, 2]. The reaction pathway in this system has a lower barrier than the barriers reported for the reduction of $CO_2$ on traditional heterogeneous catalysts [3]. MOFs have a similar structure topology to zeolites, where the effects of confinement on the transition states and intermediates are well documented [4]. Here, we focus on how the effects of pore size could be applied to improving catalysis in MOFs. We do so by comparing and contrasting results from UiO-66 (largest pore), MIL-140C, and MIL-140B (smallest pore).

We have also investigated the reduction of $CO_2$ to formic acid via a formate intermediate on chelated metal centers. The pathway for Cu active centers are formic acid desorption limited, with low (<0.2 eV) barriers for the bond breaking/formation steps. This system allows for further tuning of the activity by changing the identity of the metal atom, as well as the electron affinities of the chelating ligands.

1. Ye, J.; Johnson, J. K. *ACS Catal.* 5, 6219-6229 (2015)
2. Ye, J.; Johnson, J. K. *Catal. Sci. Technol.* 6, 8392-8405 (2016)
3. Ye, J.; Johnson, J. K. *ACS Catal.* 5, 2921-2928 (2015)
4. Jones, A. J.; Zones, S. I.; Iglesia, E. *J. Phys. Chem. C* 118, 17787-17800 (2014)

# 491022 Inorganometallic Catalyst Design: Alkane Metathesis Catalysis in Nu-1000 MOFs Functionalized with Transition Metals

**Abstract:**

Mid-weight ($C_3$-$C_8$) *n*-alkanes are petrochemical raw materials that are available in large quantities from oil and gas reserves. They are also produced in Fischer-Tropsch catalysis, which is the major route for the conversion of natural gas and coal to liquid fuels. It is desirable to convert mid-weight *n*-alkanes into higher-value molecules. Alkane metathesis, in which alkane molecules are transformed into higher and lower homologs, is a potential method for achieving such conversion.

The majority of single-site alkane metathesis catalysts are metal hydrides on amorphous $SiO_2$ and $Al_2O_3$. In each case, the metal plays dual roles: first, dehydrogenation of alkanes to alkenes, and second, perform alkene metathesis to form C–C bonds. The present work considers using a metal–organic framework (MOF), in particular NU-1000 (molecular formula $[Zr_6(\mu_3\text{–}O)_4(\mu_3\text{–}OH)_4(OH)_4(OH_2)_4](TBAPy)_2$, $H_4TBAPy$ = 1,3,6,8-tetrakis(p-benzoic-acid)pyrene), as a support for new catalyst development. Because of their tunable surface structures, MOFs are ideal for rational heterogeneous catalyst design based on structure–reactivity relationships.

We used Kohn–Sham density functional calculations with the M06-L exchange-correlation functional for computational screening of NU-1000 functionalized with various transition metals, in particular screening for propane metathesis. We considered eight transition metals (V, Nb, Ta, Cr, Mo, W, Mn, Re) along with three variations of the NU-1000 framework: fully hydrated NU-1000, partially dehydrated NU-1000, and NU-1000 decorated with yttrium oxide. The activation energies for C–H bond activation and [2+2] cycloaddition were identified as descriptors of catalytic activity of different catalysts. Using the Sabatier principle, we constructed volcano plots to evaluate the expected efficacy of the trial catalysts. This identified vanadium-functionalized, yttrium oxide-decorated NU-1000 ($VYO_x$@NU-1000) as the most suitable catalyst among the trial catalysts for the propane metathesis. The complete reaction cycle for $VYO_x$@NU-1000 catalyst was computed and the free energy of activation is predicted to be 37.8 kcal/mol. This relatively high activation energy obtained by using a dual-role single-site catalyst indicates that the overall performance of the catalyst is compromised in order to achieve a balance between different reactions. We conclude that future developments in alkane metathesis catalysis should focus on dual-site catalysts and/or tandem systems.

# 501081 Automated Tests for Physical Validity in Molecular Dynamics and Monte Carlo Simulations

**Abstract:**

Advances in recent years have made molecular dynamics (MD) and Monte Carlo (MC) simulations a powerful tool in molecular-level research, allowing the prediction of experimental observables in the study of systems such as proteins, membranes, and polymeric materials. The quality of any prediction based on molecular dynamics results will strongly depend on the validity of underlying physical assumptions.

Based on a number of examples, in which (sometimes hard-to-spot) unphysical behavior of simulations had statistically important influence on their results, we propose a two-fold approach to increase the robustness of molecular simulations. The first aspect involves tests which can be performed by the users of MD programs on their respective systems and setups. We present a number of tests of different complexity, ranging from simple post-processing analysis to more involved tests requiring additional simulations. These tests are shown to significantly increase the reliability of MD simulations by catching a number of common simulation errors violating physical assumptions, such as non-conservative integrators, deviating from the Boltzmann ensemble, or lack of ergodicity between degrees of freedom. To render the usage as easy as possible, an open-source and platform-independent Python library containing the presented tests is available.

The second aspect involves testing for code correctness. While unphysical behavior can be due to poor or incompatible choices of parameters by the user, it can just as well originate in coding errors within the program. We therefore propose to include physical validation tests in the code-checking mechanism of MD software packages. We have implemented such a validation for the GROMACS software package, ensuring that every major releases passes a number of physical sanity checks performed on selected representative systems before shipping. It is, to our knowledge, the first major molecular mechanics software package to run such validation routinely. The tests are, as the rest of the package, open source software, and can be adapted for other software packages.

# 501314 Using Graphs to Quantify Energetic and Structural (dis)Order in Organic Thin Films

**Abstract:**

The nanomorphology of polymer blend thin films critically affects performance especially in organic solar cells. However, many aspects of the underlying physics linking morphology to performance are still poorly understood. At the same time, there is increasing evidence that atomic organization can hold the key to efficient charge transport within organic electronic devices. In order to fully capitalize on these recent evidence, there is a need to quantify the atomistic and energy features of morphologies with respect to basic steps of photovoltaic process to enable the quantitative structure property relationship.

In this work, we take advantage of recent advances in molecular dynamic simulations and quantify atomic-scale morphological aspects of the thin films. Specifically, we present a graph-based technique that allows quantifying the point-cloud data (MD). In our approach, we first convert the point cloud data from atomistic simulation into a labelled, weighted, undirected graph and then use standard graph-based algorithms to calculate and quantify morphology features. The conversion of the CGMD-data into a graph preserves all the topological and geometric information about the internal structure, and local connectivity between individual atoms/beads (along and across the polymer chains). More importantly, the edges between individual beads on CGMD can be labelled by Euclidean distance, energy difference or hopping rate. Our method provides hierarchical information about the charge paths that a hole/electron needs to take to reach the electrode (path length, fraction of intra-molecular hops, path balance).

We showcase capabilities of our approach by analyzing coarse grained molecular simulations of several oligothiophene blends. We present how graph-based method allows to provide quantitative insight into the origins of few orders of magnitude difference in mobility. Taking into account the comparable dimensionality of samples, it is hard to rationalize such difference based solely on the physical distances. We combine the information from structural (Molecular Dynamics) and energy (Density Functional Theory) levels to explain large differences in mobility for these polymer blends.

# 500954 Interactive Extensible Molecular Simulations with Etomica

**Abstract:**

Etomica is an object-oriented molecular simulation programming framework [1]. It is written in Java, and was developed with a strong emphasis on modularity and extensibility. Etomica can accommodate almost any type of particle-based interaction model and any type of law governing the particles' behavior. It can allow both hard and soft potentials, any spatial dimension, on- or off-lattice, pairwise-additive or multibody potentials, Monte Carlo or molecular dynamics (or other) simulations, and so on. It also is suitable for both batch and interactive use, allowing simulation parameters, integration method, measured quantities, and other features to be modified while the simulation proceeds. This capability has led us to use Etomica to develop a suite of interactive molecular simulation modules targeted for instructional purposes [2]. However, it is also the code base we use for many of our research applications. In addition to standard molecular simulation capabilities, Etomica has a variety of tools for calculation of virial coefficients and other cluster integrals using Mayer sampling Monte Carlo.

Recently we have performed development to expand the usefulness of the Etomica framework. In particular, we have introduced capabilities for simulations to be specified using input files defined by standard packages (e.g., LAMMPS, GROMACS). We are also developing features to enhance the interactive capabilities, such that Etomica may be configured and operated interactively in a client-server configuration. In this manner we aim to offset the increasing problems associated with interactive use when based on Java running in the browser. We describe and demonstrate these developments, and discuss our plans for further expansion of Etomica's features and capabilities.

[1] A. J. Schultz and D. A. Kofke, "Etomica: An object-oriented framework for molecular simulation" J Comput. Chem. **36**, 573 (2015).

[2] http://modules.etomica.org

# 496385 A General Algorithm for Efficient Reverse-Mapping of Coarse-Grained Configurations to the Atomistic Scale

**Abstract:**

Despite the ever-growing computational resources, atomistic models (AM) are often intractable for the length and time-scales required for molecular simulations. Coarse-grained models (CG), that integrated out many degrees of freedom, have been successful at "unlocking" the length and time scales needed to perform meaningful computer experiments. The main drawback to CG models is that some quantities of interest like charge transfer in electronically conducting polymers, require atomistic resolution. As such, CG models are used to quickly achieve equilibrium at coarser length scales, and then the molecues "reverse-mapped" (or reverse coarse-grained, RCG) into AM so that the molecules can be relaxed at atomistic length scales. This RCG step is often hard-coded and can be difficult to modify for arbitrary chemistries. With the prevalence of generalized software for various computational techniques such as LAMMPS (molecular dynamics), Cassandra (Monte Carlo), and ABINIT (density functional theory), we present a generalized software that requires only two initial inputs, an atomistic input file containing the molecular topology (e.g., pdb, lammps data file), and a corresponding input file containing the CG configuration and topology. The initial RCG procedure self-consistently determines the coarse-graining scheme, and creates coarse-grain bead files, which can be used in subsequent iterations of the program to reduce program runtime. We will present the current implementation of the program, sample applications, as well as propose potential future extensions.

# 501283 Mosdef: Molecular Simulation and Design Framework

**Abstract:**

We present a collection of tools designed to enable the facile setup, initialization, atom-typing, and molecular dynamics (MD) simulations, termed MoSDeF [1]. In particular, MoSDeF has been developed to enable computational MD-based screening of soft materials [2]. The core tool, mBuild [2,3], generates starting configurations by minimizing or even eliminating the need to explicitly translate and orient components when building systems - users simply state which components to connect. mBuild supports parameterized structures through generative modeling. This allows for declaratively expressing repetitive structures, e.g. polymer chains, crystal structures, planar or spatial tiling, as well as for parameterizing the affine transformations applied to subcomponents of a composite component. For end users, this minimizes and often eliminates the need to explicitly rotate and translate components when assembling systems. Additionally, the approach enables users to programmatically vary parameters for a family of systems (e.g., polymer chain length) or interchange individual components (e.g., polymer type), while still employing the same general framework. mBuild integrates with the Foyer tool for cataloging and applying force fields to molecular systems (i.e., atom-typing). Foyer provides a force field and simulator agnostic method for defining parameter usage that relies upon SMARTS [5] based annotations of chemical context, providing both human and machine readable documentation of parameter usage, which also aids in the dissemination of force fields. These tools are coupled with the Signac-flow workflow manager [6], allowing large scale screening over parameter space. Building upon prior work [2], we demonstrate the utility and flexibility of our tool chain by screening the frictional properties of several families of monolayers.

[1] https://github.com/mosdef-hub

[2] C. Klein, J. Sallai, T. J. Jones, C. R. Iacovella, C. McCabe, and P. T. Cummings, "A Hierarchical, Component Based Approach to Screening Properties of Soft Matter," in Foundations of Molecular Modeling and Simulation, 2016, pp. 79–92.

[3] https://github.com/mosdef-hub/mbuild

[4] https://github.com/mosdef-hub/foyer

[5] http://www.daylight.com/dayhtml/doc/theory/theory.smarts.html

[6] C. S. Adorf and P. Dodd, "Signac-Flow." [Online]. Available: https://bitbucket.org/glotzer/signac-flow.

# 487730 Massively-Parallel Mesoscale Hydrodynamics on Graphics Processing Units

**Abstract:**

Multiparticle collision dynamics (MPCD) is a particle-based mesoscale method for resolving fluctuating hydrodynamics in computer simulations. A fluid is effectively represented by point particles that periodically undergo stochastic, momentum-conserving multiparticle collisions that impart long-ranged hydrodynamic interactions. MPCD is particularly useful for soft matter, including colloid-polymer mixtures, where it is important to resolve microscopic structural details and thermal fluctuations. In principle, MPCD simulations can be significantly more computationally efficient than conventional explicit-solvent molecular dynamics, allowing for large length-scales and long time-scales with fully resolved hydrodynamics. However, typical MPCD simulations may still require tens of millions of MPCD particles or more, necessitating a parallel computational approach. We present our development of the first massively-parallel, open-source implementation of the MPCD method for graphics processing units in HOOMD-blue. Our implementation fully exploits computational resources from desktop workstations up to leadership-class supercomputers, and will significantly accelerate simulations of soft matter systems where hydrodynamic interactions are important.

# 501195 Foyer: A Framework for Defining Force Field Usage Semantics and Atom-Typing Molecular Systems

**Abstract:**

The availability of forcefields for molecular simulation has reduced the effort researchers must devote to the task of determining the interactions between species, allowing them to instead focus on the motivating scientific questions. However, determining which parameters in a forcefield to use is still often a very tedious and error prone task, given that forcefields may contain hundreds of unique parameters for a single element, differentiated by the chemical context for which they apply. While tools for performing atom-typing exist (i.e., determining which parameters describe an atom's interaction potential with its environment based on its chemical context), they are often tied to a specific force field or simulation software and typically employ a rigid hierarchy where rules that identify more specialized atom types must be called ahead of more general atom types. More importantly, many of these atom-typing tools are not directly tied to the files containing the definitions of the force field parameters, which may introduce ambiguities as to appropriate usage and the source of parameters. The strong dependence of force field parameters on the chemical context of the atoms demands a codified, logic-based expression tied directly to those parameters.

Here, we present a tool - Foyer [1] - that eliminates ambiguity in force field parameter usage thereby enhancing reproducibility and simplifying the process of disseminating new force fields. Foyer can serve as a standalone tool and accepts several commonly used chemical topology formats but was also designed to facilitate automated system initialization using mBuild [2]. mBuild constructs the initial configurations for systems while Foyer seamlessly parametrizes those systems upon saving to a simulation input file. To address the issues arising from rigid rule hierarchies and the segregation of force field parameters from their atom-typing tools, Foyer provides a force field agnostic method for defining parameter usage that relies upon SMARTS [3] based annotations of chemical context to perform atom-typing and thereby simplifying dissemination of force fields. In Foyer, rigid rule hierarchies are replaced by an iterative process which allows rules to be embedded in any order within the actual force field files alongside the parameter definitions, thus delivering annotations that are both human and machine readable. This also allows rules to be tested for logical consistency, independent of the chemical species they actually describe.

We present several case studies, including the development of a novel force field for perfluoropolyethers [4], that demonstrate the forcefield annotation scheme and how it can be used to simplify the process of disseminating a force field and ensuring that end users correctly use the derived parameters. Foyer has been developed as a Python library, designed to integrate with the Molecular Simulation Design Framework (MoSDeF) [5] currently under development at Vanderbilt.

1. C. Klein, A. Z. Summers, T. Ma, C. R. Iacovella, and J. Sallai, "Foyer." [Online]. Available: https://github.com/mosdef-hub/foyer.
2. C. Klein, J. Sallai, T. J. Jones, C. R. Iacovella, C. McCabe, and P. T. Cummings, "A Hierarchical, Component Based Approach to Screening Properties of Soft Matter," in Foundations of Molecular Modeling and Simulation, 2016, pp. 79–92.
3. http://www.daylight.com/dayhtml/doc/theory/theory.smarts.html
4. J. Black, G. Silva, C. Klein, C.R. Iacovella, P. Morgado, L. Martins, E. Filipe, C. McCabe, " Perfluoropolyethers: Development of an All-Atom Force Field for Molecular Simulations and Validation with New Experimental Vapor Pressures and Liquid Densities", under review
5. https://github.com/mosdef-hub

# 494747 Using the k-d Tree Data Structure to Accelerate Monte Carlo Simulations

**Abstract:**

The k-d tree data structure is implemented in a Monte Carlo (MC) molecular simulation program to accelerate the range search for particles or interaction sites within the cutoff distance when Lennard-Jones and Coulomb interactions are computed. MC simulations are performed for different molecules in various ensembles to assess the efficiency enhancements due to the k-d tree data structure. It is found that the use of k-d trees accelerates significantly simulations for Lennard-Jones particles in the NVT and NVT-Gibbs ensembles and for n-butane and 2,4,6,8,10,12,14,16,18,20,22-undecamethylpentacosane represented by the TraPPE--UA force field in the NpT ensemble. Simulations for TraPPE--UA ethanol in the NpT ensemble and for the rigid TIP4P water model in the Gibbs ensemble gain slightly in efficiency with the k-d tree, whereas simulations for TIP4P water in the NpT ensemble do not benefit from the use of the k-d tree. The speed-up can be attributed to the reduction in the number of distance calculations in the range search from scaling as O(N) to O(logN). In addition, these tests suggest that the efficiency gain from the use of the k-d tree data structure depends on the flexibility of the molecular model (requiring configurational-bias MC moves to sample changes in conformation), on the ensemble (with open ensembles requiring special MC moves to aid particle transfers), and on the number of interaction sites per molecule (with compact multi-site models not seeing an efficiency gain). Overall, the use of the k-d tree data structure can substantially enhance MC simulation efficiency for a variety of systems, and it will enable simulations for larger system sizes in the future.

# 486849 Machine Learning for Autonomous Crystal Structure Identification

**Abstract:**

We present a machine learning technique to discover relevant structures from molecular simulation snapshots or particle tracking data. Unlike other popular methods for structural identification, our technique requires no a priori description of the target structures. Instead, we use nonlinear manifold learning to infer structural relationships between particles according to their local topology. This graph-based approach yields unbiased structural information which allows us to quantify the crystalline character of particles near defects, grain boundaries, and interfaces. We demonstrate the method by identifying relevant structural features in a simulation of colloidal crystallization, some of which are missed by standard techniques.

# 499380 Intelligent, Autonomous Exploration of Self Assembly Simulation Parameter Space

**Abstract:**

Computational "experiments" using coarse-grained models are immensely popular and have helped us understand the behavior of systems with length and time scales ranging from those of atoms to proteins. For exploratory studies in particular, researchers often spend large amounts of computational resources simulating not-so-interesting behavior of these models, such as large regions of parameter space that form trivial structures or kinetically arrested states. It is usually difficult to know which areas in parameter space should be sampled or how systems should be annealed in order to yield interesting or desirable phase behavior. Without the guidance of prior simulations, this leads to an iterative "trial-and-error" approach in many exploratory studies. Here we discuss strategies to incorporate machine learning into the experimental design loop in order to optimize computational time spent simulating interesting phenomena in the area of self assembly.

# 498035 Efficient Phase Diagram Sampling By Active Learning


**Abstract:**

We propose and implement an active learning-based efficient phase diagram sampling method, which can save upwards of an order of magnitude of state point exploration time. This adaptive sampling method is inspired by advances in machine learning research. It works by adaptively calculating the potentially most informative state points by considering their proximity to explored state points and the estimated phase boundary location. It incorporates little computational overhead and is easily generalizable to cases of high state space dimensionality. Batch sampling is also supported to take advantage of modern parallel test settings, such as supercomputing clusters. We demonstrate its use, and compare its results to traditional fixed-grid sampling. This proposed method can assist our research community by saving time and computational resources.

# 485440 A Path Entropy-Based Approach to Predict Transition Rates from Limited Information

**Abstract:**

We are interested inferring rate processes on networks. In particular, given a network's topology, the stationary populations on its nodes, and a few global dynamical observables, can we infer all the transition rates between nodes? We draw inferences using the principle of maximum path entropy. We maximize the entropy of the distribution over all possible paths a dynamical system can take subject to stationary and dynamical constraints. The present work leads to a particularly important analytical result: namely, that when the network is constrained only by a mean jump rate, the rate matrix is given by a square-root dependence of the rate, $k_{ab}$ $(\pi_b/\pi_a)^{1/2}$, on $\pi_a$ and $\pi_b$, the stationary-state populations at nodes *a* and *b*. This leads to a fast way to estimate all of the microscopic rates in the system. As an illustration, we show that the method accurately predicts transition probabilities among the metastable states of a small peptides as well as proteins at equilibrium.

# 499861 Iterative Manifold Extension for Efficient Discovery of Transition Pathways


**Abstract:**

Our understanding of high-dimensional dynamical systems, including many found in molecular modeling, can often be improved through effective free energy surfaces (FES). One of the main challenges is to find the right coarse variables upon which to construct our FES when exploration on the manifold that defines it is computationally restricted by high transition barriers. We present an algorithm to more efficiently cross transition barriers and discover new stable states by relying on the smoothness of the manifold in the original space. We first establish the effective low-dimensionality from a relatively short initial simulation burst, which gives us a local picture of the underlying manifold. After identifying the boundary of the explored region, we can extrapolate geometrically outwards to find new initial conditions to run new simulation bursts, relying on the time scale separation to correct for extrapolation errors that result in off-manifold initializations. This approach biases the simulator to cross transition barriers on the effective FES, iteratively expanding our knowledge about the underlying manifold. We demonstrate our approach on several illustrative systems.

# 495537 Learning Free Energy Landscapes Using Artificial Neural Networks

**Abstract:**

The use of adaptive biasing potential methods to estimate free energies of relevant chemical processes from molecular simulation have gained popularity in recent years. Although metadynamics and its well-tempered variant remain the most widely used, new methods such as basis function sampling, variationally enhanced sampling and Greenâ€™s function sampling have been developed which improve upon some of the shortcomings of metadynamics. In particular, the use of a basis expansion to represent the biasing potential overcomes the limitations of a Gaussian-shaped kernel and offer more flexibility in representing the often nuanced and highly nonlinear free energy landscapes. Though said approaches can outperform traditional metadynamics under many circumstances, they remain limited by the functional form or basis set chosen to represent the underlying free energy surface (FES) of interest. Without *a priori* knowledge of the FES, the choice of parameters, such as the number of terms in a basis expansion, can significantly affect the accuracy and convergence behavior of the algorithm. FESs that have sharp curvature can also introduce characteristic oscillations due to the Runge or Gibbs phenomenon. This limitation can lead to oscillations which lead to dramatic under-sampling and a non-convergent bias.

Here we develop a powerful method wherein artificial neural networks (ANNs) are used to obtain the adaptive biasing potential, and thus learn free energy landscapes. As ANNs typically represent a form of *supervised learning*, we develop an iterative scheme which refines an unbiased estimator of a system's partition function. We demonstrate that this method is capable of rapidly adapting to complex free energy landscapes and is not prone to boundary or oscillation problems. The method offers a substantial degree of flexibility to the end-user in specifying the network architecture when the topological features of the FES of interest are not known. Importantly, because the bias learned by the ANN obtains the best continuous approximation of the free energy, we see a dramatic improvement in convergence, especially for poorly sampled states over currently available and broadly used techniques.

# 501192 QM/ML: A Hybrid Quantum-Mechanics/Machine-Learning Scheme

**Abstract:**

Hybrid quantum-mechanics / molecular-mechanics (QM/MM) simulations are popular tools for the simulation of extended atomistic systems, in which the atoms in a core region of interest are treated with a QM calculator and the surrounding atoms are treated with an empirical potential. Recently, a number of atomistic machine-learning (ML) tools have emerged that provide functional forms capable of reproducing the output of more expensive electronic-structure calculations. Such ML tools are intriguing candidates for the MM calculator in QM/MM schemes, and have some obvious advantages such as the avoidance of mis-matches (e.g., between lattice constants, mechanical responses, or liquid structures) between the QM and MM calculators, and the removal of the need for the researcher to gain expertise in the MM literature. However, accurate ML models generally require a substantial amount of training data that anticipates atomic configurations that will be encountered during the course of a simulation, provided on the basis of QM methods. Here, we present a simple quantum-mechanics / machine-learning algorithm created with the goal of minimizing the amount of QM training data that needs to be anticipated and generated. Importantly, by re-training the ML calculator to the new QM information at each step, we can identically cancel the overlapping QM and MM energy and forces in a subtractive scheme, producing a simple algorithm for treating boundary interactions between the regions. The approach is analogous to the subtractive framework of the QM/MM methodology, but is not dependent on empirical parameterization, and allows the simulation of arbitrary atomic configurations while keeping QM precision, within certain physical limitations. In this talk, we outline the basic theory and algorithm and also demonstrate the application of QM/ML on the basis of simple examples.

# 500646 Automation of an Energy Renormalization Approach for the Temperature Transferable Coarse-Graining of Glass-Forming Polymers

**Abstract:**

A key objective of the Materials Genome Initiative (MGI) for soft materials is to create a coherent framework to perform coarse-grained (CG) simulations in a streamlined and efficient manner to reduce the time and cost of novel material discovery and design. Such an effort involves both advancement in coarse-graining theory, and development of appropriate machinery to give effective practical implementation. Recently, we proposed a coarse-graining strategy for glass forming (GF) liquids based on an energy renormalization (ER) scheme in which cohesive interactions are rescaled using the altered configurational entropy [1,2]. This approach yields a temperature dependent coarse-grained potential with simple form that preserves atomistic dynamics and GF activation energy from the high-temperature melt state to the low-temperature glassy state.

In this talk, we describe the implementation of the above algorithm in a computational â€œworkbenchâ€ we are developing to provide an integrated computational and data environment to support the development of coarse-grained force-fields using quantitative, progressive methods. The workbench environment has three essential elements: a modular program structure that supports the addition of new functionality through Python scripting ; a hierarchical data structure which enables unified representation of materials at different levels of granularity; an ontology based database environment. The starting point for the procedure is a set of all-atom (AA) simulation data at state points that span from the melt state to the glassy regime that are stored in the database as reference data. Starting with this upstream data, we then develop a workflow using the workbench environment to automated the parametrization of the ER potential using multiple properties derived from the AA simulation set. The theory and procedure are illustrated using a number of examples.

References

1. Wenjie Xia, Jake Song, Cheol Jeong, David D. Hsu, , Frederick R. Phelan Jr., Jack F. Douglas, and Sinan Keten, "Energy-Renormalization for Achieving Temperature Transferable Coarse-Graining of Polymer Dynamics," submitted to Proceedings of the National Academy of Sciences (PNAS) 2017.

2. Jake Song, David D. Hsu, Kenneth R. Shull, Frederick R. Phelan Jr., Jack F. Douglas, Wenjie Xia, Sinan Keten, "Energy Renormalization Method for the Coarse-Graining of Polymer Viscoelasticity," in preparation, (2017).

# 501681 Developing Theory and Data-Driven Benchmarks for General Coarse-Grained Mapping Operators

**Abstract:**

Coare-graining (CG) is a dimensionality reduction technique used to more efficiently simulate multiscale systems. CG methods rely on grouping similar particles together (mapping) and then redefining the potential energy in terms of the reduced particle number. There is a mature theory about generating potentials for CG models. There is not a rigorous theory for generating mappings from the fine-grain (FG) system to the CG system. This missing component is essential because past CG work shows that many mappings lead to homogeneous, weakly interacting, gas-like CG models. Identifying a mapping operator that preserves relevant physics is usually the most heuristic but essential step in developing CG models. The number of possible mappings is exponential with respect to the number of atoms, which is especially a problem since CG mapping are desired when the number of atoms is large.

In this talk I will outline a new graphical rigorous theory for CG mappings with techniques from the field of computer vision. Due to the ill-defined problem of evaluating specific mapping operators, the talk will emphasize data-driven benchmarks for general mapping strategies. Resulting mappings for solute-solvent and liquids will be explored.

# 495174 New Coarse-Grained Models of Hydrocarbons

**Abstract:**

New coarse-grained models for hydrocarbons are developed by utilizing particle swarm optimization (PSO) method. In this model, the bonded potentials are based on all-atom simulations of alkane, while the non-bonded potentials are optimized by the PSO method to reproduce experimental properties. Specifically, all-atom simulation trajectories were utilized to estimate the average bond length and angle. The PSO method was employed to optimize the 12-6 Lennard-Jones Potentials. New CG force-field parameters were optimized to reproduce the experimental properties of hydrocarbons including density (*Ï*), self-diffusion coefficient (*D*), heat of vaporization (*H*v) and surface tension (*Î³*). The potentials are transferable over a range of hydrocarbons with different chain lengths .

# 494168 Constrained Best Subset Selection Methodology for the Regression of Helmholtz Energy Equations

**Abstract:**

Accurate thermodynamic properties are important for the development of new processes and technologies and simulating an optimized flowsheet. Unfortunately, equations of state typically used, such as Peng Robinson and Soave-Redlich Kwong, are inaccurate in critical regions to be able to provide accurate models. A recent development in equations of state uses first principles to determine one unifying equation explicit in Helmholtz energy to overcome these inaccuracies. These equations have been developed for over 100 pure substances and have been expanded to mixtures [1, 2, 3, 4].

A major challenge in the development of Helmholtz energy equations is the fitting of large datasets containing correlated data to the main unifying equation [3]. The current fitting procedures depend on cycling between linear and nonlinear regression techniques that are restricted to equality constraints and result in local solutions requiring multiple starts or an experienced user manually selecting an initial starting point [4]. Advancing these techniques to eliminate the need for multistart heuristics and enforce inequality constraints on the resulting models would allow us to control the thermodynamic behavior, improve extrapolation behavior, and maintain thermodynamically feasible limitations on the regressed equation while simultaneously fitting the data [5].

In order to address these challenges associated with fitting data to Helmholtz energy equations, we have developed a new model-building procedure from thermodynamic data. The main idea is to apply best subset selection to allow for the controlling of thermodynamic slopes with inequality constraints while fitting all the data simultaneously. A bank of terms is chosen to best represent the different phase regions and thermodynamic behavior. In order to avoid overfitting, our procedure systematically selects a subset of these terms to optimally fit the multiple thermodynamic property data sets. We rely on a global optimization solver [6] to find an optimal solution that optimizes the fit as well as the number of terms in the model according to an information criterion.

The proposed approach was used to fit a new Helmholtz energy equation for toluene. The resulting models were validated against results from the literature [7]. With the same data and bank of terms, the new methodology results in accuracy within the range of the experimental error of the thermodynamic properties incorporated. Our methodology can be applied to other pure and pseudo-pure chemical data sets to automatically generate new Helmholtz energy equations for use in simulations and optimization. **References cited**

[1] Lemmon, E. W.; Huber, M. L.; McLinden, M. O. NIST Standard Reference Database 23: Reference Fluid Thermodynamic and Transport Properties-REFPROP, Version 9.1, National Institute of Standards and Technology. 2013; https://www.nist.gov/srd/refprop.

[2] Span, R. Multiparameter equations of state: An accurate source of thermodynamic property data; Springer-Verlag, 2000.

[3] Span, R.; Wagner, W.; Lemmon, E. W.; Jacobsen, R. T. Multiparameter equations of state - recent trends and future challenges. Fluid Phase Equilibria 2001, 183-184, 1-20.

[4] Wagner, W.; PruÃŸ, A. The IAPWS formulation 1995 for the thermodynamic properties of ordinary water substance for general and scientific use. *J. Phys. Che. Ref.* Data 2002, 387.

[5] Lemmon, E. W.; Jacobsen, R. T. A New Functional Form and New Fitting Techniques for Equations of State with Application to Pentafluoroethane (HFC-125). J. Phys. Chem. Ref. Data 2005, 34, 69â'' 108.

[6] Tawarmalani, M.;Sahinidis, N. V. Global optimization of mixed-integer nonlinear programs: A theoretical and computational study, Mathematical Programming, 99, 563-591, 2004.

[7] E. W. Lemmon. Correlations short fundamental equations of state for 20 industrial fluids. J. Chem. Eng. Data, (51):785{850, 2006.

# 501374 Identifying Equilibrated Simulation Trajectories with Artificial Neural Networks


**Abstract:**

Determining which microstates to include in an ensemble average is a ubiquitous simulation task whose reproducibility is hampered by being underspecified. For a single simulation, the task is not so bad: Including or omitting one microstate from a sufficiently large set won't change the answer, so heuristics like "The potential energy is higher here, so we'll cut those states out" will do. The problem arises when thousands of simulation trajectories need to be analyzed, demanding efficient heuristics that are automatable and deterministic. Artificial neural networks are one type of machine learning algorithm that provides a reproducible path for applying pattern recognition heuristics. Here we use the TensorFlow machine learning library to train an artificial neural network on the problem of distinguishing "equilibrated" from "not equilibrated" hypothetical observation sequences. We generate training populations and test populations of normally-distributed observation sequences with embedded linear and exponential correlations. We train a two-neuron artificial network to distinguish the correlated and uncorrelated sequences. We characterize the predictive capabilities of the network as a function of training population size, the number of points in each sequence, and the number of training rounds performed. We find diminishing returns on predictive accuracy with sample sizes over 512 points, training populations of 8000 sequences, and 2000 training rounds, which takes about three minutes to train. We find that this simple network is powerful enough for over 98% accuracy in identifying exponentially-decaying observation sequences in microseconds. These observations suggest neural networks as an effective tool for automating this and other heuristic bottlenecks in high-throughput simulation pipelines.

# Deep Cavity Cavitand/Alkane Assembly State Switching

Tulane University

Du Tang, Wes Barnett, Bruce Gibb, Hank Ashbaugh

## Biology Routinely Utilizes Confinement

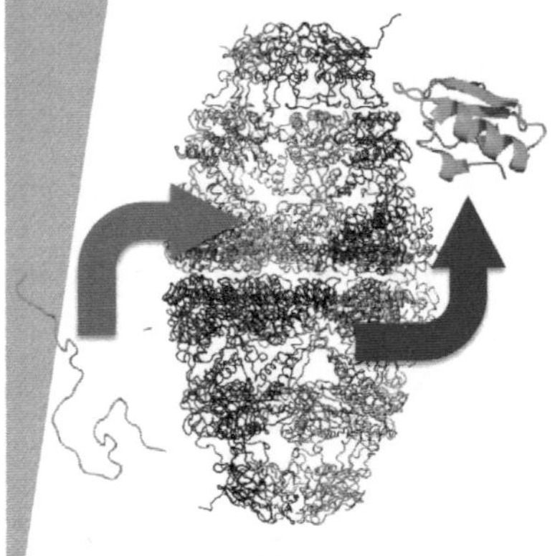

chaperone complex ~35 nm

viral capsid ~100s nm

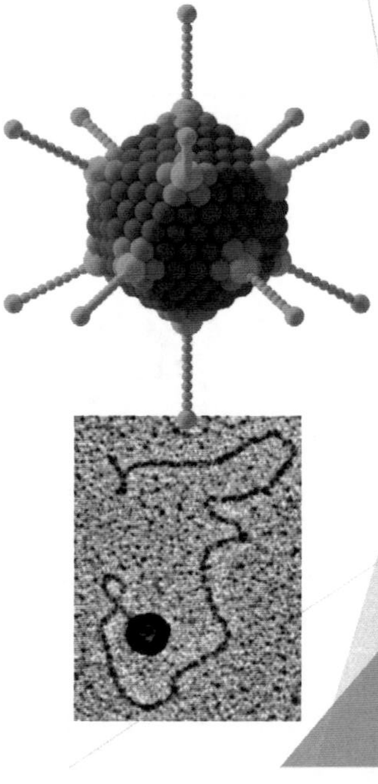

## Deep-cavity Cavitand Structure

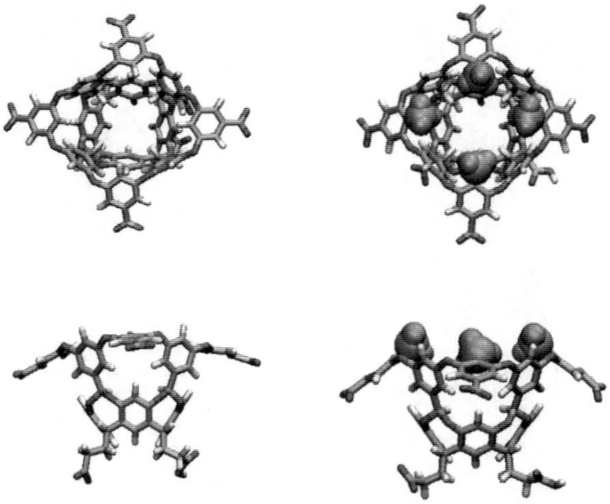

OA (octa-acid) TEMOA (tetra-endo-methyl-octa-acid)

3

## So Many Ways to Assemble …

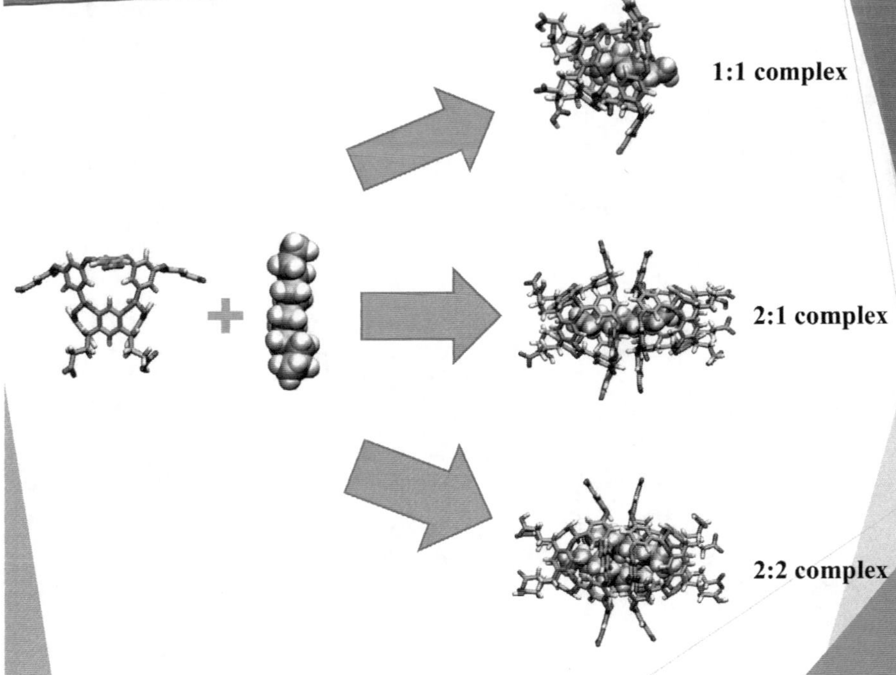

## OA versus TEMOA dimerization

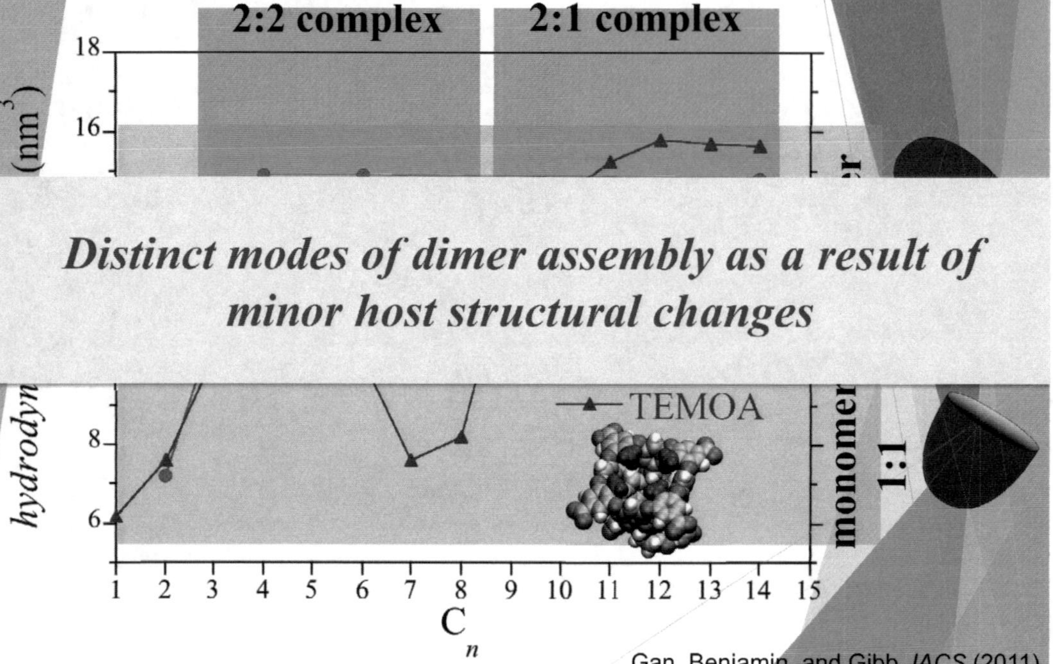


***Distinct modes of dimer assembly as a result of minor host structural changes***

Gan, Benjamin, and Gibb *JACS* (2011)

## Complex Simulation Methods

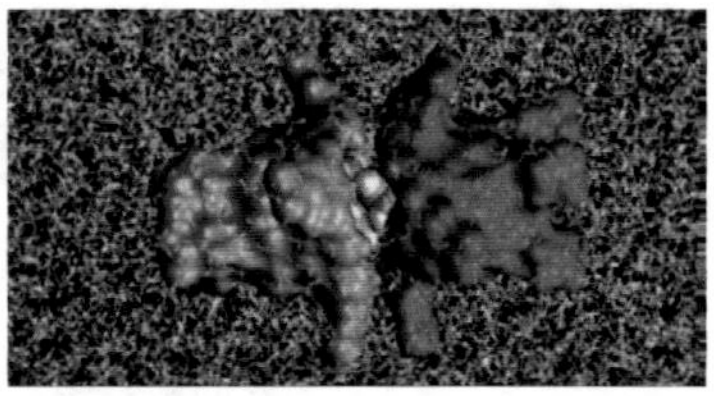

- GROMACS 5.1 MD simulation package
- Water: TIP4PEW
- Cavitand: GAFF
- Alkanes: LOPLS-AA
- -6e net charge, 298.15K,1bar

Potential of mean force

- Umbrella sampling
- 29 to 41 windows
- 0.5 Å-spaced windows
- k = 15000 kJ/ (mol $nm^2$)
- 15 ns at each window

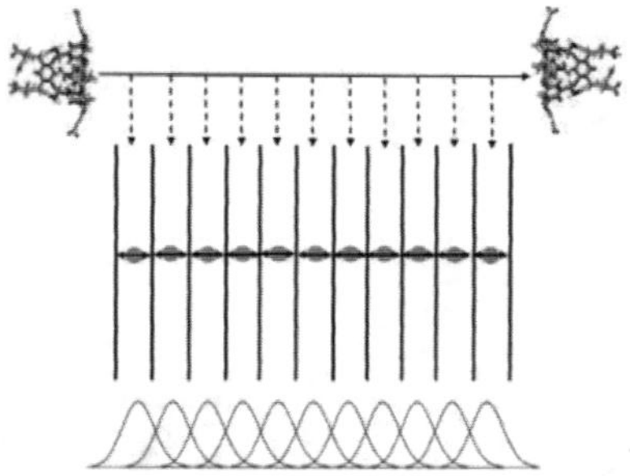

## OA Complex Formation Simulations

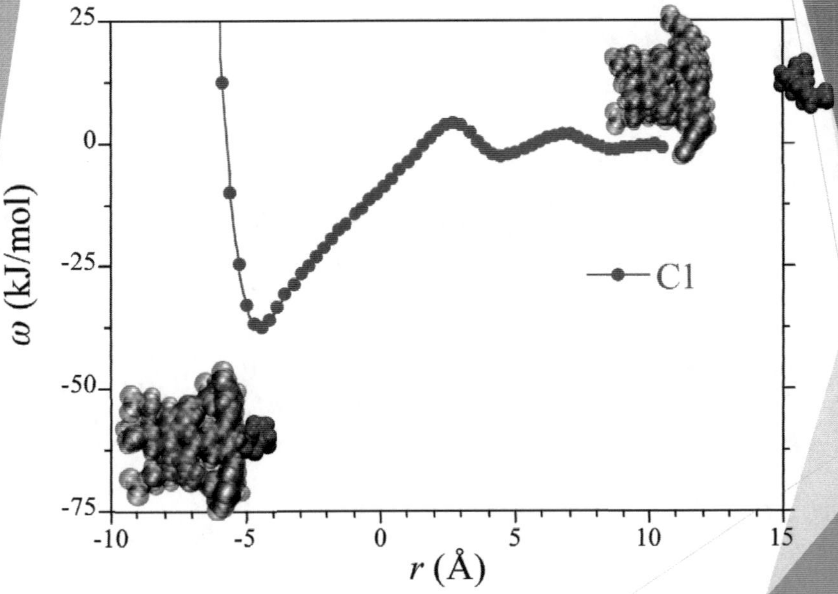


## Minima of Association Free Energy

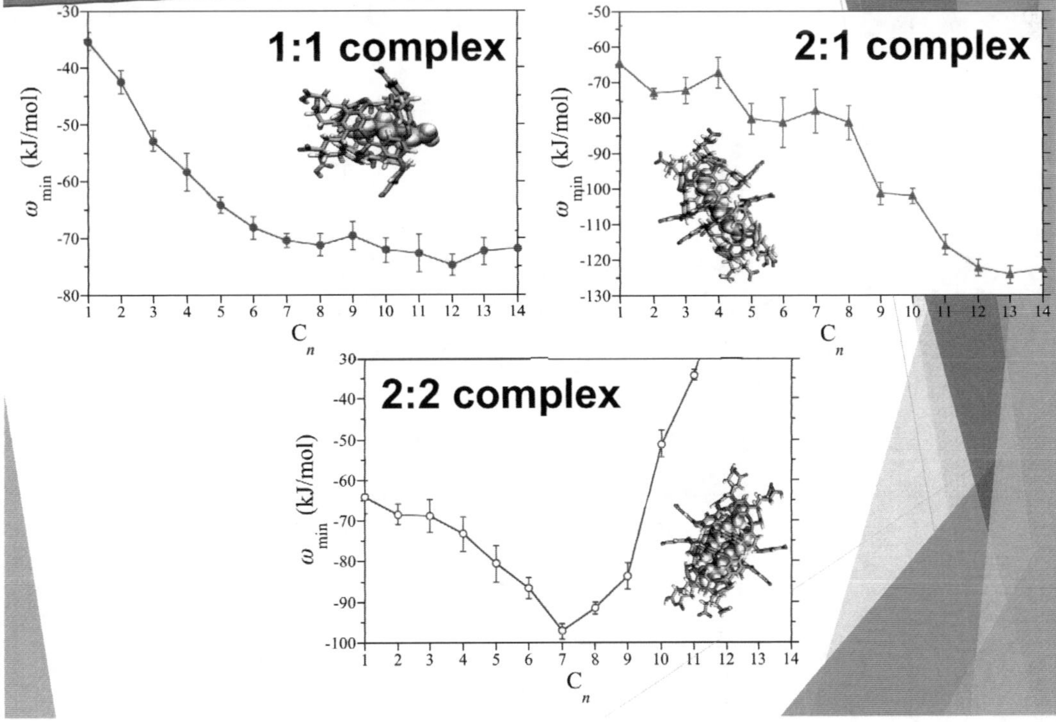

## Assembly Reaction Equilibrium

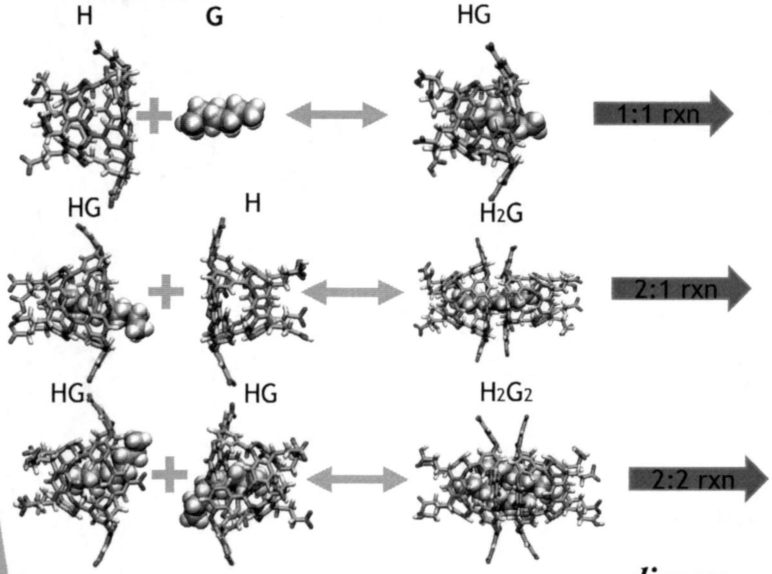


$$\frac{[HG]}{[H][G]} = K_{1:1}$$

$$\frac{[H_2G]}{[HG][H]} = K_{2:1}$$

$$\frac{[H_2G_2]}{[HG]^2} = K_{2:2}$$

***monomers*** ***dimers***

$$K_{1:1} = \alpha \exp(-\omega_{1:1}/RT)$$

$$K_{2:1} = \beta \exp(-\omega_{2:1}/RT)$$

$$K_{2:2} = \beta \exp(-\omega_{2:2}/RT)$$

## Different Cavitand/Alkane Complexes Distributions

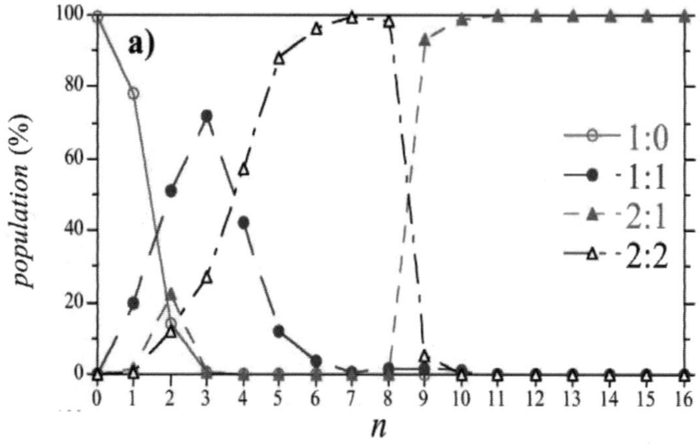


**OA**

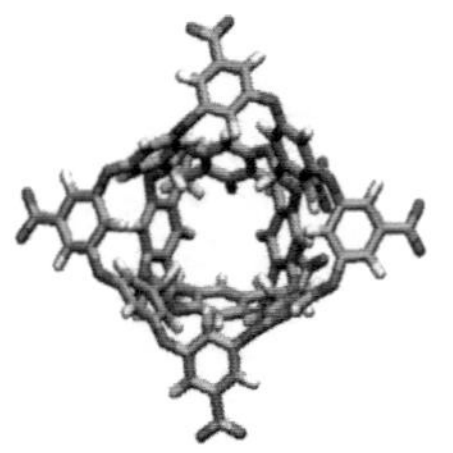

## Different Cavitand/Alkane Complexes Distributions

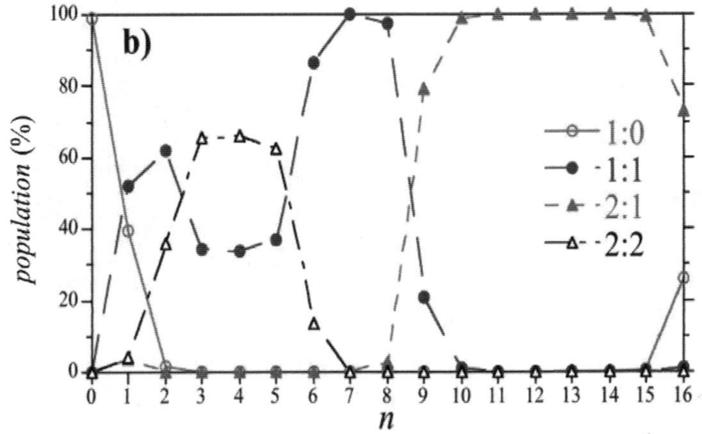


**TEMOA**

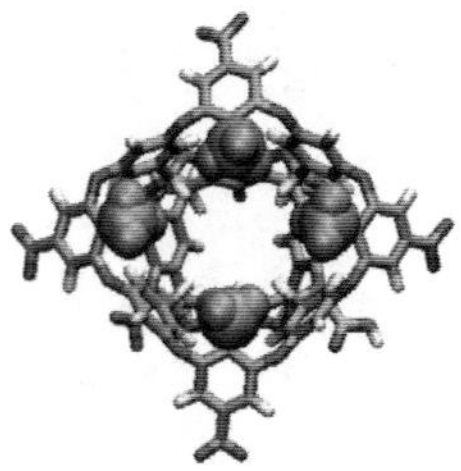

## Cavitand/Alkane Complex Volumes

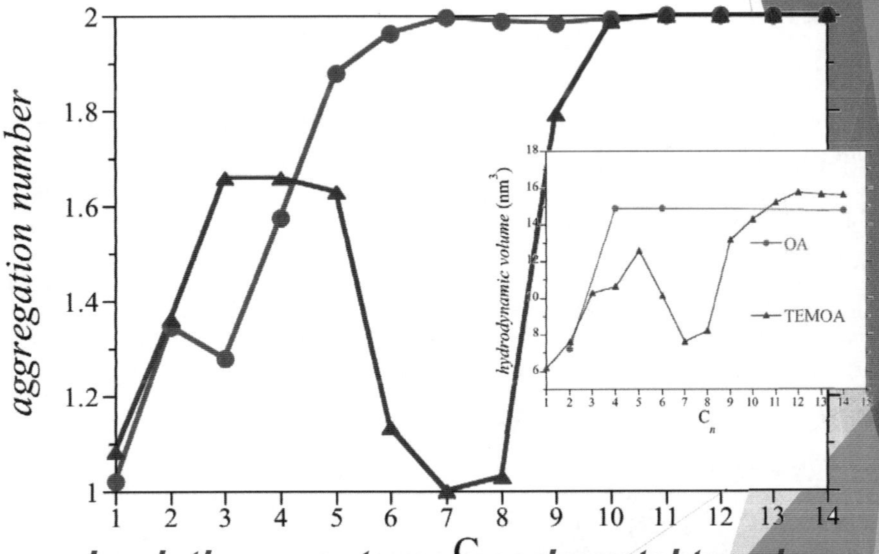


***simulations capture experimental trends***

Gan, Benjamin, and Gibb *JACS* (2011)

## OA versus TEMOA 2:2 Complexation

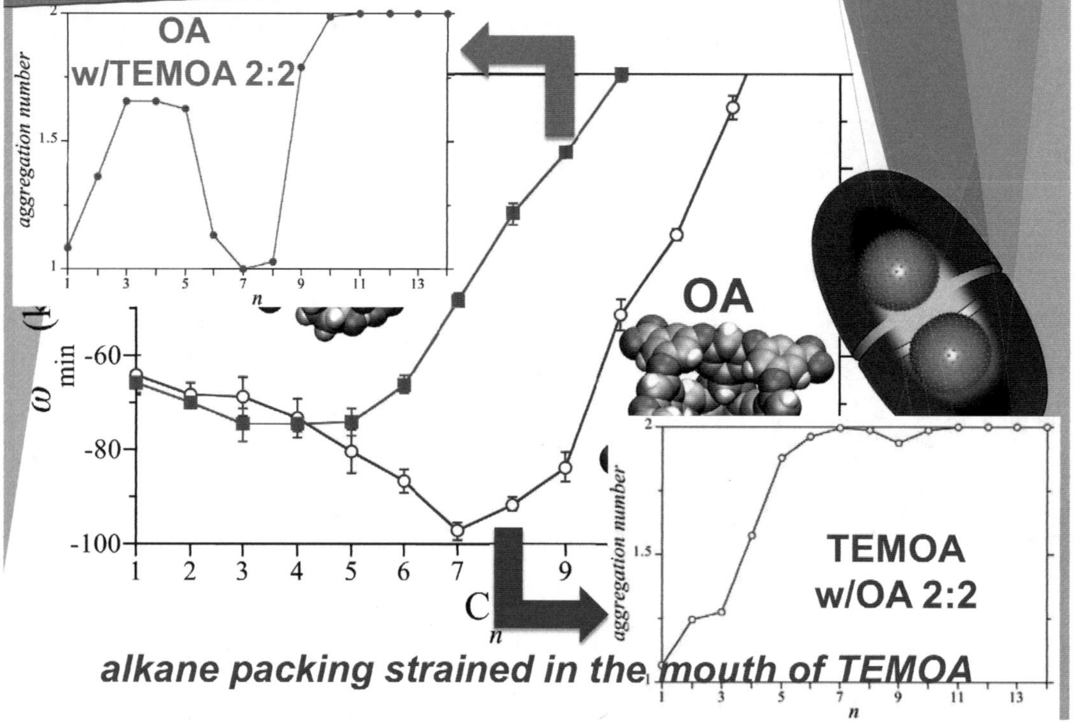


## In Summary

- Simulation accurately reproduce OA's monotonic and TEMOA's non-monotonic assembly patterns
- Conformational freedom of guests within TEMOA capsules are severely limited. In turn, it has repulsive contributions to TEMOA 2:2 and 2:1 complex association free energies.
- The balance between monomeric and dimeric capsules are therefore shifted.
- Small chemical modifications to constituents of a supramolecular assembly can dramatically impact on their structure and properties.

**Acknowledgement**

Tulane University NSF LONI

# 501193 Elucidating Protein (Folding) Kinetics Near Organic Surfaces As a Function of Surface Hydrophobicity

**Abstract:**

Protein-surface interactions are ubiquitous in biology and in modern biosensing devices, but how such interactions impact protein stability and dynamics is not well understood. Surface hydrophobicity characterized by water contact angle is often used to describe surface-induced modulations in the protein stability. The role of surface hydrophobicity in protein diffusion and folding kinetics is much less well-understood. Here, we use peptide loop-closure (end to end contact formation) as a proxy for measuring protein folding kinetics and how it is impacted by protein-surface interactions. We use self-assembled monolayers (SAMs) for which hydrophobicity can be controlled by changing the headgroup as these surfaces are routinely employed in laboratory experiments. We find that the rate of contact formation between peptide ends is significantly altered by its interactions with the SAM surface. Furthermore, the change with respect to bulk is a function of surface hydrophobicity. Interestingly, we do not find a significant impact of protein-surface interactions on protein diffusion for these small peptides diffusing around on defect-free SAM surfaces.

# 501288 Serum Albumin Interactions with Doxorubicin-Loaded Graphene Oxide in an Aqueous Environment with Blood pH Level: A Molecular Dynamics Simulation Study

**Abstract:**

Graphene oxide (GO) as a nanocarrier has been shown to possess desirable delivery and targeting properties for doxorubicin (DOX) anticancer drug [1]. In general, an effective drug delivery system should be able to load the drug, carry it in the blood stream, and release it at the action site. In this study, we use molecular dynamics (MD) simulation to investigate the interactions between albumin, which has a crucial role in the pharmacokinetics and pharmacodynamics of nanoparticulate drug delivery systems, and DOX-loaded GO in an aqueous environment mimicking blood with a pH level of 7.4. Our objective is to elucidate the effect of serum albumin on the stability of the drug delivery system during intravenous delivery. Understanding these fundamental molecular-level interactions will help in designing improved nanocarriers for an efficient delivery of DOX to the action site without any appreciable premature release of the drug in the blood stream. We are performing a series of MD simulations of the DOX/GO/albumin system in the NAMD software using the CHARMM force field at the blood pH level and body temperature using an implicit solvent method. DOX-albumin, albumin-GO, and DOX-GO interactions are calculated and compared. Preliminary results will be given and a direction for future work will be outlined.

[1] M. Mahdavi, F. Rahmani, S. Nouranian, Molecular simulation of pH-dependent diffusion, loading, and release of doxorubicin in graphene and graphene oxide drug delivery systems, *Journal of Materials Chemistry B* 4(46) (2016) 7441-7451.

# 502084 Elucidating the Interaction Mechanisms of Thermo-Responsive Ligand with Proteins

**Abstract:**

Poly (N-vinylcaprolactam) (PVCL) is a prototype thermo- and salt responsive polymer. PVCL is a widely investigated biomaterial that has potential applications in drug delivery and protein purifications as responsive hydrophobic ligands for protein adsorption and elution during down stream processing. Here classical molecular dynamics simulations were conducted to investigate the energetics and interaction mechanisms between the PVCL polymer chains and bovine serum albumin (BSA) in aqueous solutions. Water structures surrounding the polymer chains and BSA as well as their interactions including hydrogen bonding, electrostatic and van der Waals interactions were determined. Significant insights were obtained on the effects of polymer hydration state (i.e. collapsed or hydrated), polymer chain length as well as the presence of salt ions on the protein-ligand binding interactions.

# 487177 Study of Interaction and Transpassing of Human Beta Defensin-3 with Popg and Pops Membrane

**Abstract:**

Human beta defensin type 3 (hBD-3) is a cysteine rich small peptide. It has 45 residues and a charge density of +11. hBD-3 can form dimer or higher order oligomer. It has the potent antimicrobial activity even at high salt concentrations, which is believed to depend on its interaction with biological membrane. In this study, the interaction of hBD-3 dimer withÂ 1 -Palmitoyl-2-oleoyl-sn-glycero-3 â€"phosphatidylserine (POPS) andÂ 1 -Palmitoyl-2-oleoyl-sn-glycero-3 â€"phosphatidylglycerol (POPG) membrane bilayer is investigated using all- atom molecular dynamic simulation. Factors influencing the structure and dynamics of hBD-3 in lipids under different conditions are investigated, including altering-disulfide bonds between the cysteine residues of two hBD-3 monomers, protonation state of hBD-3, and NaCl concentration. Protonation and NaCl concentration does not influence notably the structure and dynamics of hBD-3 but hBD-3 structure is found to be more flexible in absence of disulfide bond. hBD-3 dimer showed a bit instability in POPS at protonated state other than that it was quite stable in both POPG and POPS in all conditions.

In order to understand the trans-passing process of hBD-3 dimer through the POPG/POPS lipid bilayer, the Potential Mean Force (PMF) is calculated using Umbrella Sampling method. Results are compared between different lipid bilayers.

JCTC Journal of Chemical Theory and Computation

Article

pubs.acs.org/JCTC

# Free Energy Calculations of Microcin J25 Variants Binding to the FhuA Receptor

Pin-Kuang Lai and Yiannis N. Kaznessis*

Department of Chemical Engineering and Materials Science, University of Minnesota, 421 Washington Avenue SE, Minneapolis, Minnesota 55455, United States

Supporting Information

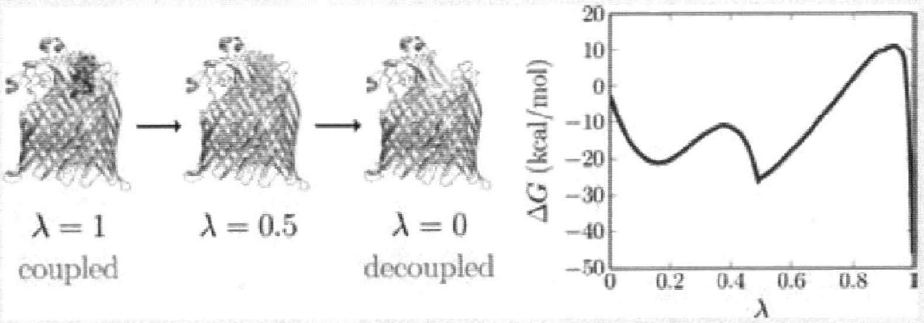


ABSTRACT: Computer simulations were performed to study the antimicrobial peptide microcin J25 (MJ25), a 21-mer peptide with an unusual lasso structure and high activity against Gram-negative bacteria. MJ25 has intracellular targets. The initial step of MJ25 acquisition in bacterial cells is binding to the outer-membrane receptor FhuA. Molecular dynamics simulations were implemented to study the binding mechanism of MJ25 to FhuA and to search for important binding residues. The absolute binding free energy calculated from combined free energy perturbation and thermodynamic integration methods agrees well with experimental data. In addition, computational mutation analysis revealed that His5 is the key residue responsible for MJ25 and FhuA association. We found that the number of hydrogen bonds is essential for binding of MJ25 to FhuA. This atomistic, quantitative insight sheds light on the mechanism of action of MJ25 and may pave a path for designing active MJ25 analogues.

## 1. INTRODUCTION

Antibiotics are vital to modern medicine, but resistance to available antibiotics has become a pressing issue over the last two decades. Antimicrobial peptides (AMPs) are a promising alternative treatment to kill bacteria.[1,2] Scientists have tried to engineer new AMPs starting with natural analogues to enhance their activity.[3−5] Despite the efforts to design new AMPs, the underlying mechanism of action at the atomistic level is still poorly understood. This lack of understanding hampers efforts to develop AMP-based therapeutics.

Microcins are gene-encoded AMPs with molecular weights less than 10 kDa. They are produced mostly by Gram-negative Enterobacteria.[6] Microcins have receptor-mediated and metabolic inhibition mechanisms of antimicrobial activity. Among all of the inhibitory microcins discovered, microcin J25 (MJ25) has attracted much attention in recent years.[6−10] MJ25 is a 21-mer peptide with an unusual lasso ring structure, and it exhibits high activity against Gram-negative bacteria. MJ25 binds to the iron transporter FhuA located on the outer bacterial membrane.[11] It is then transported through FhuA via the energy transduction complex TonB−ExbB−ExbD and then binds to the inner-membrane protein SbmA.[12,13] The final intracellular target is the secondary channel of RNA polymerase, where MJ25 blocks the diffusion of nucleoside triphosphate substrate to the enzyme catalytic site.[14,15] A second intracellular target of MJ25 is likely the respiratory chain through the production of reactive oxygen species.[16]

Despite numerous experimental studies on MJ25, previous computational work is limited. Ferguson et al.[17] used molecular dynamics (MD) simulations of uncyclized MJ25 in an aqueous phase. However, since the ring structure has been shown to be crucial to antimicrobial activity, it is useful to simulate MJ25 in its active form. Pan et al.[18] performed a computational redesign of the MJ25 peptide using a two-stage sequence selection procedure based on both energy minimization and fold specificity. Hegemann and co-workers optimized and incorporated integrin motifs on MJ25.[19] The application of computational methods to study binding of MJ25 to its cognate receptors has been limited by a lack of information about the structures of the binding complexes.

Recently the structure of the MJ25−FhuA complex was determined by X-ray crystallography at a high resolution of 2.3 Å.[9] This encouraged a first computational study of the MJ25−FhuA complex. In this work, we focused on the binding free energy for the MJ25−FhuA complex and a mutation analysis to identify important binding sites. The goal was to understand the nature of the interaction between MJ25 and FhuA in terms

Received: April 21, 2017
Published: June 16, 2017

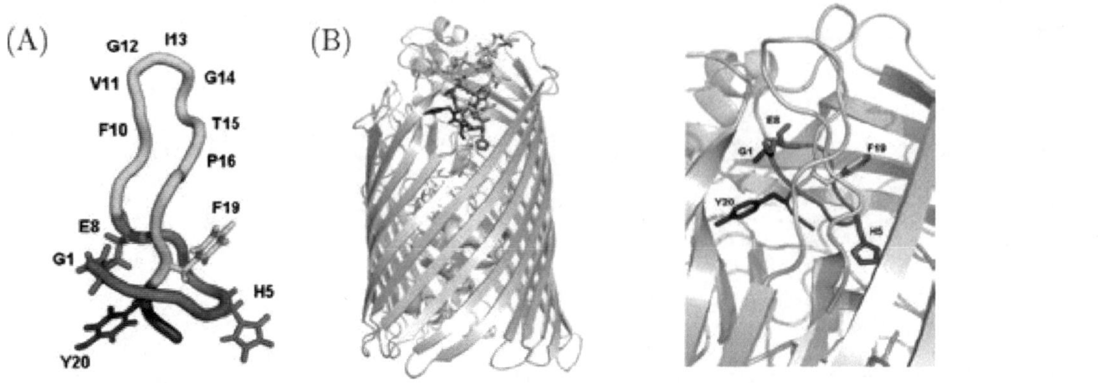


**Figure 1.** (A) Lasso structure of MJ25. The loop region (G1−E8) is shown in red, the upper tail region (Y9−F19) in green, and the lower tail region (Y20−G21) in blue. (B) Structure of the MJ25−FhuA complex (PDB entry 4CU4).

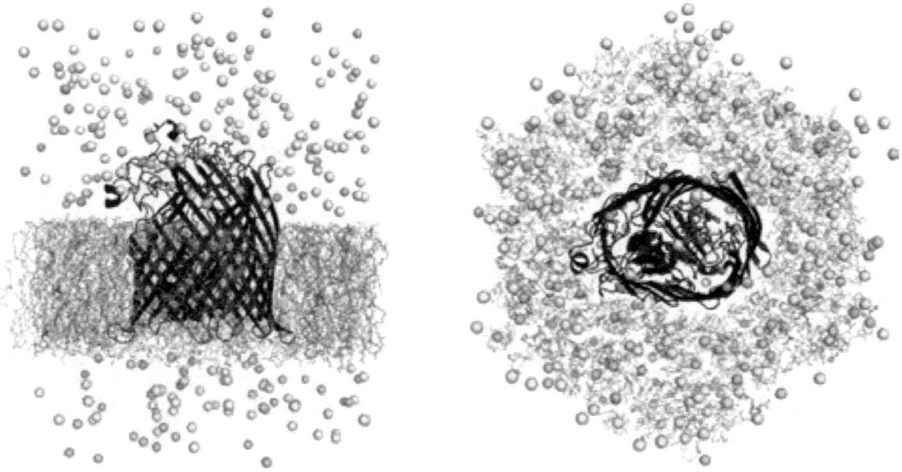

**Figure 2.** MD system for MJ25 and the FhuA receptor. MJ25 is shown in red, the FhuA receptor in blue, POPE lipids in gray, POPG lipids in purple, chloride ions in green, and potassium ions in yellow.

of thermodynamic driving forces. In what follows, we present the methods we employed.

## 2. METHODS AND THEORY

**2.1. Modeling of MJ25 and FhuA.** The structures of MJ25 and FhuA receptor were initially taken from the available structure (PDB entry 4CU4).[9] The lasso structure of MJ25 is formed when the N-terminal amide forms an isopeptide bond with the carboxylic acid of glutamic acid (E8). The C-terminus tail then threads back into the ring to form the lasso-like structure (Figure 1A). Figure 1B illustrates the structure of the MJ25−FhuA complex. FhuA is a 79 kDa outer-membrane siderophore receptor. It is a monomeric $\beta$-barrel protein consisting of 22 antiparallel $\beta$-strands. The N-terminus folds inside the $\beta$-barrel, forming a plug domain (residues 20−160). The side-chain ring of histidine (His5) of MJ25 inserts deeply into the plug region of FhuA in what has been reported to be a vital binding site.[9]

Residues 405 to 417 of FhuA are missing from the experimentally determined structure. We used another crystal structure of FhuA (PDB entry 1QFG) to derive the structure of the missing part. This was done by means of a least-squares fit of the positions of the $C_\alpha$ atoms on the $\beta$-barrel of the FhuA receptor using common residues of both structures.

**2.2. System Construction.** The CHARMM-GUI membrane builder was used to embed the MJ25−FhuA complex inside a lipid bilayer, solvate it with water, and add ions (Figure 2).[20−23] The lipid bilayer consists of a random mixture of the lipids palmitoyloleoylphosphatidylethanolamine (POPE) and palmitoyloleoylphosphatidylglycerol (POPG). The proportion of POPE to POPG is 3:1, which mimics the components of the membrane of Gram-negative bacteria.[24]

We note that FhuA resides on the outer membrane of bacteria,[11] whereas POPE/POPG mixtures mimic the inner one. The outer membrane is asymmetric, composed of phospholipids on the inner leaflet and lipopolysaccharides (LPSs) on the outer leaflet. There have been a number of recent efforts to build and model bacterial outer membranes.[25−27] The area of outer-membrane simulations is indeed becoming an active one. However, the simulation of LPSs on the outer leaflet is still challenging mainly because of the structural complexity of LPSs, the difficulties of large, asymmetric system construction for LPS−membrane environments, and importantly, a relative absence of proper force fields, especially for the vast repertoire of mono- and polysaccharides.[25]

We chose to model FhuA in a symmetric lipid bilayer for three reasons. First, in a previous study[28] a lipid bilayer was used to simulate the membrane environment of FhuA, with the protein structure remaining relatively stable during the simulation. Second, the simulation of lipid bilayers as membrane mimics is a mature technique with well-established building and simulation protocols and accurate, well-tested force fields. Third, on the basis of the structure of FhuA, we assumed that the binding site of MJ25 on the FhuA receptor is in contact with the water subphase, not close to LPSs. We could thus hypothesize with reasonable confidence that the actual membrane composition may have little effect on the binding affinity.

We also note that the dissociation constant for the MJ25–FhuA complex was measured in detergents. As discussed in ref 11, in vitro results can be considered to reliably mimic the actual in vivo behavior because in vitro and in vivo measurements of inhibition values matched well with increasing MJ25 concentration.

TIP3P waters were added on each side of the membrane with a thickness of approximately 15 Å for the binding free energy simulation. PROPKA was used to assign the protonation states of amino acids at pH 7.[29] Histidine on MJ25 (His5) was protonated, which is consistent with experimental observations.[13] The positive charge is balanced by the negative charge on the C-terminus, and therefore, MJ25 is neutral. The net charge of the FhuA receptor is −13. There were 216 POPE molecules and 72 POPG molecules in the system. POPE is neutral while POPG has a net charge of −1. Potassium ions were added to neutralize the charges on the pore and lipid bilayer. Additional potassium and chloride ions were placed to reach an ionic concentration of 0.15 M. The entire system was built in a hexagonal cell with the $z$ axis perpendicular to the hexagonal face.

**2.3. MD Simulation Details.** All of the MD simulations in this work were performed with NAMD2.11[30] with a time step of 2 fs. The CHARMM36[31] force field was used for proteins and lipids, and the TIP3P model[32] was used for waters. Parameter modifications were required for the unusual ring structure. The details are presented in Figure S1 in the Supporting Information. Energy minimization was performed for 2000 steps. The system was then heated to 300 K at a rate of 1 K/ps at 1 atm. Harmonic constraints were enforced on the protein main-chain atoms and lipid headgroups during heating. An additional 100 ps run at 300 K with the same constraints was then performed. The last equilibration step was for 1 ns with all of the constraints relaxed. The production run was performed in the *NPT* ensemble for 50 ns. Semi-isotropic pressure coupling at 1 atm was achieved using the Langevin piston,[33,34] while temperature control was accomplished via the Langevin thermostat.[35] The oscillation period and damping time for pressure control were 50 and 25 ps, respectively. The damping coefficient for temperature control was set to 1 ps$^{-1}$. The electrostatic interactions were calculated using the particle-mesh Ewald (PME) method.[36] The grid spacing for the PME method was set to 1 Å. The van der Waals interactions were smoothly turned off between 10 and 12 Å with the force switching method. The neighbor pair-list distance was 16 Å. The RATTLE algorithm[37] was used to constrain all of the hydrogen-containing bonds.

**2.4. Standard Binding Free Energy.** Methods for the calculation of binding free energy by computer simulation have been developed and applied to numerous ligand–protein systems.[38−53] In this study, the protocol for calculation of the binding free energy from MD simulations developed by Gumbart et al.[51] was adopted. The double-decoupling scheme was implemented.[47] The calculation was divided into several stages as shown in the thermodynamic cycle depicted in Figure 3A. The ligands in both the aqueous solution (bulk phase) and

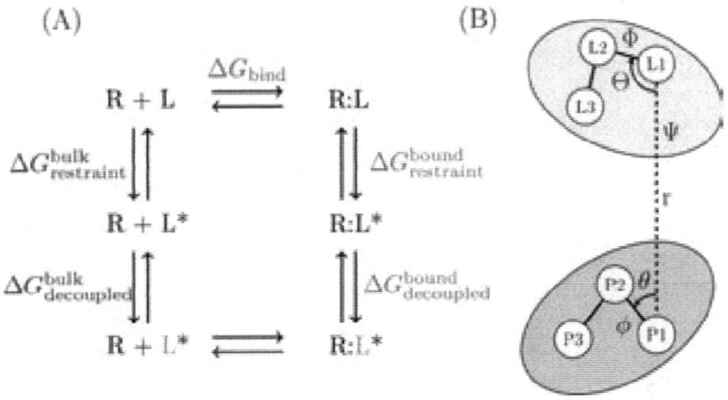


**Figure 3.** (A) Thermodynamic cycle for calculation of the binding free energy: $\Delta G_{\text{bind}} = \Delta G^{\text{bulk}}_{\text{restraint}} + \Delta G^{\text{bulk}}_{\text{decoupled}} - \Delta G^{\text{bound}}_{\text{restraint}} - \Delta G^{\text{bound}}_{\text{decoupled}}$. (B) Illustration of the reference frame of a ligand–receptor system. The triplet {P1, P2, P3} are three groups chosen from a receptor protein. The triplet {L1, L2, L3} are three groups chosen from a ligand. The positional degrees of freedom are defined by the P1–L1 separation distance ($r$), the L1–P1–P2 angle ($\theta$), and the L1–P1–P2–P3 dihedral angle ($\phi$). The orientational degrees of freedom for the ligand are defined by the P1–L1–L2 angle (Θ) and two dihedral angles, P1–L1–L2–L3 (Φ) and P2–P1–L1–L2 (Ψ).

the binding site (bound phase) were first restrained and decoupled from their environments. The absolute binding free energy is the summation of all of the contributions: $\Delta G_{\text{bind}} = \Delta G^{\text{bulk}}_{\text{restraint}} + \Delta G^{\text{bulk}}_{\text{decoupled}} - \Delta G^{\text{bound}}_{\text{restraint}} - \Delta G^{\text{bound}}_{\text{decoupled}}$. Figure 3B presents the definition of geometric constraints.

Positional, orientational, and conformational restraints are required to circumvent the "wandering ligand" issue.[52−54] Three sites are selected for ligands (L1, L2, L3) and for receptors (P1, P2, P3). The positional degrees of freedom are defined by the separation distance $r$ and the two angles $\theta$ and $\phi$. The orientational degrees of freedoms are defined by an angle Θ and two dihedral angles, Φ and Ψ. The choice of these sites is arbitrary, with the requirements that any four points are not on the same plane and any three points are not on the same line. In this work, the sites for MJ25 were L1 = Y20, L2 = V6, and L3 = I17. For FhuA the three sites were P1 = E130, P2 = T73, and P3 = R93. For MJ25, one residue from each structure region was selected, namely, the ring (V6), the upper loop (I17), and the lower tail (Y20). For FhuA, E130 was chosen because it is aligned with Y20 on MJ25 along the $z$ axis, a natural choice for calculating the separation distance $r$. In addition, T73 and R93 were selected because they are close to E130 and also in the plug domain.

The free energy calculations were based on thermodynamic integration (TI), which was performed by coupling the force constant of each harmonic restraint potential to a 14-grid-point order parameter Λ with the values [0 0.1 0.2 0.3 0.4 0.5 0.6 0.7 0.8 0.9 0.99 0.999 0.9999 1]. The energy gradient increases rapidly close to Λ = 1, and therefore, a finer grid was applied in that region. The potential gradient with respect to each individual collective variable at each Λ was calculated from MD simulations with 50 000 steps for equilibration (0.1 ns) followed by 200 000 steps for production (0.4 ns). The restraint free energy change was obtained by numerical integration.[41,51] The scaling of the force constant was performed bidirectionally, and the free energy value was calculated by averaging the results.

The calculation of the free energy of coupling/decoupling of MJ25 relied on the free energy perturbation (FEP) method. The system was divided by the order parameter $\lambda$ unevenly between the initial state ($\lambda = 0$) and the final state ($\lambda = 1$). An MD simulation consisting of 200 000 production steps after 50 000 equilibration steps was performed for each window. The simulation was carried out bidirectionally (forward and backward) for MJ25 in both aqueous solution and the binding site. The free energy change was calculated by means of the Bennett acceptance ratio (BAR),[46] combining statistical data assessed from the forward and backward directions. The convergence of the FEP simulation was assessed by analyzing the overlap of the energy changes ($\Delta U$) in both directions using the ParseFEP tool.[55]

**2.5. Calculation of the Relative Free Energy upon Mutation.** Techniques such as alanine scanning mutagenesis are commonly applied to determine the free energy change between wild-type and alanine mutant variants for each residue. In principle, this protocol can be extended to mutation with any amino acid besides alanine. In practice, this results in a large library, and rapid and accurate determination of free energy changes upon mutation is required even for short peptides. Perhaps the simplest method is the docking and scoring method.[56,57] Another intermediate approach is based on molecular mechanics with Poisson–Boltzmann or generalized Born and surface area continuum solvation methods (MM/PBSA or MM/GBSA).[42] The most sophisticated and accurate approaches are free energy perturbation (FEP) and thermodynamic integration (TI), which rely on alchemical transformation from one residue to another in MD simulations.[40,50,51] The first two methods are computationally inexpensive and are usually applied in the high-throughout screening method. However, the accuracy has been found to be system-dependent. In this study, we focused on the FEP method to analyze the free energy changes upon mutation of each residue to alanine. We also investigated the mutation to other amino acids for selected positions of MJ25. The theoretical background and computational details are discussed below.

For a point mutation, the relative binding free energy ($\Delta\Delta G_{\mathrm{bind}}$) between the wild type and the mutant can be computed. The calculation is based on the thermodynamic cycle depicted in Figure 4. Instead of calculating the absolute binding free energies for the wild type and for mutant directly and independently, the free energy change due to a mutation is estimated in the bulk and bound phases ($\Delta G^{\mathrm{bulk}}_{\mathrm{Mut}}$ and $\Delta G^{\mathrm{bound}}_{\mathrm{Mut}}$, respectively). The alchemical transformation for a mutation is carried out with the dual-topology approach.[39,58] Both the initial state ($\lambda = 0$) and the final state ($\lambda = 1$) are defined concurrently. As the MD simulation progresses, the potential energy is properly scaled on the basis of the $\lambda$ value. The two states are allowed to interact with the environment, whereas they do not see each other in the course of the simulation.

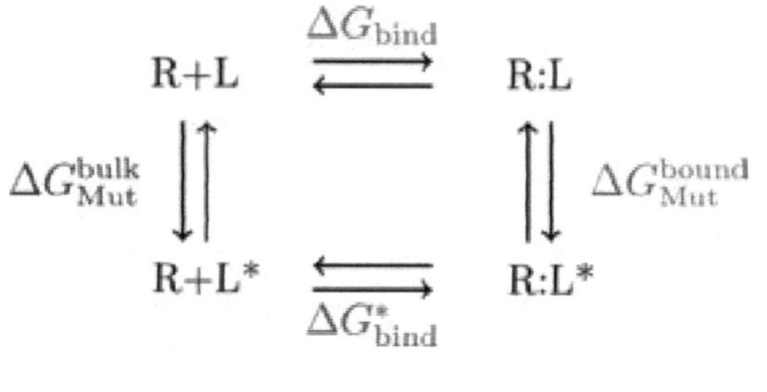


**Figure 4.** Thermodynamic cycle for the calculation of the relative binding free energy ($\Delta\Delta G_{\mathrm{bind}}$) for MJ25 mutants. The vertical arrows correspond to the point mutation in the bound and bulk phases. The horizontal arrows correspond to the binding free energy from the bulk to the binding site. R is the receptor; L is the wild-type ligand; L* is the mutant. $\Delta\Delta G_{\mathrm{bind}} = \Delta G^{*}_{\mathrm{bind}} - \Delta G_{\mathrm{bind}} = \Delta G^{\mathrm{bound}}_{\mathrm{Mut}} - \Delta G^{\mathrm{bulk}}_{\mathrm{Mut}}$.

There are two main strategies to set up the system for an alchemical transformation simulating a mutation in MD.[59] The first method is to estimate the free energy changes in the bound phase and in the bulk phase separately in two individual simulated systems. The advantage is that the system in the bulk phase is smaller, which reduces computational time. However, one issue might arise when dealing with mutation of charged residues. The charge neutrality of the whole system is not conserved, potentially resulting in unstable systems. The second method is to combine the two systems in a single simulation box, performing the forward transformation in the binding state while simultaneously performing the backward transformation in the bulk, which is well-separated from the binding pocket. This approach avoids the problems due to mutation of a charged residue. Although a combined simulation approach is proposed for mutation involving charged resides initially, it can apply to neutral residues as well. The comparison between the two approaches has not been fully discussed before. In this work, we applied both methods to calculate the relative binding free energy. We assess and discuss their utility.

The system setup for MD simulations is summarized here. For the separated simulation method, the system for the bound phase was the same as that used for the absolute binding free energy calculations. The bulk phase contained 3765 water molecules in a 50 Å × 50 Å × 50 Å cubic box. For the combined system, additional 30 Å thick water layers were added at both sides in the $z$ direction of the original bound system. There were about 40 000 water molecules in total. The dual-topology file was taken from VMD 1.93 and was modified to be compatible with the CHARMM36 force field.

The FEP transformation was evenly divided into 50 windows. An MD simulation consisting of 50 000 production steps after 5000 equilibration steps was performed for each window. The simulation was carried out bidirectionally for both the separated and combined system approaches. The free energy changes from both directions were again calculated by the Bennett acceptance ratio (BAR)[46] to combine statistical data.

## 3. RESULTS AND DISCUSSION

**3.1. Structural Analysis of MJ25 and the FhuA Receptor.** The root-mean-square displacements (RMSD) compared with the energies of the minimized structures show that the RMSD for FhuA stabilizes after 10 ns while that for MJ25 takes 30 ns to converge (see Figure S2). The equilibrium structure of MJ25 can be superimposed with the X-ray structure with an RMSD of 1.06 Å over 21 $C_\alpha$ atoms. Figure S3 illustrates the superimposed structures. The lasso structure of MJ25 remains stable during the course of the simulation. However, the side chains are relatively flexible. For example, the imidazole ring of His5 and aromatic ring of Tyr9 do interact with various FhuA residues during the simulation. These conformational changes alter the relevant hydrogen-patterns.

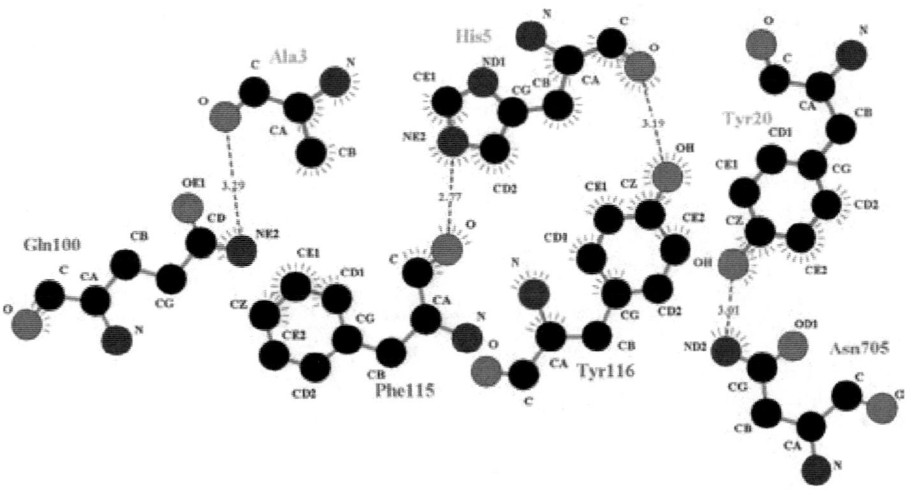


**Figure 5.** Hydrogen bonds between MJ25 and the FhuA receptor found in the X-ray structure.[9] Three hydrogen bonds are formed with the plug domain, and one hydrogen bond is formed with the $\beta$-barrel domain. This plot was generated using Ligplot.[60,61]

Figure 5 presents the hydrogen bonds in the initial X-ray structure of the binding complex.[9] The imidazole ring (NE2) and backbone carbonyl group (O) of His5 from MJ25 make hydrogen bonds with the backbone carbonyl group (O) of Phe115 and the side chain (OH) of Tyr116 on the FhuA receptor, respectively. In addition, the backbone carbonyl group (O) of Ala3 forms a hydrogen bond with the side-chain group (NE2) of Gln100. One hydrogen bond is formed on the $\beta$-barrel domain between the side chain (OH) of Tyr20 and the side chain (ND2) of Asn705.

The number of hydrogen bonds during the last 10 ns of simulation is shown in Figure 6. The cutoff for the donor–acceptor (D–A) distance was 3.5 Å, and the D–H–A angle cutoff was 30°. On average, only two hydrogen bonds existed in every frame. Numerous unique hydrogen bonds were formed; however, most of them had a short lifetime.

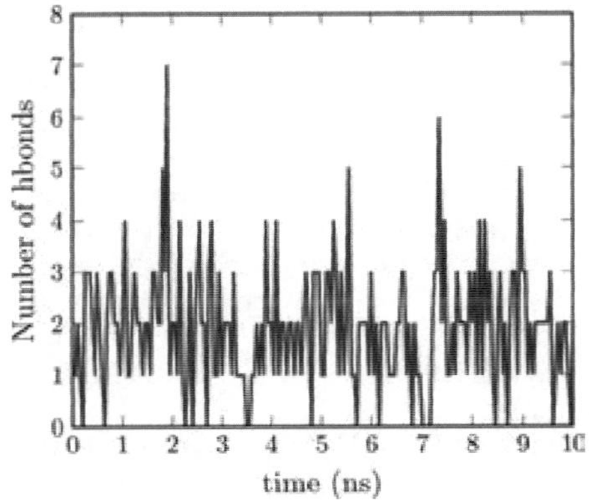


**Figure 6.** Number of hydrogen bonds between MJ25 and the FhuA receptor from the last 10 ns of MD simulation.

Figure 7 illustrates the hydrogen bonds with the highest percent occupancy, defined as the ratio of the time during which the hydrogen bond is present to the total time length of the considered trajectory. Other hydrogen-bond pairs had less than 5% occupancy. The hydrogen bonds were determined using the VMD Hbonds plugin.

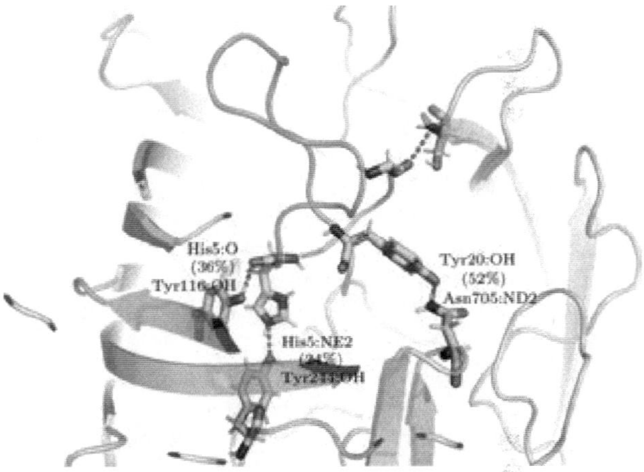


**Figure 7.** Illustration of hydrogen bonding (shown in magenta) between MJ25 and the FhuA receptor during the last 10 ns of the MD simulation. The hydrogen bonds displayed here are ones with higher occupancy (occupancies are shown in the parentheses).

The imidazole ring of His5 from MJ25 forms a hydrogen bond with the side chain of Tyr244 on the plug domain. The backbone carbonyl group of His5 from MJ25 makes a hydrogen bond with the side chain of Tyr116 on the plug domain. The side chain of Tyr20 on the MJ25 tail also makes a hydrogen bond with the side chain of Asn705 on the $\beta$-barrel domain. The hydrogen bonds between O of His5 and OH of Tyr116 and between OH of Tyr20 and ND2 of Asn705 are consistent with the X-ray structure, whereas the hydrogen bond from NE2 of His5 to O of Phe115 shifts to OH of Tyr244. It can be observed in Figure S3 that the imidazole ring flips to the other side. Furthermore, the hydrogen bond between O of Ala3 and NE2 of Gln100 was not stable during the simulation.

**3.2. Binding Free Energy.** The equilibrium values for the collective variables used to constrain the MJ25 structure are as follows: $r = 28.6$ Å, $\theta = 80°$, $\phi = -41.9°$, $\Theta = 88.2°$, $\Phi = -125°$, and $\Psi = -129°$. These values were taken at the end of the 50 ns equilibration. The individual contributions to the binding free energy are listed in Table 1. The results for the forward and backward transformations are shown in Figure S4.

**Table 1. Decomposition of Binding Free Energy Contributions Shown in the Thermodynamic Cycle**

| contribution | $\Delta G$ (kcal/mol) | $\Delta G$ (kcal/mol) | time (ns) |
|---|---|---|---|
| $\Delta G_{\Theta}^{\text{bound}}$ | 0.54 ± 0.17 | | |
| $\Delta G_{\Phi}^{\text{bound}}$ | 0.68 ± 0.21 | | |
| $\Delta G_{\Psi}^{\text{bound}}$ | 0.41 ± 0.19 | | |
| $\Delta G_{r}^{\text{bound}}$ | 0.45 ± 0.01 | | |
| $\Delta G_{\theta}^{\text{bound}}$ | 0.14 ± 0.00 | | |
| $\Delta G_{\phi}^{\text{bound}}$ | 0.32 ± 0.12 | | |
| $\Delta G_{c}^{\text{bound}}$ | 1.72 ± 0.05 | | |
| $\Delta G_{\text{restraint}}^{\text{bound}}$ | | 4.26 ± 0.36 | 11.2 |
| $\Delta G_{o}^{\text{bulk}}$ | 6.61 | | |
| $\Delta G_{a}^{\text{bulk}}$ | 2.68 | | |
| $\Delta G_{r}^{\text{bulk}}$ | 0.71 | | |
| $\Delta G_{c}^{\text{bulk}}$ | 7.81 ± 0.01 | | |
| $\Delta G_{\text{restraint}}^{\text{bulk}}$ | | 17.81 ± 0.01 | 11.2 |
| $\Delta G_{\text{decoupled}}^{\text{bound}}$ | | −46.44 ± 1.20 | 81.6 |
| $\Delta G_{\text{decoupled}}^{\text{bulk}}$ | | −68.88 ± 0.42 | 63.2 |
| $\Delta G_{\text{bind}}$ | | −8.89[a] ± 1.32[b] | 167.2 |
| $\Delta G_{\text{bind}}^{\text{exp}}$ | | −8.13[c] | |

[a] $\Delta G_{\text{bind}} = \Delta G_{\text{restraint}}^{\text{bulk}} + \Delta G_{\text{decoupled}}^{\text{bulk}} - \Delta G_{\text{restraint}}^{\text{bound}} - \Delta G_{\text{decoupled}}^{\text{bound}}$ [b] The total error was calculated as the square root of the sum of the individual squared errors. [c] From the dissociation constant:[11] $\Delta G_{\text{bind}}^{\text{exp}} = RT \ln K_d$.

A value of −8.89 ± 1.32 kcal/mol for the absolute binding free energy was obtained from the MD simulations. This agrees well with the experimentally determined value of −8.13 kcal/mol. The experimental binding free energy was obtained from the dissociation constant ($\Delta G_{\text{bind}}^{\text{exp}} = RT \ln K_d$), which was found to be 1.2 $\mu$M.[11]

The restraint free energy in the bound state is 4.26 kcal/mol. This was obtained by numerical integration of the energy gradient with respect to each collective variable (position, orientation, and conformation). All of the gradients could be calculated in one simulation, requiring 11.2 ns to complete bidirectional simulations. The contributions from the position $(r, \theta, \phi)$ and orientation $(\Theta, \Phi, \Psi)$ constraints are relatively small (about 0.5 kcal/mol each). This suggests that MJ25 is very stable at its equilibrium position.

The restraint free energy for conformation at the binding site $\Delta G_c^{\text{bound}}$ is 1.72 kcal/mol. This is about 3−4 times larger than that of other constraints in the bound state. The collective variable for conformation is based on the RMSD of the equilibrium structure at the binding site as a reference. The heavy atoms in the main chain (N, $C_\alpha$, C, O) and the side-chain $C_\beta$ atoms wee included in the RMSD calculation. As a reference, we present the RMSD versus time plots for MJ25 with constraints in both the bound and unbound states (see Figure S5). Compared with the conformation restraint free energy for the bulk phase ($\Delta G_c^{\text{bulk}}$), 7.81 kcal/mol, the loss of conformation free energy upon binding is 6.09 kcal/mol. This is a significant loss. The conformation change of MJ25 in going from the bulk phase to the binding site might be crucial for the recognition by the FhuA receptor.

The position ($\Delta G_r^{\text{bulk}} + \Delta G_a^{\text{bulk}}$) and orientation ($\Delta G_o^{\text{bulk}}$) restraint free energies for MJ25 in the bulk phase are 3.39 and 6.61 kcal/mol, respectively. These can be derived analytically as follows:

$$e^{-\beta(\Delta G_r^{\text{bulk}} + \Delta G_a^{\text{bulk}})} = \frac{1}{F_t} \int_0^\infty dr\, r^2 e^{-\beta u_r} \int_0^\pi d\theta \sin\theta \int_0^{2\pi} d\phi\, e^{-\beta(u_\theta + u_\phi)} \quad (1)$$

$$e^{-\beta \Delta G_o^{\text{bulk}}} = \frac{1}{8\pi^2} \int_0^\pi d\Theta \sin\Theta \int_0^{2\pi} d\Phi \int_0^{2\pi} d\Psi\, e^{-\beta(u_\Theta + u_\Phi + u_\Psi)} \quad (2)$$

where $u_q$ is the restraint potential for each degree of freedom $q$ and $F_t$ denotes the translational factor with units of volume (1661 Å$^3$). Combining these values with $\Delta G_c^{\text{bulk}}$ yields 17.81 kcal/mol for the total restraint free energy in the bulk phase, which is about twice as large as the binding free energy ($\Delta G_{\text{bind}}$).

The accuracy of decoupled free energy calculations is key in calculating the binding free energy. Over 85% of the computational time was spent on them. The decoupled free energy was obtained by calculating the free energy difference between the initial, fully coupled state ($\lambda = 0$) and final, fully decoupled state ($\lambda = 1$) for MJ25. The two states were divided into multiple intermediate steps or windows. The free energy differences for successive windows were calculated and summed to obtain the final results. The decoupling process becomes reversible if sufficient windows are defined. In this work, uneven window sizes were used to save computational time. The smallest window size was $\Delta\lambda = 0.00125$, and the largest window size was $\Delta\lambda = 0.02$. These window sizes were chosen so that the difference between the forward and backward free energies at each window was smaller than 0.5 kcal/mol. There were 79 and 102 windows for the bulk and bound phases, respectively.

The decoupled free energies in the bound and bulk phases are −46.44 and −68.88 kcal/mol, respectively. The average absolute deviations between forward and backward steps in each window are 0.1 kcal/mol for simulations in both phases. The standard deviations are 1.2 and 0.42 kcal/mol, respectively, as derived from the bidirectional analysis using the ParseFEP tool.[55] The agreement between forward and backward simulations indicates that the process is approximately reversible during the time frame examined. Finally, the binding free energy was obtained by adding all of the contributions from the restrained and decoupled simulations ($\Delta G_{\text{bind}} = \Delta G_{\text{restraint}}^{\text{bulk}} + \Delta G_{\text{decoupled}}^{\text{bulk}} - \Delta G_{\text{restraint}}^{\text{bound}} - \Delta G_{\text{decoupled}}^{\text{bound}}$).

**3.3. Computational Mutation Analysis.** The relative binding free energies upon mutation were estimated to identify the key residues for MJ25 binding.

*Alanine Scanning Mutagenesis.* The alanine scanning simulation was performed for 17 residues of MJ25 to identify hot spots. Mutation of Ala3 to Ala3 was excluded. In addition, Gly1, Glu8, and Gly2 were ruled out from mutation because of their importance in production, maturation, export, and stability as reported by Pavlova et al.[62] Figure 8 presents the relative binding free energy ($\Delta\Delta G$) at each position from the separated and combined system approaches. Overall, the results from the two methods agree well, with the average absolute deviation being 0.62 kcal/mol. The ring (G1−E8) and lower tail (Y20−G21) regions have a significant free energy increase upon mutation compared with the upper tail region (Y9−F19). This indicates that the important binding sites are in the ring and lower tail regions.

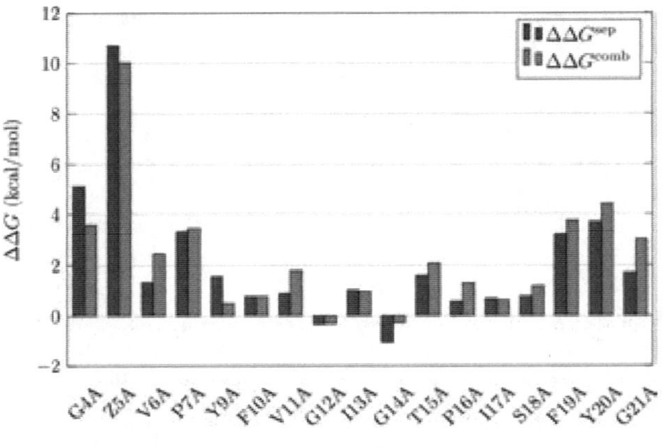


**Figure 8.** Relative binding free energies from alanine scanning mutagenesis. The blue and red bars present results from the separated system and combined system methods, respectively. The standard deviation for each bar is around 0.1 kcal/mol (not shown). The symbol Z indicates a protonated histidine at position 5.

Figure 9 demonstrates the relative free energy results for the mutation from protonated histidine (Z) to alanine (A) from the

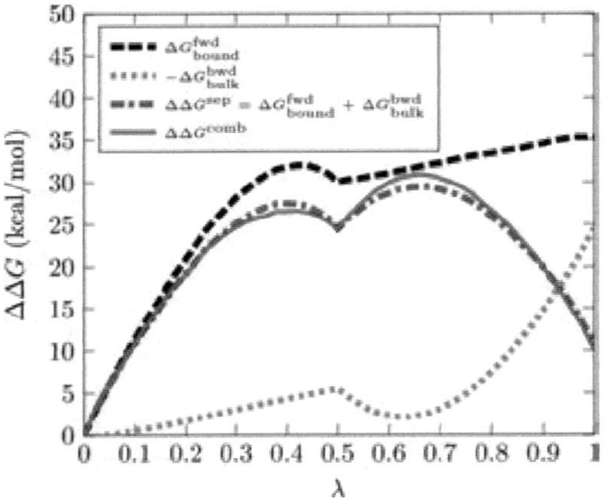


**Figure 9.** Relative binding free energies of mutation from protonated histidine (Z) to alanine (A) for the separated and combined system approaches. The dashed, dotted, and dash-dotted lines indicate results from the separated system approach. The black line is for the Z → A transformation. The gray line is for the A → Z transformation plotted with $-\Delta G_{\text{bulk}}^{\text{bwd}}$ as a function of $1 - \lambda$. The sum of these two contributions (shown in green) is the relative binding free energy. The red solid line is the result computed directly from the combined system method.

separated and the combined system approaches. The combined method transforms the wild type to mutants in the bound state, and the reverse transformation is done in the bulk phase. For one FEP simulation with equal size (0.02) in between the initial and the final states, $\Delta\Delta G^{\text{comb}}$ ($\lambda = 0 \rightarrow \lambda = 0.02$) = $\Delta G_{\text{bound}}^{\text{fwd}}$ ($\lambda = 0 \rightarrow \lambda = 0.02$) + $\Delta G_{\text{bulk}}^{\text{bwd}}$ ($\lambda = 1 \rightarrow \lambda = 0.98$). This is the same for other windows. The agreement between the two methods for all of the intermediate steps is satisfying. The simulations were started from different initial structures, and the results indicate that the FEP method for alanine transformation was reproducible and that sufficient sampling was obtained. The comparison between the two approaches for each mutation facilitates addressing some of the known pitfalls in free energy calculations. This will be further explained in the following section.

Pavlova et al.[62] showed that Gly2, Gly4, and Pro7 are essential for bacterial growth inhibition. Pan et al.[63] also indicated that variations in the ring structure likely disrupt import of MJ25 to the cytoplasm, which is consistent with our alanine scanning results. The relative free energy changes of G4A and P7A are both 4 kcal/mol. Furthermore, Z5A has a free energy change of about 10 kcal/mol, which is much higher than those for alanine substitutions at other positions.

Previously, His5 has been demonstrated to be crucial for SbmA-mediated transport of MJ25 into the cytoplasm of susceptible cells.[13] We also find that His5 plays an important role in FhuA-mediated transport, which is consistent with the experimental results that His5 is essential for recognition and binding to the plug domain of FhuA and subsequently for transport.[9] The importance of His5 is further studied in the next section.

The role of the upper loop region for FhuA recognition remained controversial in previous studies. The Val11 to Pro16 region was proposed to be critical for MJ25 recognition at the bacterial membrane.[11] Moreover, Phe10 was found to be essential for permeation.[62] However, Pan et al.[63] indicated that multiple concurrent amino acid substitutions can be tolerated in this region of MJ25 without loss of recognition by FhuA. The alanine scanning results from our work provide support for the latter viewpoint that the upper loop region can tolerate alanine substitutions. The calculated relative free energies are mostly less than 2 kcal/mol.

In the study by Pavlova et al.,[62] the aromatic residues that straddle the ring, Phe19 and Tyr20, have no tolerance for non-wild-type side chains. Our work reveals that these two residues have high relative free energies when mutated to alanine, about 4 kcal/mol, a result that is compatible with the experimental results.

In summary, computational alanine scanning provides results consistent with experiments showing the importance of the ring and lower tail regions for the transport of MJ25. In addition, we provide some evidence that the loop region from Phe10 to Ser18 may tolerate substitutions.

*Analysis of Histidine Substitutions.* The relative binding free energy of Z5A is over 10 kcal/mol (see Figure 8). Considering the value of the total binding free energy (−8.89 kcal/mol), this mutation would impact the binding affinity significantly. This motivated us to perform additional substitutions at this position. The positively charged histidine (denoted as Z) was mutated to two charge-equivalent residues, lysine (K) and arginine (R). In addition, it was mutated to a neutral histidine counterpart to evaluate the role of the positive charge in binding.

In principle, the dual-topology approach applies to any type of point mutation. Nevertheless, a severe convergence challenge became apparent for the direct transformation from histidine to other residues with bulky side chains, especially at the binding site. The position and conformation of the emerging side chains are challenging to predict because of the negligible interactions with the environment initially. As the interactions gradually turn on, the bulky side chains might be trapped by an energy barrier. This prevents efficient sampling of stable conformations. For smaller side chains like alanine, the energy barrier is relatively small. The improved convergence can be observed from the good agreement between the two different methods, as shown in Figure 8. Therefore, we propose another strategy to calculate the relative binding free energy between two bulky side chains.

To take advantage of the better convergence of alanine mutations, the double-alanine-transformation scheme is proposed. For instance, the relative binding free energy between protonated histidine (Z) and lysine (K) in this scheme is equivalent to $\Delta\Delta G(Z \rightarrow K) = \Delta\Delta G(Z \rightarrow A) + \Delta\Delta G(A \rightarrow K) = \Delta\Delta G(Z \rightarrow A) - \Delta\Delta G(K \rightarrow A)$. Instead of transforming from protonated histidine to lysine by FEP, we transform histidine to alanine as well as lysine to alanine. We start with switching the original histidine to a lysine directly (single topology). After minimization and equilibration, the FEP transformation was performed from lysine to an alanine side chain (dual topology) as described in Alanine Scanning Mutagenesis. It was observed that the convergence improved considerably.

Figure 10 presents the relative binding free energies from protonated histidine (Z) to other residues for the separated and combined system approaches. The neutral histidine counterpart (Z5H) has a free energy change as large as that for alanine. On the contrary, the charge-equivalent substitutions (Z5K and Z5R) show almost no difference in binding association compared to the protonated histidine. Pan et al.[63] reported that among 15 amino acid substitutions at position 5, only positively charged histidine and arginine retain the antimicrobial function. As inferred from our simulation, the decreased antimicrobial ability of other substitutions is likely due to the inability of these variants to be transported across the FhuA receptor. Although lysine was not included in the library, the lysine variant is expected to be potent on the basis of our results. In fact, Mathavan et al.[9] reported that the charge-conserved lysine mutant has antimicrobial activity against *Escherichia coli* W3110.

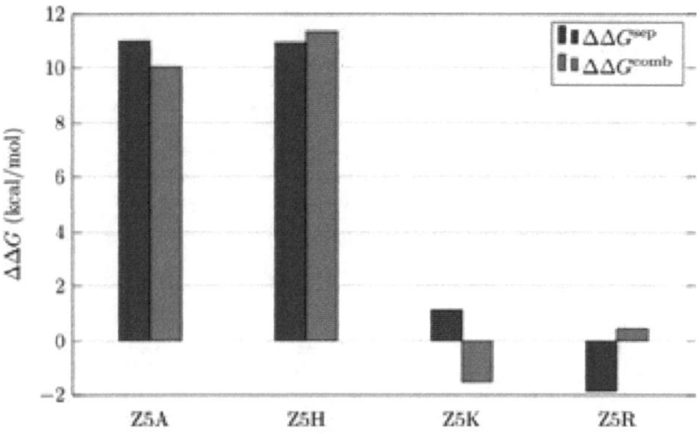


**Figure 10.** Relative binding free energies of mutation from protonated histidine (Z) to alanine (A), neutral histidine (H), lysine (K), and arginine (R) at position 5. The blue and red bars present results from the separated system and combined system methods, respectively. The standard deviation for each bar is around 0.1 kcal/mol (not shown).

The hydrogen bonding for the protonated histidine is crucial for binding, as shown in Figures 5 and 7. It is of interest to see whether hydrogen bonds still form for the mutants. Figure 11 displays the hydrogen bonds for the Z5K and Z5R mutants. For the lysine variant, the NZ atom is capable of making hydrogen bonds with O of Phe115 and O of Gly99. In addition, for the arginine variant, the NH1 and NH2 atoms can also form hydrogen bonds with O of Gly99 and O of Phe115, respectively. Although the hydrogen-bonding pairs change after the mutation, the number of hydrogen bonds formed is preserved for the lysine and arginine mutants. On the contrary, no stable hydrogen bonds were observed for the Z5A and Z5H mutants. The energy of the hydrogen bond is about 2 to 5 kcal/mol. Losing two hydrogen bonds can result in a reduction of up to 10 kcal/mol in the binding free energy. This might explain the differences in relative free energy shown in Figure 10.

In addition to lysine and arginine, Pavlova et al.[62] reported that analogues with substitutions of histidine with three other neutral residues, tryptophan, tyrosine, and valine, can also permeate through bacterial cells. It is expected that a tryptophan side chain can serve as a hydrogen-bond donor and a tyrosine side chain can serve as both a donor and acceptor. The observation for valine is probably due to the oxygen atom on the main chain serving as a donor; however, we suspect that the binding affinity for the valine mutant is weaker than those for other polar or charged variants. From the experimental data and our computational results, it may be inferred that the positive charge at position 5 is preferable but not absolutely necessary for MJ25 permeation as long as the number of hydrogen bonds between the mutant and the FhuA receptor is preserved.

One critical issue in computational mutagenesis studies by the FEP method is the conservation of the binding mode after a transformation.[49,59] If a mutation results in significant changes in the structure of the molecule and the binding mode, it is arguably an intractable challenge to resolve during a short FEP simulation. Figure S6 illustrates the overlaps of the structure of the initial protonated histidine with those of the final states of the lysine and arginine variants. The binding mode appears to be conserved, which ensures that mutation simulation is feasible and the results are pertinent. The conservation of binding modes for other substitutions of alanine scanning were also confirmed (results not shown).

*Improvement of Efficient and Accurate Sampling.* One common issue in FEP simulations is whether the sampling is sufficient. If the molecular conformation is trapped by an

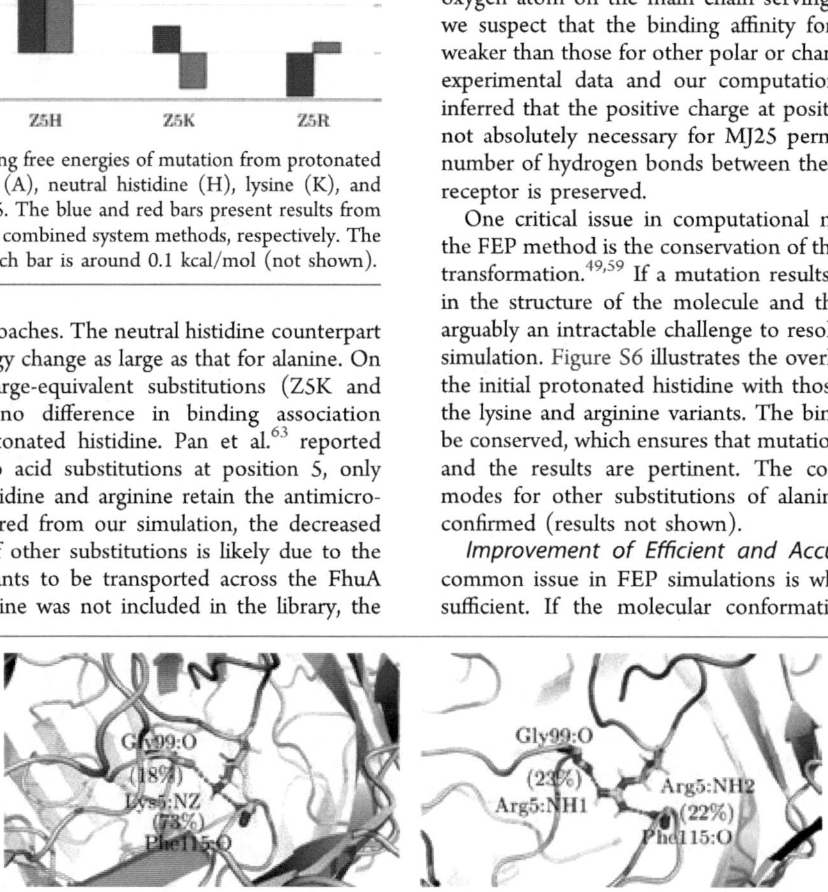


**Figure 11.** Illustration of hydrogen bonding (shown in magenta) between MJ25 variants and the FhuA receptor with protonated histidine mutated to Lys (left) and Arg (right). The values in parentheses are hydrogen-bond occupancies.

energy barrier, correct sampling may not be attained even with long simulation times. In this work, one problem was encountered when a smaller residue was transformed into a larger residue around the lasso ring. Figure 12 illustrates two

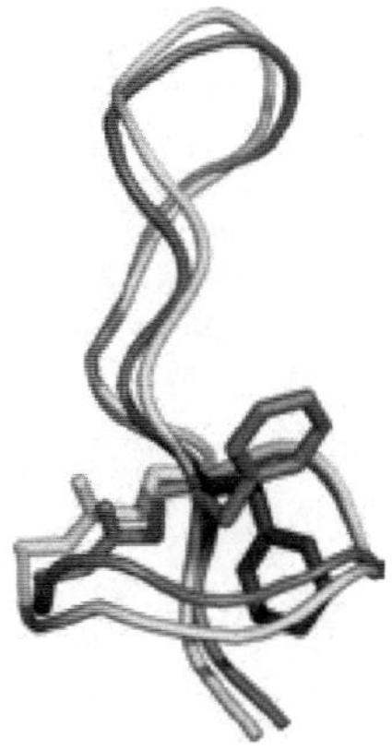

**Figure 12.** Schematic representation of the phenylalanine side chain (F19) in different conformations. The green color shows the initial structure in which the aromatic ring is above the loop. The blue color shows the structure with the aromatic ring trapped under the loop during the mutation transformation.

conformations with one phenylalanine side chain above the ring and one below the ring. The former is taken from a snapshot from an equilibrium structure. When a backward transformation simulation from the alanine residue to the phenylalanine residue was performed, the interactions between phenylalanine and the environment were negligible initially. The phenylalanine side chain experienced little repulsive forces when the aromatic ring flipped to the other side of the lasso ring. When the interactions were gradually switched on as a function of the coupling parameter $\lambda$, the strong repulsive forces prevented the aromatic ring from flipping back. This is a case where adequate sampling of equilibrium states is difficult to achieve.

For proteins with simple topology, the mutated residue may overcome energy barriers in the course of simulation. On the contrary, for proteins with complex topology such as lasso structures, care must be taken to avoid simulations starting from improper initial structures. Some strategies are useful to avoid this issue. This is not limited to lasso structures.

First, using the same atom coordinates for similar scaffolds in the dual-topology file in VMD may help. For instance, if the residues from the wild type and the mutants both have $C_\alpha$, $C_\beta$, and $C_\gamma$ atoms, applying the same positions to the mutant atoms may help. The structure generated by available internal coordinates is generated in a linear manner without considering the adjacent environment. This may engender unwanted overlap with other residues. Some of the overlap might be relaxed in minimization or equilibrium steps; however, for complicated proteins, a large energy barrier may prevent the simulation from returning to stable conformations.

Second, slightly turning on the interactions between the emerging residues and the environment during the equilibrium stage may help. Even a small repulsive force can prevent atoms from jumping into and being trapped in a high energy valley such as was the case depicted in Figure 12. For example, this tactic was used for the backward simulation from alanine to phenylalanine.

Third, mutating from larger residues to smaller residues is preferable. The reverse transformation involves growing or inserting a large number of atoms, which requires a longer time to equilibrate. As described in Analysis of Histidine Substitutions, direct transformation from histidine to lysine or arginine may encounter serious convergence issues. The double-alanine-transformation scheme proposed in this work effectively solves the issue.

## 4. CONCLUSIONS

This is the first computational study of MJ25 with the FhuA receptor. The binding free energy from MD simulation agrees with the experimental data. Computational mutation analysis revealed that the ring and lower tail regions of MJ25 are essential for binding recognition. The His5 residue is particularly important. Two charge-equivalent substitutions (Lys and Arg) were found to have similar association affinities with the FhuA receptor. The preservation of hydrogen bonds with the FhuA receptor is critical for binding of the mutants. Several strategies to improve sampling for free energy calculations are provided. The double-alanine-transformation scheme can be extended to any type of point mutation as long as the binding mode is conserved. MJ25 is a potent antimicrobial peptide, and a better understanding of the binding mechanism may facilitate the design of new drugs that relieve some of the pressure exerted on public health by the antibiotic resistance challenge.

## ■ ASSOCIATED CONTENT

### Supporting Information

The Supporting Information is available free of charge on the ACS Publications website at DOI: 10.1021/acs.jctc.7b00417.

Force field parameters for the isopeptide bond (Figure S1), plots of backbone-atom RMSD versus time for MJ25 and the FhuA receptor (Figure S2), superimposition of the X-ray and MD equilibrium structures of MJ25 (Figure S3), detailed information on the FEP calculation for each collective variable (Figure S4), plots of backbone-atom RMSD versus time for MJ25 with constraints in the bound and unbound states (Figure S5), and superimposition of the structures of wild-type MJ25 and MJ25 with charge-equivalent substitutions of Lys and Arg at position 5 (Figure S6) (PDF)

## ■ AUTHOR INFORMATION

### Corresponding Author

*E-mail: yiannis@umn.edu. Phone: 612-624-4197. Fax: 612-626-7246.

### ORCID

Pin-Kuang Lai: 0000-0003-2894-3900

Yiannis N. Kaznessis: 0000-0002-5088-1104

### Funding

This work was supported by the National Institutes of Health (Grant GM111358) and the National Science Foundation (Grant CBET-1412283). We acknowledge computational support from the Minnesota Supercomputing Institute (MSI) and the Extreme Science and Engineering Discovery Environment (XSEDE), which is supported by the National Science Foundation (Grant ACI-10535753). Support from the University of Minnesota Digital Technology Center and the

University of Minnesota Institute for Engineering in Medicine is also acknowledged.

## Notes

The authors declare no competing financial interest.

## REFERENCES


(1) Zasloff, M. Antimicrobial peptides of multicellular organisms. *Nature* **2002**, *415*, 389−395.

(2) Brogden, K. A. Antimicrobial peptides: pore formers or metabolic inhibitors in bacteria? *Nat. Rev. Microbiol.* **2005**, *3*, 238−250.

(3) Thaker, H. D.; Som, A.; Ayaz, F.; Lui, D.; Pan, W.; Scott, R. W.; Anguita, J.; Tew, G. N. Synthetic mimics of antimicrobial peptides with immunomodulatory responses. *J. Am. Chem. Soc.* **2012**, *134*, 11088−11091.

(4) Qu, X.-D.; Harwig, S. S. L.; Shafer, W. M.; Lehrer, R. I. Protegrin structure and activity against *Neisseria gonorrhoeae*. *Infect. Immun.* **1997**, *65*, 636−639.

(5) He, J.; Eckert, R.; Pharm, T.; Simanian, M. D.; Hu, C.; Yarbrough, D. K.; Qi, F.; Anderson, M. H.; Shi, W. Novel synthetic antimicrobial peptides against *Streptococcus* mutants. *Antimicrob. Agents Chemother.* **2007**, *51*, 1351−1358.

(6) *Handbook of Biologically Active Peptides*, 2nd ed.; Kastin, A., Ed.; Academic Press: San Diego, CA, 2013.

(7) Hegemann, J. D.; Zimmermann, M.; Xie, X.; Marahiel, M. A. Lasso peptides: an intriguing class of bacterial natural products. *Acc. Chem. Res.* **2015**, *48*, 1909−1919.

(8) Li, Y.; Zirah, S.; Rebuffat, S. *Lasso Peptides: Bacterial Strategies To Make and Maintain Bioactive Entangled Scaffolds*; Springer: New York, 2014.

(9) Mathavan, I.; Zirah, S.; Mehmood, S.; Choudhury, H. G.; Goulard, C.; Li, Y.; Robinson, C. V.; Rebuffat, S.; Beis, K. Structural basis for hijacking siderophore receptors by antimicrobial lasso peptides. *Nat. Chem. Biol.* **2014**, *10*, 340−342.

(10) Forkus, B.; Ritter, S.; Vlysidis, M.; Geldart, K.; Kaznessis, Y. N. Antimicrobial probiotics reduce *Salmonella enterica* in Turkey gastrointestinal tracts. *Sci. Rep.* **2017**, *7*, 40695.

(11) Destoumieux-Garzón, D.; Duquesne, S.; Peduzzi, J.; Goulard, C.; Desmadril, M.; Letellier, L.; Rebuffat, S.; Boulanger, P. The iron-siderophore transporter FhuA is the receptor for the antimicrobial peptide microcin J25: role of the microcin Val11-Pro16 $\beta$-hairpin region in the recognition mechanism. *Biochem. J.* **2005**, *389*, 869−876.

(12) Salomón, R. A.; Farías, R. N. The peptide antibiotic microcin 25 is imported through the TonB pathway and the SbmA protein. *J. Bacteriol.* **1995**, *177*, 3323−3325.

(13) de Cristóbal, R. E.; Solbiati, J. O.; Zenoff, A. M.; Vincent, P. A.; Salomón, R. A.; Yuzenkova, J.; Severinov, K.; Farías, R. N. Microcin J25 uptake: His5 of the MccJ25 lariat ring is involved in interaction with the inner membrane MccJ25 transporter protein SbmA. *J. Bacteriol.* **2006**, *188*, 3324−3328.

(14) Adelman, K.; Yuzenkova, J.; La Porta, A.; Zenkin, N.; Lee, J.; Lis, J. T.; Borukhov, S.; Wang, M. D.; Severinov, K. Molecular Mechanism of Transcription Inhibition by Peptide Antibiotic Microcin J25. *Mol. Cell* **2004**, *14*, 753−762.

(15) Mukhopadhyay, J.; Sineva, E.; Knight, J.; Levy, R. M.; Ebright, R. H. Antibacterial peptide microcin J25 inhibits transcription by binding within and obstructing the RNA polymerase secondary channel. *Mol. Cell* **2004**, *14*, 739−751.

(16) Bellomio, A.; Vincent, P. A.; de Arcuri, B. F.; Farías, R. N.; Morero, R. D. Microcin J25 Has Dual and Independent Mechanisms of Action in *Escherichia coli*: RNA Polymerase Inhibition and Increased Superoxide Production. *J. Bacteriol.* **2007**, *189*, 4180−4186.

(17) Ferguson, A. L.; Zhang, S.; Dikiy, I.; Panagiotopoulos, A. Z.; Debenedetti, P. G.; Link, A. J. An experimental and computational investigation of spontaneous lasso formation in microcin J25. *Biophys. J.* **2010**, *99*, 3056−3065.

(18) Pan, S. J.; Cheung, W. L.; Fung, H. K.; Floudas, C. A.; Link, A. J. Computational design of the lasso peptide antibiotic microcin J25. *Protein Eng., Des. Sel.* **2011**, *24*, 275−282.

(19) Hegemann, J. D.; De Simone, M.; Zimmermann, M.; Knappe, T. A.; Xie, X.; Di Leva, F. S.; Marinelli, L.; Novellino, E.; Zahler, S.; Kessler, H.; Marahiel, M. A. Rational improvement of the affinity and selectivity of integrin binding of grafted lasso peptides. *J. Med. Chem.* **2014**, *57*, 5829−5834.

(20) Wu, E. L.; Cheng, X.; Jo, S.; Rui, H.; Song, K. C.; Dávila-Contreras, E. M.; Qi, Y.; Lee, J.; Monje-Galvan, V.; Venable, R. M.; et al. CHARMM-GUI Membrane Builder toward realistic biological membrane simulations. *J. Comput. Chem.* **2014**, *35*, 1997−2004.

(21) Lee, J.; Cheng, X.; Swails, J. M.; Yeom, M. S.; Eastman, P. K.; Lemkul, J. A.; Wei, S.; Buckner, J.; Jeong, J. C.; Qi, Y.; et al. CHARMM-GUI Input Generator for NAMD, GROMACS, AMBER, OpenMM, and CHARMM/OpenMM Simulations Using the CHARMM36 Additive Force Field. *J. Chem. Theory Comput.* **2016**, *12*, 405−413.

(22) Jo, S.; Kim, T.; Iyer, V. G.; Im, W. CHARMM-GUI: a web-based graphical user interface for CHARMM. *J. Comput. Chem.* **2008**, *29*, 1859−1865.

(23) Brooks, B. R.; Brooks, C. L.; MacKerell, A. D.; Nilsson, L.; Petrella, R. J.; Roux, B.; Won, Y.; Archontis, G.; Bartels, C.; Boresch, S.; et al. CHARMM: the biomolecular simulation program. *J. Comput. Chem.* **2009**, *30*, 1545−1614.

(24) Murzyn, K.; Róg, T.; Pasenkiewicz-Gierula, M. Phosphatidylethanolamine phosphatidylglycerol bilayer as a model of the inner bacterial membrane. *Biophys. J.* **2005**, *88*, 1091−1103.

(25) Jo, S.; Wu, E. L.; Stuhlsatz, D.; Klauda, J. B.; MacKerell, A. D.; Widmalm, G.; Im, W. Lipopolysaccharide membrane building and simulation. *Methods Mol. Biol.* **2015**, *1273*, 391−406.

(26) Ma, H.; Irudayanathan, F. J.; Jiang, W.; Nangia, S. Simulating Gram-negative bacterial outer membrane: a coarse grain model. *J. Phys. Chem. B* **2015**, *119*, 14668−14682.

(27) Patel, D. S.; Qi, Y.; Im, W. Modeling and simulation of bacterial outer membranes and interactions with membrane proteins. *Curr. Opin. Struct. Biol.* **2017**, *43*, 131−140.

(28) Faraldo-Gómez, J. D.; Smith, G. R.; Sansom, M. S. Molecular dynamics simulations of the bacterial outer membrane protein FhuA: a comparative study of the ferrichrome-free and bound states. *Biophys. J.* **2003**, *85*, 1406−1420.

(29) Li, H.; Robertson, A. D.; Jensen, J. H. Very fast empirical prediction and rationalization of protein pKa values. *Proteins: Struct., Funct., Genet.* **2005**, *61*, 704−721.

(30) Phillips, J. C.; Braun, R.; Wang, W.; Gumbart, J.; Tajkhorshid, E.; Villa, E.; Chipot, C.; Skeel, R. D.; Kale, L.; Schulten, K. Scalable molecular dynamics with NAMD. *J. Comput. Chem.* **2005**, *26*, 1781−1802.

(31) Klauda, J. B.; Venable, R. M.; Freites, J. A.; O'Connor, J. W.; Tobias, D. J.; Mondragon-Ramirez, C.; Vorobyov, I.; MacKerell, A. D., Jr.; Pastor, R. W. Update of the CHARMM all-atom additive force field for lipids: validation on six lipid types. *J. Phys. Chem. B* **2010**, *114*, 7830−7843.

(32) Jorgensen, W. L.; Chandrasekhar, J.; Madura, J. D.; Impey, R. W.; Klein, M. L. Comparison of simple potential functions for simulating liquid water. *J. Chem. Phys.* **1983**, *79*, 926−935.

(33) Martyna, G. J.; Tobias, D. J.; Klein, M. L. Constant pressure molecular dynamics algorithms. *J. Chem. Phys.* **1994**, *101*, 4177−4189.

(34) Feller, S. E.; Zhang, Y.; Pastor, R. W.; Brooks, B. R. Constant pressure molecular dynamics simulation: the Langevin piston method. *J. Chem. Phys.* **1995**, *103*, 4613−4621.

(35) Izaguirre, J. A.; Catarello, D. P.; Wozniak, J. M.; Skeel, R. D. Langevin stabilization of molecular dynamics. *J. Chem. Phys.* **2001**, *114*, 2090−2098.

(36) Essmann, U.; Perera, L.; Berkowitz, M. L.; Darden, T.; Lee, H.; Pedersen, L. G. A smooth particle mesh Ewald method. *J. Chem. Phys.* **1995**, *103*, 8577−8593.

(37) Andersen, H. C. RATTLE: A Velocity version of the SHAKE algorithm for molecular dynamics calculations. *J. Comput. Phys.* **1983**, *52*, 24−34.

(38) Zhao, C.; Caplan, D. A.; Noskov, S. Y. Evaluations of the absolute and relative free energies for antidepressant binding to the

amino acid membrane transporter LeuT with free energy simulations. *J. Chem. Theory Comput.* **2010**, *6*, 1900−1914.

(39) Pearlman, D. A. A Comparison of Alternative Approaches to Free Energy Calculations. *J. Phys. Chem.* **1994**, *98*, 1487−1493.

(40) Chodera, J. D.; Mobley, D. L.; Shirts, M. R.; Dixon, R. W.; Branson, K.; Pande, V. S. Alchemical free energy methods for drug discovery: progress and challenges. *Curr. Opin. Struct. Biol.* **2011**, *21*, 150−160.

(41) Darre, L.; Domene, C. Binding of capsaicin to the TRPV1 ion channel. *Mol. Pharmaceutics* **2015**, *12*, 4454−4465.

(42) Kollman, P. A.; Massova, I.; Reyes, C.; Kuhn, B.; Huo, S.; Chong, L.; Lee, M.; Lee, T.; Duan, Y.; Wang, W.; Donini, O.; Cieplak, P.; Srinivasan, J.; Case, D. A.; Cheatham, T. E. Calculating Structures and Free Energies of Complex Molecules: Combining Molecular Mechanics and Continuum Models. *Acc. Chem. Res.* **2000**, *33*, 889−897.

(43) Woo, H.-J.; Roux, B. Calculation of absolute protein-ligand binding free energy from computer simulations. *Proc. Natl. Acad. Sci. U. S. A.* **2005**, *102*, 6825−6830.

(44) Deng, Y.; Roux, B. Calculation of standard binding free energies: Aromatic molecules in the T4 lysozyme L99A mutant. *J. Chem. Theory Comput.* **2006**, *2*, 1255−1273.

(45) Gumbart, J. C.; Roux, B.; Chipot, C. Efficient determination of protein-protein standard binding free energies from first principles. *J. Chem. Theory Comput.* **2013**, *9*, 3789−3798.

(46) Bennett, C. H. Efficient estimation of free energy differences from Monte Carlo data. *J. Comput. Phys.* **1976**, *22*, 245−268.

(47) *Free Energy Calculations: Theory and Applications in Chemistry and Biology*; Chipot, C., Pohorille, A., Eds.; Springer: Berlin, 2007.

(48) de Ruiter, A.; Oostenbrink, C. Free energy calculations of protein-ligand interactions. *Curr. Opin. Chem. Biol.* **2011**, *15*, 547−552.

(49) Mobley, D. L.; Klimovich, P. V. Perspective: Alchemical free energy calculations for drug discovery. *J. Chem. Phys.* **2012**, *137*, 230901.

(50) Michel, J.; Essex, J. W. Prediction of protein-ligand binding affinity by free energy simulations: assumptions, pitfalls and expectations. *J. Comput.-Aided Mol. Des.* **2010**, *24*, 639−658.

(51) Gumbart, J. C.; Roux, B.; Chipot, C. Standard binding free energies from computer simulations: What is the best strategy? *J. Chem. Theory Comput.* **2013**, *9*, 794−802.

(52) Hermans, J.; Shankar, S. The free energy of xenon binding to myoglobin from molecular dynamics simulation. *Isr. J. Chem.* **1986**, *27*, 225−227.

(53) Gilson, M. K.; Given, J. A.; Bush, B. L.; McCammon, J. A. The statistical-thermodynamic basis for computation of binding affinities: a critical review. *Biophys. J.* **1997**, *72*, 1047.

(54) Roux, B.; Nina, M.; Pomès, R.; Smith, J. C. Thermodynamic stability of water molecules in the bacteriorhodopsin proton channel: a molecular dynamics free energy perturbation study. *Biophys. J.* **1996**, *71*, 670.

(55) Liu, P.; Dehez, F. o.; Cai, W.; Chipot, C. A toolkit for the analysis of free-energy perturbation calculations. *J. Chem. Theory Comput.* **2012**, *8*, 2606−2616.

(56) Brooijmans, N.; Kuntz, I. D. Molecular recognition and docking algorithms. *Annu. Rev. Biophys. Biomol. Struct.* **2003**, *32*, 335−373.

(57) Halperin, I.; Ma, B.; Wolfson, H.; Nussinov, R. Principles of docking: An overview of search algorithms and a guide to scoring functions. *Proteins: Struct., Funct., Genet.* **2002**, *47*, 409−443.

(58) Gao, J.; Kuczera, K.; Tidor, B.; Karplus, M. Hidden thermodynamics of mutant proteins: a molecular dynamics analysis. *Science* **1989**, *244*, 1069−1072.

(59) Rashid, M. H.; Heinzelmann, G.; Kuyucak, S. Calculation of free energy changes due to mutations from alchemical free energy simulations. *J. Theor. Comput. Chem.* **2015**, *14*, 1550023.

(60) Laskowski, R. A.; Swindells, M. B. LigPlot+: multiple ligand-protein interaction diagrams for drug discovery. *J. Chem. Inf. Model.* **2011**, *51*, 2778−2786.

(61) Wallace, A. C.; Laskowski, R. A.; Thornton, J. M. LIGPLOT: a program to generate schematic diagrams of protein-ligand interactions. *Protein Eng., Des. Sel.* **1995**, *8*, 127−134.

(62) Pavlova, O.; Mukhopadhyay, J.; Sineva, E.; Ebright, R. H.; Severinov, K. Systematic structure-activity analysis of microcin J25. *J. Biol. Chem.* **2008**, *283*, 25589−25595.

(63) Pan, S. J.; Link, A. J. Sequence Diversity in the Lasso Peptide Framework: Discovery of Functional Microcin J25 Variants with Multiple Amino Acid Substitutions. *J. Am. Chem. Soc.* **2011**, *133*, 5016−5023.

# 488632 Predicting the Dimer Structure of Defensins Using a Combined Simulation Strategy

**Abstract:**

Defensins are cationic cysteine-rich small molecules with molecular masses ranging from 2 to 6 kDa. They can be classified into alpha, beta and theta categories. They are important to the human innate immune system, and have a broad spectrum of antibacterial activity against both Gram-positive and Gram-negative bacteria. Defensins have been shown to form oligomers, and human defensins have been shown to function in a dimer or even higher ordered oligomer form.

Human beta defensin 3 (hBD-3) is believed to function as a dimer, as Schibli et al [i]in 2002 predicted a symmetric dimer structure of hBD-3 using NMR. On the other hand, the noncovalent electrostatic interaction contributes to the diverse binding interfaces of hBD-3 dimer as predicted from docking software. In order to solve the discrepancy and predict a trustable dimer structure of hBD-3, a combined simulation strategy was developed in this project, which uses replica-exchange implicit solvent CHARMM simulation to predict the initial dimer structure, then uses explicit solvent all atom molecular dynamics simulations to refine the structure. The method was initially tested on both human alpha defensin 5 (HD-5) and human beta defensin 2 (hBD-2), and was proven to work well. The same strategy was applied on hBD-3 to predict its trustable dimer structure, which agrees with experimental findings.

---

[i]. DJ. Schibli, HN. Hunter, V. Aseyev, TD. Starner, JM. Wiencek, PB. McCray Jr, BF. Tack, HJ. Vogel. â€œThe solution structures of the human beta-defensins lead to a better understanding of the potent bactericidal activity of HBD3 against Staphylococcus aureusâ€. J Biol Chem.277 (2002), pp. 8279â€“89.

---

# 489746 Full-Atom Molecular Simulations of Lysozyme Confined in Realistic Silica Mesopores â€“ Insights in Conformation and Accessibility of Active Sites

**Abstract:**

Confinement of biomolecules in ordered mesoporous materials is receiving growing attention, because of potential applications in drug delivery systems, as well as in bio-separations and bio-catalysis. Experiments with mesoporous silica SBA-15 demonstrated that this material can be used to stabilize enzymes thanks to spatial and electrostatic confinement effects, similar to what is seen in the GroEL/GroES chaperonin complex [1]. In particular, experimental results of hen egg white lysozyme immobilized on SBA-15 show that the pore surface properties (curvature, electric charge and hydrophobicity), as well as the solute properties (pH and ionic strength) play a significant role in the stability of the confined enzyme, as well as its associated catalytic activity [2].

Because of this, it is of interest to systematically investigate the interactions between proteins and the pore surface of mesoporous materials, to explain how charge and surface curvature, related to pore size, dictate the proteinâ€™s orientation and affect the accessibility of its active site. A recent series of theoretical studies has shown that protein adsorption on a surface, modelled to represent silica, is mainly driven by electrostatic forces with positively charged residues helping to anchor the enzyme to the silica wall [3-7].

We have carried out new studies to understand the complex process of lysozyme adsorption on a realistic, curved silica mesopore surface. In these studies, we investigate the behavior of lysozyme in contact with a realistic representation of a pore of SBA-15, as used in complementary experimental work [1,2]. This representation includes pore roughness. We perform restrained all-atom molecular dynamics (MD) simulations, and we discuss our insights concerning how electrostatic and steric effects of the silica boundary induce conformational changes, and affect the accessibility of the active site of a confined lysozyme molecule.

**REFERENCES**

[1] J. Siefker, P. Karande, M-O Coppens; *Expert Opinion in Drug Delivery* **11**, 1781-1793, 2014

[2] M. Lynch, J. Liu, M. Nigra, and M-O Coppens; *Langmuir* **32**, 9604- 9610, 2016

[3] K. Kubiak-Ossowska, P.A. Mulheran, *J. Phys. Chem. B* **133**, 12189-12200, 2009

[4] K. Kubiak-Ossowska, P.A. Mulheran, *Langmuir* **26**, 7690-7694, 2010

[5] K. Kubiak-Ossowska, P.A. Mulheran, *Langmuir* **26**, 15954-15965, 2010

[6] K. Kubiak-Ossowska and P. A. Mulheran, *Langmuir* **28**, 15577-15585, 2012

[7] N. Hildebrand, S. Koppen, L. Derr, K. Li, M. Koleini, K. Rezwan, L.C. Ciacchi, *J Phys Chem C.* **119**, 7295-7307, 2015

# 492227 Aggregation and Self-Assembly of Biomimetic Polymers at Interfaces

**Abstract:**

Peptoids, or N-glycine substitutes, have been replacing conventional peptides for use as scaffolds, anti-fouling coatings, and functionalized membranes. Their ease of synthesis, thermal, and enzymatic stability and large combinatorial library of sequences make them attractive replacements for peptides[1]. In this talk, we will highlight the driving forces of aggregation and self-assembly of peptoids using molecular dynamics simulations and metadynamics[2–6] enhanced sampling.

We discuss how peptoids differ from peptides with simulations of model systems (sarcosine and alanine), using a forcefield that is re-parameterized for peptoids[7]. We examine the origins of peptoid hydrophilicity, and the differences between the conformational space explored by peptoids and peptides. Considering these results, we try to understand how peptoids form nanostructures on surfaces. For a comprehensive examination of the assembly process, it is essential to understand both solution behavior (to understand nanostructure seed formation), and interfacial behavior (to understand complex formation for nanostructures). Thus, we investigate how peptoids form clusters in solution and the effect of ions on these clusters. We find that pi-stacking and backbone flexibility allow peptoids to retain elongated conformations. Further, we discuss how peptoids may exhibit crystal facet selectivity (for gold) and how their assembly may be templated by the surface structure (for mica). These simulations highlight molecular-level origins of assembly of peptoids in different environments, which are poorly understood, using a multi-scale modeling approach.

Reference:

(1) Zuckermann, R. N. *Biopolymers* 2011, *96*(5), 545–555.

(2) Pfaendtner, J.; Bonomi, M. *J. Chem. Theory Comput.* 2015, *11*(11), 5062–5067.

(3) Barducci, A.; Bussi, G.; Parrinello, M. *Phys. Rev. Lett.* 2008, *100*(2), 20603.

(4) Valsson, O.; Tiwary, P.; Parrinello, M. *Annu. Rev. Phys. Chem.* 2016, *67*(1), 159–184.

(5) Raiteri, P.; Laio, A.; Gervasio, F. L.; Micheletti, C.; Parrinello, M. *J. Phys. Chem. B* 2006, *110*(8), 3533–3539.

(6) Tiwary, P.; Parrinello, M. *J. Phys. Chem. B* 2015, *119*(3), 736–742.

(7) Mirijanian, D. T.; Mannige, R. V.; Zuckermann, R. N.; Whitelam, S. *J. Comput. Chem.* 2014, *35* (5), 360–370.

# AUTHOR INDEX

# AUTHOR INDEX

# AUTHOR INDEX

# AUTHOR INDEX

# AUTHOR INDEX

# AUTHOR INDEX

**AIChE**
120 Wall Street, FL 23
New York, NY 10005-4020

ISBN 978-1-5108-5807-7

CURRAN ASSOCIATES INC.
proceedings.com